DATE DUE

OCT 2 1 1984			
SEP 2 3 1985			
OCT 1 4 1988			
	201-6503		Printed in USA

Elementary
Statistical
Methods

INTERNATIONAL SERIES IN DECISION PROCESSES

INGRAM OLKIN, Consulting Editor

Elementary Statistical Methods

Third Edition

Helen M. Walker

Claremont Graduate School and University Center

Joseph Lev

New York State Education Department

HOLT, RINEHART AND WINSTON, INC.

New York · *Chicago* · *San Francisco* · *Atlanta* · *Dallas*
Montreal · *Toronto* · *London* · *Sydney*

PREFACE TO THE
THIRD EDITION

Like the preceding edition, this book provides an introduction to both descriptive statistics and statistical inference. Basic concepts are presented by the use of intuitive approaches, and applications are stressed. By these devices an understanding of the underlying meanings and assumptions is developed without resort to intricate mathematics.

An effort has been made to present the various topics with such simplicity and clarity as to make the book almost self-instructive. Formulas are not derived but are presented as economical descriptions of concepts or procedures. The assumptions underlying a formula are stated when these limit the conditions under which the formula may be used.

The major purpose of the present revision was to expand the contents to embrace such topics as the chi-square and F distributions, tests of hypotheses about variances, analysis of variance in its simplest form, tests of hypotheses about the comparison of categorical populations, multiple regression and correlation, and sampling from a finite population with applications to survey sampling.

Because of the continually expanding role of statistical methods in the social and behavioral sciences, these topics, which were considered "advanced" a decade ago, must now be brought within the general understanding of beginning students. For this reason, inclusion of these topics in an elementary text is amply justified.

Rewriting the text to include new material offered an opportunity to make certain improvements in those sections which have been carried over from the 1958 edition (Chapters 1 to 14 of the present text). The symbolism has been changed to conform to what appears to be the best modern usage. By changing the order of presentation so that hypothesis testing precedes regression and correlation, the latter subjects have been given a more satisfactory treatment consistent with the development of multiple regression and correlation in

v

Chapter 18. Numerous small inconsistencies and infelicities of expression have been corrected.

The preceding edition was intended as a text for a one-semester introductory course. This edition will usually require two semesters. However, it can be used in an introductory one-semester course if any or all of these conditions exist: (a) the class is able to move rapidly, (b) some topics toward the end of the book are omitted, and (c) the availability of an electric computer makes it feasible to eliminate some of the work on computational methods.

The exercises in this text are intended to serve two purposes. Some are the usual type of practice exercise, presenting applications of methods under discussion. Other exercises are intended to clarify meanings, to encourage the student in an active search for relationships, and to extend the development of concepts already presented. Because the sets of exercises are timed to come at particular stages in the expository treatment, it seems desirable to place them immediately after the exposition of particular topics rather than at the end of chapters or in a separate manual. Answers to questions in the exercises appear in the Appendix.

In preparing exercises the authors have made repeated use of the same two sets of data, so that the same data may appear in exercises in tabulation; graphing; computation of percentiles, means, and standard deviations; and determination of relationships between pairs of variables. The decision to use the same sets of data in many types of problems is based on the belief that it is easier for a student to comprehend a statistical concept if he is familiar with the subject matter of the data. It should be pointed out, however, that a great variety of additional data are included in the book to enrich the reader's experience. In the interest of emphasizing statistical principles, a new idea is often introduced with a miniature problem in which the arithmetic is so simple that computation requires little attention. These short problems have been specifically planned to facilitate the development of meanings. Short problems are also used occasionally to test the student's grasp of principles.

We would like to express our thanks to Mrs. Marjorie Royle for her help in proofreading and her assistance in preparing the answer key.

To Harvard University Press, Iowa State University Press, G. W. Snedecor, Wm. G. Cochran, and *Biometrika,* we are indebted for permission to reprint their tables in the Appendix of this book.

We are indebted also to the Literary Executor of the late Sir Ronald A. Fisher, F.R.S., to Dr. Frank Yates, F.R.S., and to Oliver & Boyd Ltd., Edinburgh, for permission to reprint Appendix Tables III, IV, VI, and VII from their book *Statistical Tables for Biological, Agricultural and Medical Research.*

Claremont, California   H. M. W.

Albany, New York   J. L.

May 1969

HOW TO USE
THIS BOOK

This section is addressed in part to teachers and in part to students. However, since the students in this case are adults, no important distinction need be made.

Physical Features of the Book. Before studying any text the wise student familiarizes himself with its physical structure. Among the features of this book which help economize student time and energy are the Lists of Tables, Figures, and Formulas, and a Glossary of Symbols in the back of the book. If the text is read selectively rather than sequentially, unfamiliar terms or symbols may be encountered. The page on which these are introduced and explained can be discovered by consulting the Glossary of Symbols or the Index, or both. The manner of numbering tables, figures, and formulas identifies the chapter in which they are placed.

Reading Habits. Some students who have consciously trained themselves to read very rapidly and are not accustomed to the study of scientific or mathematical material may find it necessary to revise their patterns of study. Until they come to understand that in a work of this nature one cannot read a page at a gulp, they may feel a sense of frustration and think that progress is slow because the number of pages covered in a given period of time is far below their average. Students must learn to gear themselves according to the subject matter. While some material may appropriately be covered at high speed, other material calls for thoughtful perusal—for continual comparison of the section under consideration with earlier sections. As Chrystal, a great English teacher of algebra in the nineteenth century, once said, "Every mathematical book that is worth anything must be read 'backwards and forwards'." His advice is still valuable.

The following pattern of study is suggested as likely to produce the clearest understanding for a given expenditure of time. (1) Read a chapter rapidly to get a general impression of the material contained and the purpose of the discussion.

In this rapid preliminary reading try to keep a chapter or two ahead of the class discussion. (2) Just before a section is taken up in class, go over it in detail, comparing, rereading, solving problems, verifying numerical statements, and making note of all points that are not fully clear. (3) As soon as possible after a section has been studied in class, read it once more to consolidate the understanding gained from the class discussion. (4) Several weeks later, reread the chapter once more. The practice of rereading material studied previously not only clarifies one's thinking but tends to improve morale. Points that seemed completely obscure on a first reading now seem so obvious that one experiences a real sense of growth. On the other hand, some points which seemed quite simple on a first reading now, in the light of further knowledge, have new implications which raise new queries. That, too, is a mark of growth.

Students unaccustomed to scientific or mathematical work are often discouraged by their need to read statistical material many times. They should realize that this is a characteristic of this type of material.

Abstract Imagination.   The purpose of statistical method is to abstract from a mass of individual observations certain information which will enable one to comprehend the whole. Therefore, statistical measures are essentially *abstractions*. An understanding of such abstract concepts has to be built up by experience, just as one builds up the concepts suggested by the words "quaint," "patroitism," "irresponsibility," "nevertheless," and "implication," for which no photograph or diagram can suffice. These concepts must grow in the imagination. To stimulate such imagination is the purpose of certain sets of artificial data used throughout the text and of certain developmental exercises, both verbal and numerical. The building of concepts is a more important undertaking than any other aspect of the course, because without a clear comprehension of concepts all skills are useless.

The beginner does not know which concepts can be clarified graphically and which cannot, and sometimes he is needlessly troubled by his inability to "see" a statistic (for example, the variance) for which he can obtain no satisfactory visual image. If a student is uncertain about the existence of a useful visual symbol, he might well ask his teacher before he indulges in too much worry.

Mathematical Preparation.   The extreme diversity of mathematical background usually found in a beginning class creates problems for both the teacher and the student. This text is written primarily for the person whose study of mathematics ended when he left high school. At points where such persons are likely to experience difficulty, either special explanations are provided or references are made to Walker's *Mathematics Essential for Elementary Statistics* (Holt, Rev. Ed., 1951). Teachers who use this book may want to require each student to take the self-scoring tests with which each chapter begins. Students who make no errors on a test need do nothing further with that chapter. Those who make errors should study the explanatory material and the practice exercises, then take a second self-scoring test parallel to the first one. Each student can go through the book at his own speed, reporting each week on the chapter he has studied and the number of errors made on each test. A student with good mathematical background can finish the book in a few hours. The student who cannot needs the help he will secure from its use. By this method practically no class time is taken up with difficulties that relate to the simpler aspects of mathematics.

Order of Topics.   The sequence of development followed in this book is not the only one which might successfully have been adopted. However, if a person should attempt to study *this* book in a different sequence, rearranging the topics in some new order, he will almost certainly suffer the frustration of encountering the use of technical terms and concepts before he has studied the basic explanation clarifying them. The authors have devoted a great deal of thought to the arrangement of topics and have consistently used the earlier sections to prepare for later ones.   Most of the accruing advantages will be lost if the sequence of topics is drastically revised. However, the ensuing confusion can be somewhat mitigated by consistent reference to the Index and Glossary of Symbols.

CONTENTS

Transforming ordered categories into scaled scores   Normalizing a
distribution of scores   Profile chart   Transforming ranks to normal
deviates   Discovery of the normal curve

Situations in which samples are employed   Population   Randomness
Sample   Parameter and statistic   Statistical inference   Law of large
numbers   Sampling variability   Sampling distributions   The
normal distribution for statistics   The standard error   Estimation
by a single value   Unbiased estimate   Efficient estimate   The
standard score for means   The confidence interval   Confidence
interval for the mean when the sample is large   Student's
distribution   Confidence interval for the mean when the sample is
small

A problem involving a hypothesis about means   A statistical
hypothesis   Test of hypothesis that $\mu = 0$ Level of significance
One-sided test   Comparing the means of two populations
Effectiveness of a test of hypothesis   The power of a test   Power
function   Computing the power of a test

Meaning of proportion   Conversion of dichotomous data to
measurement data   Symbolism for proportions   Statistical inference
about proportions   Population distribution   Sampling distribution
of a proportion   Point estimation of a proportion   Interval
estimation of a proportion   Test of a hypothesis about a proportion
The sign test   A table for testing hypotheses about a proportion
Hypothesis that P is the same for two populations

Estimation (or prediction) of one variable from another   The exact
straight-line relationship   The formula for an exact straight-line
relationship   Independent and dependent variables   Inexact
relationship   Bivariate frequency distribution   Mathematical model
The problem in miniature   A general routine of computation for the
regression formula   Checks on computation   Computation by hand
or by electronic computer   Estimation of the conditional standard
deviation   Variation about the regression line   Standard error of
regression coefficient   Applications of the regression formula
Graphic representation of the regression coefficient   The assumption

of linearity   Origin of the term "regression"   Routine for
computing regression from a scatter diagram

II. Tests of independence in contingency tables

Nature of problem   Symbolism   Degrees of freedom for an $r \times c$
contingency table   Illustration of a 3×3 contingency table
Contingency table in which one trait is dichotomous   Contingency
table in which both traits are dichotomous   Relation of χ^2 to z
Relation of the ϕ coefficient to χ^2 with 1 degree of freedom   The
2×2 contingency table with small frequencies   Comparison of two
proportions based on the same individuals   Misuses of the chi-
square test

D. Additional Comments on Survey Sampling

Further applications   Additional methods

E. References on Survey Sampling

1

THE ROLE
OF STATISTICS

Statistical method is one of the devices by which men try to understand
the generality of life. Out of the welter of single events, human beings
seek endlessly for general trends; out of the vast and confusing variety
of individual characters, they continually search for underlying group
characters, for some picture of the group to which the individual belongs.
This group picture is not merely a summary of the individuals who form
the group. It transcends the individuals. It has meaning of its own not
to be discovered through the most intense contemplation of any single
individual. Such group trends are sometimes apprehended subjectively
and almost intuitively by a person of penetrating insight, who observes,
without the aid of numerical computations, "this group is more variable
than that" or "these two characteristics are not likely to occur in the
same individual." Such statements are fundamentally statistical in
nature, because they relate to tendencies of a group as a whole which
either have no meaning with reference to a single individual, or may not be
true of a particular selected individual. That such a statement is vague,
subjective, debatable, quite possibly false, and inefficiently arrived at does
not keep it from being statistical in nature. Controlled, objective methods
by which group trends are abstracted from observations on many
separate individuals are called statistical methods.

Statistics in Modern Life.   To a very striking extent our culture has
become a statistical culture. Even on the most elementary level it is
impossible to understand psychology, sociology, economics, finance, or the
physical sciences without some general idea of the meaning of an average,
of variation, of relationship, of sampling, or of how to read charts and

1

tables. Even a person who has never heard of an index number is affected in the intimate details of his daily living by the gyrations of those index numbers which describe the cost of living. Perhaps the clearest indication of the extent to which statistical ideas pervade modern life is provided by the use which advertising firms make of such ideas. Quoting statistics in favor of almost any product they wish to popularize, even if those statistics have little relation to the quality of the product, and making liberal use of graphs, they tacitly assume that the American people are accustomed to thinking in statistical terms or that if they do not understand they may at least be impressed by a statistical argument.

Let us look briefly at some of the ways in which statistical facts and statistical ideas impinge on the life of an ordinary person. The process begins early. The parents of a newborn baby are soon asking how his rate of development compares with that of the "average" child and are perhaps quite unaware of the large amount of statistical research required to furnish the norms needed to provide an answer for their queries. For the baby's ills as well as their own, they expect the doctor to have available tested remedies. But such testing is the outcome of the statistical analysis of data obtained from large-scale experiments. They expect to be able to buy clothing and shoes accurately sized for children of various ages. But how could a manufacturer get the information which makes such sizing possible unless measurements had been made and statistically analyzed for large numbers of children at each age level?

Perhaps the parents take out an insurance policy to provide for the eventual college education of the new baby. An insurance policy actually represents a gamble in which the individual bets that he will not live a specified number of years or that some stated disaster will happen to him, and the company bets that the disaster will not occur. Rates that will permit the company to stay in business but not to make excessive profits can be set only by a very large-scale continuing analysis of mortality data, accident data, and the like. It is impossible to foretell or control the chance factors in life, but it is possible statistically to estimate their group impact and by spreading costs over many people to mitigate the effect of calamity on any one person.

In our "average" family, money is probably not plentiful, and they watch the cost-of-living index anxiously, although they may have no idea of how the Bureau of Labor Statistics works year in and year out to gather the statistical data on which that index is based. Or they may watch apprehensively the published figures on unemployment, without knowing anything about the enormous sampling project conducted by the Census Bureau and the Bureau of Labor Statistics in order to obtain current information on the extent of unemployment in various industries and various regions.

When the erstwhile baby is of school age, his parents expect the school to obtain and make proper use of information about his aptitudes, abilities, and progress. Yet without the psychological and statistical labor of many people over many years, there would be no tests of reading readiness, of intelligence, of achievement, or of vocational preference.

Just as parents expect the medical profession to have reliable information about the effects of all kinds of drugs and treatments, old and new (information that can be secured only through large-scale experiments and statistically analyzed), so do they take it for granted that the teaching profession will have tested information concerning how children learn; how they develop physically, mentally, and emotionally; which school practices are beneficial to children's development. Such knowledge requires the observation of *many* children, with statistical analysis to extract meaning from the observations.

Population increases or population shifts will affect the living arrangements and personal comfort of our "average" family. Statistical information concerning population changes is essential to the long-range planning of such agencies as housing and traffic authorities, public utilities, and school boards. Thus the "average" person reaps the benefit of statistical studies or suffers when they are not properly made.

The legislator is faced with many decisions which are essentially statistical in nature, relating to such matters as proposals to revise the social security program, to authorize programs of slum clearance and public housing, to extend or withhold foreign aid, to change the pattern of unemployment compensation, or to encourage or discourage prepayment for health services. To reach a sound decision on such matters requires a thorough statistical analysis similar to the analyses which permit the successful operation of insurance companies, but all too often the debate on such bills appeals more to emotion than to statistical fact.

Persons who hold managerial positions in industry must be able to interpret statistical studies of the quality of the manufactured product and the efficacy of the manufacturing process—studies which involve drawing inferences from samples. Businessmen are often concerned with studies of consumer preference and the effectiveness of an advertising campaign, which are also based on samples.

The teacher who does not understand certain basic statistical ideas is in a position to do real harm to pupils because of his ignorance. If he is not aware of the universality of human variability and of ways of measuring it, he may use a test norm as a standard to which he tries to make all pupils conform. If he does not understand the approximate nature of all measurement, the conventional measures of test reliability, the qualities to be sought in selecting standardized tests, and methods of determining the reliability of teacher-made tests, he may place too much faith in unreliable measures of pupils. If he has never studied correlation and regression, he may confuse correlation with causation and thus propose irrelevant action, or he may assume that the child with the highest intelligence quotient in his class should be expected to stand highest in other desirable traits and therefore be censured if he does not excel, thus demanding the impossible of some pupils and giving insufficient challenge to others.

The student majoring in education, psychology, sociology, or economics soon discovers that, unless he has at least a minimum acquaintance with statistical vocabulary and statistical method, much of the important literature of his field is incomprehensible to him. Often he discovers that he himself has become interested in working on some project to which statistical methods are basic.

Everyone who reads the newspapers is now familiar with the various opinion polls and so has some vague idea about generalizations based on samples. The educated person needs to understand something about the practice of drawing samples and making inferences from sample data and to become sensitive to sources of bias in any sample on which he is depending for important information; he needs to know that sampling has become a highly technical matter and that, if he plans to make an important investigation utilizing sampling, he must either study the literature on sampling methods or consult an expert, perferably both.

Persons who do not feel confident of their own ability to interpret a statistical statement usually resort to one of two extreme positions, each unsatisfactory to the well-educated man. One extreme is the uncritical acceptance of any statement buttressed by statistical data, however fantastic. The other extreme is uncritical suspicion of *all* statistical reasoning. Aside from these two extremes, the only alternatives appear to be either reliance on authority or some comprehension of what statistical reasoning is and on what principles it depends for validity and cogency. One cannot count on having a satisfactory authority always at hand, however, and the person who wishes to make judicious decisions in regard to personal matters, business matters, or public policy will often need to base those decisions on information that is essentially statistical in nature.

Choice of Action. A very important use of statistical information is to furnish a basis for choice between two or more courses of action. A statistical study can throw light on the probable consequences of each of the alternative actions, but it does not pretend to be a substitute for value judgments. The person or group responsible for choosing a course of action can make a more informed decision when they have more light on what the results of each action are likely to be, but they still have to decide which results they prefer.

For this purpose a person or a group sometimes utilizes statistical information already published or instruments already constructed by statistical methods. For example, a young person trying to choose a vocation is not likely to have the resources for gathering his own data on the income, the living conditions, the hours of work, or the demand for new entrants in several vocations, but there may be such data already in print. Suppose he learns from such data that there is great demand for new workers in vocation A—salaries are high, hours of work long—whereas in vocation B salaries are lower, competition for jobs is more intense, but hours are shorter and living conditions pleasanter. He must still choose which he prefers, but his choice is no longer blind.

Sometimes a statistical study is initiated to provide the basis for a specific, immediate decision. The field of market research provides a great many examples of this situation. A merchant employs interviewers to ask housewives whether they would use a certain product if it were put on the market, whether they prefer one type of product or another, and so forth and uses the tabulated returns to make merchandising decisions. The sponsor of a program on radio or television analyzes data on program ratings to help him decide whether to continue a particular kind of advertising. Large manufacturing plants keep up a continuous statistical study of their products in order to detect flaws that might indicate the need to overhaul the machinery, in order to choose between alternate processes, and the like.

Sometimes a large investigation provides the basis for widespread, general decisions. This is the situation, for example, in most medical research. The immediate decision may be the choice between saying "yes" or saying, "The evidence is not convincing" to such a question as, "Do cigarette smokers have, in general, a shorter life-span than non-smokers?" or "Does this vaccine protect against the common cold?" But that decision may cause many individuals and groups to make other related decisions as to a course of action.

Action Suggested by Statistical Inquiry. A statistical investigation is not limited to providing a basis for choosing between alternatives which are known prior to the investigation. The purpose of the inquiry may be to suggest a new course of action. Some of the basic problems confronting our society, such as delinquency and alcoholism, do not yet have well-defined solutions even as possibilities. It is hoped that statistical studies underway on factors related to these problems will suggest remedial courses of action.

Basic Statistics. A large part of statistical enterprise concerns itself with compilation of statistical data—its summarization and presentation as a basis for a variety of decisions. In business, statistical data deal with such matters as items produced, labor used in production, sales, and profits or losses. In government, statistical data are concerned with such matters as size and composition of population, and wealth of the nation and its welfare.

The purposes to be served by the basic data are not always well defined at the time of compilation. Data obtained from a census may serve as a basis for revising congressional districts; they may help a chain store in establishing new branches; or they may point out the need for new facilities for higher education ten or fifteen years later.

In the presentation of basic statistics, consideration needs to be given to the uses to which they will be put in making decisions leading to action. A great deal of data, such as population or wealth of many localities, is presented in the form in which they were obtained. The users may then relate and summarize these data in ways which are most helpful to them.

Data for the nation as a whole, or for states, are often of sufficient interest so that governmental agencies present summaries of them in a variety of ways. As an illustration of such a summarization, consider the consumers' price index prepared by the United States Bureau of Labor Statistics. This is a weighted average of retail prices expressed as a percent. This index of prices is very important in interpreting wages and costs under the conditions of varying prices. Other well-known summaries of this sort are indices of stock prices and of production.

Summary of Purposes. Statistical investigations are justified by their usefulness in leading to improved courses of action. Some investigations are directly related to a specific purpose and are planned accordingly. Other investigations do not have clearly defined purposes, but it is expected that the information obtained in these investigations will serve a variety of purposes.

Issues Preliminary to Any Statistical Inquiry. At this point the reader may be mistakenly assuming that by studying a statistics text he will learn certain statistical computations and a kind of universal routine for attacking a statistical problem and that he can then apply that routine and perform those operations in any situation he wishes to explore statistically. This is a serious misapprehension of the way a statistical study should be approached. Before any statistical enterprise is begun, whether it be on a large scale or a small scale, whether its pattern be simple or complex, certain general issues must be faced. Decisions must be reached on the following points:

1. What concrete questions is the study designed to answer? The ability to ask good questions is one of the marks of an effective, imaginative, and creative research worker. Help at this point must come primarily from the content field rather than from methodological consideration.

2. To what class of individuals are the answers intended to apply? To those individuals examined only or to some larger group?

3. If the answers are to have some general import beyond the cases actually observed, how is the sample to be chosen? How large should it be? Are there any elements of bias in the proposed plan for selecting the sample of individuals to be examined?

4. What observations or measurements are to be made on these individuals and how? Each content field has its own special methodology for making observations and measurements, and this methodology must be learned by the specialist in that field. In this text only general matters related to a wide variety of fields can be mentioned.

5. How shall data be gathered for analysis? Some useful methods are described in Chapter 2.

6. What summary measures shall be computed? This is the main theme of the book.

7. How shall results be presented? Some aspects of this problem are discussed in Chapter 3.

2

GATHERING
AND RECORDING
DATA

Once the purpose of a statistical investigation has been defined, the problem is to collect data which are relevant to that purpose, to analyze these data, and to present them in a meaningful manner. This chapter will deal somewhat sketchily with problems of gathering and recording data in order to facilitate their analysis and interpretation.

Subjects of the Investigation. Statistical data may be considered broadly as observations made on subjects. Subjects differ vastly from one inquiry to another. Sometimes subjects are individual persons, sometimes they are families or institutions or communities, sometimes they are physical objects such as buildings or machines or samples of a manufactured product. The variety of things which may be made the subjects of inquiry is practically endless.

Before the investigation begins, it is necessary to *define the units* about which investigations are to be made. For example, in planning a study of low-rent public housing developments in a city, it would be necessary to decide whether observations are to be made on families, on persons, on dwelling units, or on some other type of individual. If the family is selected as the unit, any of the following definitions might be selected, as well as many others: (a) a family moving into one of these developments during the period covered by the study; (b) a family living in one of these developments for any portion of the period covered by the study; (c) a family applying for entrance to the development during the period covered by the study. It would also be necessary to define "family." If single persons are accepted, does a single person constitute a "family" for the purpose of the fnquiry? Do two or three single persons living in the same apartment constitute one family? If so, does that family have a "head" to be enumerated when the heads of families are classified? If the person who signs the lease has relatives other than wife and children living in the same apartment with him, is there one family or two? Using an apartment in one of the developments as the unit would permit making observations on all the persons living in it during a designated period, thus securing information about turnover which could not be obtained from a study of family units. Almost every study presents similar perplexities of definition which must be resolved before the data are gathered. Each unit or subject is also called *an individual* and sometimes *an element*. Obviously the term *individual* does not necessarily mean a person.

Variables.   For each subject of an investigation, one or more characteristics are observed. These characteristics are selected in the light of the purposes of the investigation. For example, the observations made on a family might include the number of persons in the family, the number of children, the number of adults, the number of children of school age, the number of persons employed, the total family income, the occupation of each parent, the country of birth of each parent, the language customarily spoken in the home, the number of persons per room, the number of years of education for each adult in the family, and so on and on.

Observations on a characteristic of individuals permit those individuals to be grouped in different classes or categories. All the individuals in one class are then said to have the same *value* of that characteristic. Thus, if applicants for a position are classified by sex, the characteristic sex is said to have the two values, male and female.

If a characteristic has the same value for all the individuals in an investigation, that characteristic is called *a constant*. If it has different values for different individuals, it is called *a variable*. Thus, for the children in a public elementary school, such characteristics as sex, age, intelligence quotient, or score on any standardized test would be variables. In a boys' school, sex would be a constant and the other characteristics variables.

A variable for which there are two possible values is called *a dichotomy*. Dichotomies are very familiar—people are classified as male or female, as living or dead, as employed or not employed, as native born or foreign born, as literate or illiterate; answers are classified as correct or incorrect; schools are classified as public or private, as sectarian or nonsectarian; plants are classified as exogenous or endogenous; candidates for an office are classified as winners or losers; and so on.

The statistical treatment of dichotomous variables is particularly simple because dichotomous variables can be ordered arbitrarily, and they present no problems of size of interval. It is often convenient to code the two classes as 0 and 1 as will be explained later.

If a variable can assume three or more values, these values may or may not have a necessary order. The cars in a parking lot might be classified as to make, state in which they are currently registered, and company with which they are insured. Each of these variables has values which identify distinct classes, and these classes have no natural order. Students in a graduate school might be classified according to place of birth, type of college from which they obtained their bachelor's degree, department in which they are majoring, favorite type of recreation, and so on. For any two elements the values of the variable are "same" or "different" but never "more than" or "less than."

These variables are usually termed either categorical variables or nominal variables. Sometimes they are called qualitative variables to distinguish them from variables whose values have quantitative meaning. Whether the classes are designated by names, by letters, by numerals, or by other symbols, the designations are only labels with no numerical significance whatever. The only statistics which can be computed for nominal variables are based on the number or proportion of elements in the various classes. Nominal variables are sometimes called "attributes."

The categories defined by the values of a variable may be a set of distinct but ordered classes. Then if two elements are not in the same class, one of them is "higher" or "greater" or "more than" the other. Thus the condition of the tires of an automobile might be described as (a) perfect, (b) very good, (c) slightly worn, (d) badly worn, or (e) beyond all usefulness. These categories stand in an obvious order yet do not necessarily represent equally spaced points on a scale. Such an ordered variable is susceptible to forms of statistical analysis which could not be applied if classes were unordered.

For some variables it may be possible to arrange all the individuals under consideration in meaningful order, so that no two fall into the same class. Then the individuals are said to be in rank order, and the ordinal number corresponding to an individual's position in the series is his rank. Here only relations of "more than" and "less than" are involved. It is necessary to know that A surpasses B, but not to know by what amount. If in an ordered set of classes some contain more than one individual, the individuals are said to be in a ranking with ties. Many very useful statistics are developed around the idea of rank order. Some of these are introduced in Chapters 5, 8, 13, and 17.

When the intervals between successive values of a variable are all equal, the variable is said to be scaled. Then and only then is it legitimate to perform the operations of addition and subtraction on these scale values. Then and only then can we subtract the score of element B from that of element A to find the amount by which A exceeds B. Without equality of intervals on the scale, true measurement is impossible. Much that is called "measurement" in educational work does not deserve that name because equality of intervals has not been provided for.

Unless the real zero point of the scale is known, the ratio of one value to another cannot be found. An arbitrary origin does not suffice for this purpose. Thus we cannot say that 40° temperature is twice as warm as 20° because whether we use the Fahrenheit or Centigrade scales, the origin we call "zero" is only arbitrary.

Scaled variables may be classified as to *whether the scale is discrete or continuous*. Thus age and height are continuous variables because they can take any value—integer or fraction—along the scale, which is continuous, without a break. On the other hand, number of siblings is a discrete variable. The values of the variable must be integers and cannot take any of the fractional values between the integers.

Sample or Entire Population.   The question as to whether observations are to be made on all the individuals with whom the study is concerned (that is, the *population*) or only on some sample of those subjects must be decided in each situation on its own merits. An example of an inquiry in which the goal is to make observations on all individuals is the United States Census. On the other hand, when norms are to be established for a mental test, it is obvious that it is neither necessary nor economical to attempt to measure all children of the appropriate age in the country. Even if that could be done, the test would be used for children born at a later time and so not included in the original study. Clearly, the group on which test norms are established must be considered a sample of a larger population.

The procedures of sampling in experiments and in surveys are treated in several excellent textbooks and are beyond the scope of this book. However, an indication of the general logic of sampling is given in Chapters 9 and 19.

Instruments for Making Observations.   If observations are to be made on subjects, instruments for making these observations must be provided. In work with physical materials or mechanical devices, the instruments are devices for measuring size, strength, chemical composition, electrical charge, and so on.

In the behavioral sciences—psychology, sociology, anthropology, and education—the instrument might be a questionnaire, a rating scale, an interview, a galvanometer, a performance test, a pencil and paper test, or an observation that certain behavior does or does not occur.

Every book on a content field—whether intended for the administrator, the research worker, the practitioner, or the layman—is directly or indirectly concerned with the collection of data. Each field has its special lore which cannot be rehearsed here where we are dealing chiefly with problems common to many fields. The choice of instruments to be used in a particular investigation depends on the purpose and the subjects of that investigation. The construction of instruments must be made by persons well trained in the subject-matter field and its particular techniques of measurement. However, there are two important general requirements which must always be met if observations are to furnish dependable information.

The first requirement is that observations should be *reliable*—in other words, that repetition of the observation on the same subjects should produce approximately the same results. The second requirement is that observations should be *valid*—in other words, that they should provide a true measure of the characteristic they purport to measure.

Both reliability and validity are amenable to statistical investigation. Chapter 13 will discuss these important characteristics of observations.

The manner in which an instrument is to be used is also an important consideration. Precise instructions must be prepared to assure that observations are made in a uniform manner. If the instrument is a mental test with limited time, the entire test, or its component parts, must be timed accurately and uniformly for all subjects. If the instrument is a form on which information is to be recorded by an interviewer, careful training must be given to all interviewers so that practice will be uniform and records comparable.

Personnel of the Investigation.   In some enterprises carried out by a small team and in many individual studies, the same person or persons determine the goals of the study, plan its design, and execute the details. However, in a governmental agency or in a large business enterprise, the purpose of the investigation is likely to be decided by the higher administrative officials of the organization, and the conduct of the investigation is likely to be turned over to technicians who work in consultation with the administrative officials.

The planning of the investigation and the construction of instruments require skill of very high order. Specialists in one or more subject-matter fields and in statistics participate in this aspect of the investigation. The actual use of the instruments in making observations may often be left to personnel with more limited training. Such personnel may have a variety of titles, depending on the area of work. In industry, they are often called inspectors. In census work, they are called enumerators.

The statistical analysis of the data is carried on under supervision of persons with advanced training in statistics. However, tabulations and computations may be performed by clerical workers and by operators of the various mechanical devices used in statistical work. The preparation of the final report is again likely to be an enterprise carried out by analysts of the highest skill.

Source of Observations. In many statistical undertakings the observations are made specifically for the purpose at hand; in fact, the instruments are devised for the sake of the investigation. However, many other investigations use observations which have been accumulated as operating records of a public agency or a commercial orginization. The problem of devising instruments is then minimized or even eliminated, but the problems of sampling, organizing data, analyzing, and reporting still remain.

Original Records. For observations to be useful in a statistical investigation, it is necessary that they be recorded. Important considerations in the planning of forms for the original recording of data are:

Accuracy of the record
Simplicity of recording
Identification of the subject
Accuracy and ease of reading the record
Simplicity of use in further analysis
Completeness of record in respect to its later use

The time card in Figure 2-1 satisfies all the requirements here listed. Accuracy of the record is assured by an exact time mechanism which automatically stamps the time. The record is made simply by inserting the card into the mechanism. The pressure of the card operates the mechanism. The employee is clearly identified. Since the record shows the date, hour, and minute stamped in a preassigned position, it is easy to read. The tabular arrangement simplifies computing time for payroll purposes.

An answer sheet for a multiple choice test which is scorable by an electrically operated machine is shown in Figure 2-2. This answer sheet requires further processing. An electric scoring machine will give the number of correct answers either for the entire sheet or for a specified portion of it.

The examples which have been given provide a separate form for the original record of each individual, but often a single form is used for a

BROWN, JONES, SMITH & CO.

TIME CARD
Week beginning August 4, 1969

Roe, Richard
Packer
Shipping Department

	Time in	Time out	Time in	Time out
Mon.	8:30	12:00	12:55	5:00
Tues.	8:31	12:00	12:55	5:00
Wed.				
Thurs.				
Fri.				

Figure 2-1   Time Card

group of individuals. The sheets used in census enumeration have 40 lines on each side, all the data for one person being placed on a single line of that sheet. One advantage of such recording is that if an enumerator in the field handles only a few large sheets, he is less likely to lose one than if he handles many small sheets.

The original record form may also provide for recording subsequent action on the basis of the information. Bills customarily have space for endorsement as having been paid. Applications for employment may also provide space for later evaluation, as approved or disapproved for employment.

Summary Records.   Frequently data on several variables for the same individual, gathered on separate original records, must be brought together on one single summary record. A familiar example of such a summary record is a teacher's record book, where one line is allocated to each pupil and various columns to the tests. The original test records from the answer sheets of the pupils are transcribed into the appropriate spaces in the class book. At the end of the school term the various test scores are averaged. The class book is also likely to have a space for a final grade.

In order to simplify further analysis it is often convenient to enter summary data on a separate form for each subject. The chief advantage of using such separate summary forms is the convenience in sorting under a variety of classifications. Three illustrations of summary forms in common use will be described.

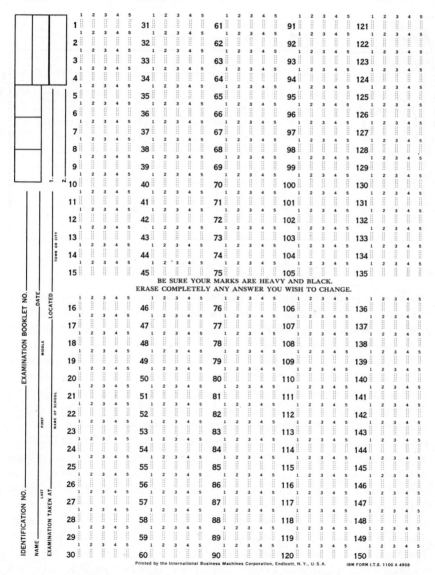

Figure 2-2   Answer Sheet for Multiple-choice Test

The Punched Card for Electrically Operated Machines. Several firms have constructed electric machines which can be used for sorting, tabulating, and making computations from data which have been recorded by punching holes in cards. Figure 2-3 shows a punched card which was used in devising a battery of tests for selecting employees.

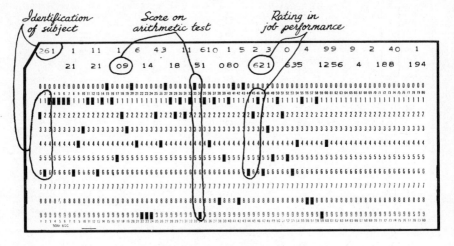

Figure 2-3   Punched Card

One card is prepared for each subject or sometimes more than one card if there are a great many variables. In each column of this card any digit from 0 to 9 can be recorded by punching a hole in the space identified with that digit. For a number consisting of several digits, several columns will be needed—one column for each digit of the number. In designing the plan for punching the cards, each variable is allotted as many columns as the maximum number of digits required for an entry on that variable. The set of columns assigned to a variable is called its *field*. A field is also allotted to identify the subject, usually by means of a code number. As each card contains 80 columns, 80 digits can be recorded. In the hands of an experienced person these 80 digits suffice for the record of a very large amount of information, and the novice will find it a great economy to submit his instruments to such an expert before he gathers his data. Without expert advice *in advance*, one may incur much avoidable expense in transcribing data from the original instruments to the punched cards. In some of the columns, letters of the alphabet can be recorded, but these are used in reporting results, not in analysis.

The electric equipment can be used to sort cards into classes, to count cards, to list some or all the entries, and to make a variety of complex computations. Such equipment is expensive to rent or to buy and is customarily used in an organization with a volume of work sufficiently great to warrant the expense. Card equipment which is much cheaper to buy will be described in the following paragraphs.

Edge Punched Cards.   Cards with a row of little holes around the edge, or sometimes two rows, can be purchased in various standard sizes, as illustrated in Figure 2-4. If a study is to involve a very large number

Figure 2-4   Examples of Edge Punched Cards

of individuals, it may be economical to have a card specially printed by one of the commercial firms handling such materials.

Data are recorded by notching the card, from a hole to the edge. A special clip must be used for this, as it is important that the notch be made smoothly and accurately. If a variable has only two values, a single hole can be assigned to that variable; then a notch may represent one value and the absence of a notch another. If a variable has more than two values, a field of several holes may have to be assigned to the variable. When cards of a certain category are to be selected, a needle is passed through the appropriate holes of the entire batch of cards. As the needle is lifted, the cards not classified in the category are drawn out with the needle, the other cards which have been notched remaining behind. More elaborate systems for complex classifications with edge punched cards are described in the technical literature on this subject.

Use of Recorded Data. Gathering and recording data are the first steps in a statistical investigation—as they are, in fact, in any scientific enterprise. To make the recorded data useful to the investigation, it is necessary to select pertinent sections of the data and to analyze them in a variety of ways. The decision as to which data are to be selected for further analysis depends on the special purposes of the study. The remaining chapters of this book will be concerned with methods of analysis of data which are common to many statistical investigations.

Calculation by Computer. Much of the statistical calculation which was formerly done by use of a desk calculator is now being carried out by electronic computers. These machines can receive exceedingly complex instructions and apply them to large sets of data. The computer then provides fast and accurate answers. As a result, analyses which might formerly have been set aside as requiring excessive labor are being conducted routinely.

Instructions to a computer must, of course, be presented in a language which is comprehensible to it. Several languages which serve as a means of communication between the analyst and the computer have been devised. By use of one of these languages, the analyst sets up a program which the computer is to follow. When the program has been developed, it is entered on punch cards or a tape and presented to the computer, which carries out the instructions faithfully.

A simple example of the use of a computer appears in Figure 2-5. The procedure presented there was carried out on a device having the form of a typewriter which is connected by ordinary telephone lines to a remote computer. The purpose of the calculation was to obtain the totals and arithmetic averages of the following two sets of numbers:

Set 1:   15,   25,   10,   30,   20
Set 2:   2.4,   1.9,   4.7,   3.2

The language which was used is called "Basic." The program representing the instructions appears in Figure 2-5 in lines numbered 10 to

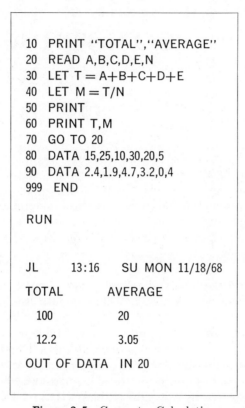

```
10   PRINT "TOTAL","AVERAGE"
20   READ A,B,C,D,E,N
30   LET T = A+B+C+D+E
40   LET M = T/N
50   PRINT
60   PRINT T,M
70   GO TO 20
80   DATA 15,25,10,30,20,5
90   DATA 2.4,1.9,4.7,3.2,0,4
999  END


RUN


JL        13:16     SU MON 11/18/68

TOTAL            AVERAGE

 100              20

 12.2             3.05

OUT OF DATA  IN 20
```

Figure 2-5 Computer Calculation

999. Line numbers indicate the sequence in which operations are to be performed. When the program is completed, the computer is told to "run." It then provides the required totals and averages in tabular form as shown at the lower part of the figure. The final statement indicates that all data which were presented have been used.

The calculations shown here are very simple and serve only to illustrate the operation of the computer. Far more complex operations can be carried out in similar fashion.

3

CONSTRUCTION
AND USE
OF TABLES

One of the simplest and most revealing devices for summarizing data and presenting them in a meaningful fashion is the statistical table. The table may be only an intermediate step leading to further analysis, or it may be an end product of the entire statistical investigation.

A well-planned table is a unified, coherent, and, in a sense, a complete story about some aspect of a set of data. Elements of the data are set up in rows and columns in order to indicate important relationships. The story is carried by the title and by suitable phrases which describe the actual numbers in the rows and columns. The elements of the table and their functions will be described and illustrated in the following pages.

The Table as a List. The simplest kind of table is a mere list of the subjects—usually in the vertical *column* at the left of the table—with certain facts about the subjects in the other columns of the table, all the facts for one subject appearing in the same horizontal line or *row*. Such listings are illustrated in Appendix Tables VIII and IX. Although the order in which the subjects are listed in these tables is of no particular importance, in many tables the subjects are placed in a meaningful order. Thus in a table giving data for the 50 states, the names of the states might be placed in alphabetical order for readers who wished to locate a particular state readily; in a historical study, the names might be placed in order of their admission to the Union; or, if the purpose of the table were to compare the states with respect to some characteristic such as density of population or per pupil expenditure for education, that characteristic might be placed in descending order. A list of candidates who have passed a civil service examination is usually arranged in the descending order of their examination scores because their relative position on this list—their rank in the group—determines their eligibility for appointment to a position.

If the data to be tabulated are on cards, with a separate card for each subject, these cards can be sorted in any desired order, and then the data can be transcribed. If the original data already appear as a list but a new list in some other order is desired, it may be helpful to prepare a set of cards, one for each subject, and then to sort these cards in the desired order prior to setting up the new list. A convenient feature of electric equipment for use with punched cards is that it makes possible an automatic listing, in tabular form, of the data on the punched cards.

Summary Tables.   A list is often a preliminary table, providing a basis for other tables which summarize the information contained in the list and present it in a way which brings forward important facts and relationships contained in the data. The remainder of this chapter will be concerned with the methodology of constructing summary tables.

The Elements of a Table.   Table 3-1 is a simple table which summarizes and makes evident information contained in the listing of Appendix Table VIII about the sizes of Sections I, II, and III. Although it is a simple table, it contains the elements which characterize tables generally.

Table 3-1   Enrollment in Three Sections of a First Course in Statistical Methods

Teachers College, Columbia University

SECTION	NUMBER OF STUDENTS
I	41
II	27
III	28
Not identified	2
Total	98

These include:

THE TABLE NUMBER	*Table 3-1.*
THE TITLE	*Enrollment in Three Sections of a First Course in Statistical Methods.* The title is a brief description of the contents of the table.
THE HEAD NOTE	*Teachers College, Columbia University February,*

The head note may be used to provide additional information about the contents of the table which would make the title undesirably long if included in it.

THE FIELD — The tabular entries, which in this table are the frequencies *41, 27, 28, 2*, and *98*.

THE COLUMN HEADS — *Section and Number of Students.* A column head describes the entries in a column.

THE STUB — The left-hand column contains the major scheme of classification of the tabular entries. In Table 3-1 the stub consists of the entries *I, II, III, Not identified*, and *Total*.

LINE CAPTION — Each entry on a line of the stub. Thus, *I, II,III, Not identified*, and *Total* are line captions.

STUB HEAD — A statement placed above the stub and used to describe the line captions in the stub. As the stub is a column, the stub head, *Section*, is also one of the column heads.

The significance of these structural elements is more clearly brought out in Table 3-2, which will be discussed in some detail with reference to its structure as a table and from the point of view of procedures needed to obtain the entries.

Table 3-2 Number of Men and Women in Each Section of a First Course in Statistical Methods

Teachers College, Columbia University

SECTION	NUMBER OF STUDENTS		
	MEN	WOMEN	TOTAL
I	28	13	41
II	24	3	27
III	18	10	28
Not identified	1	1	2
Total	71	27	98

Table 3-2 tells in greater detail the same story as Table 3-1. Because it shows another variable—namely, the sex of the students—two more columns are needed. There are now three columns in the field, and each column requires an explanatory heading. Instead of saying in this heading

"Number of Men" and "Number of Women," the phrase "Number of Students" is placed as brace heading (also called a spanner head) over all three columns. Note the use of a horizontal ruling to indicate the relation of the spanner head to the column heads.

Tallying. The entries, or frequencies, in Table 3-2 could be obtained from the entries in the first two columns of Table VIII in the Appendix by counting all the entries for a given section which are unstarred in the first column, and then counting all the entries for the same section which are starred in the first column. This process could be repeated for all three sections. The results of these counts could be checked against the complete counts for the sections made by ignoring the stars, as in Table 3-1, or by counts of all starred entries and all unstarred entries separately, ignoring sections. A check of some sort is absolutely essential for accuracy. However, this method of obtaining the frequencies is laborious and likely to produce errors because it is so easy to lose count. It is better to obtain the frequencies by *tallying.*

First, we note the nature of the breakdown desired for the final table, which, in this case, must provide for one classification into two groups (men and women) and for a second classification into four groups (three sections and a group of those "not identified"). A diagram having 2-by-4 boxes or cells will be required, and each cell must be large enough to hold all the tallies likely to be placed in it. Suitable headings should be placed on the rows and columns so that the person tabulating will not make a mistake by forgetting which is which.

Next, for each of the subjects in turn we enter in the appropriate box a short stroke called a tally (/). Thus for subject number 1, a tally is placed in the box which is in row III and the column "Men." The fifth tally in each set is drawn across the other four //// to facilitate counting the tallies. Sometimes the line for each tenth tally is slanted at a different angle to facilitate counting by tens, thus: Ж Ж Ж Ж //. The diagram and the tallies which have been inserted in it are shown in Figure 3-1.

Accuracy of tallying is considerably increased and eyestrain decreased if one person can read from the raw data while a second person enters tallies in the diagram. When an electronic computer is used, such hand tallying is of course unnecessary.

Table Containing Percents. Table 3-2 would be more informative if the percents as well as the numbers of the men and women in the three sections were exhibited as in Table 3-3. Some distinctive structural features of this table merit consideration. The words "Men," "Women," and "Both" are used as *spanner heads* (or *brace heads*), and appropriate horizontal rules indicate the relation of each to the two column heads under it.

Section	Men	Women
I	||| |||| |||| |||| |||| |||	|||| |||| |||
II	|||| |||| |||| |||| ||||	|||
III	|||| |||| |||| |||	|||| ||||
Not identified				

Figure 3-1   Diagram Showing Tallies for Frequencies in Table 3-2

Table 3-3   Percent of Men and Women in Each Section
of a First Course in Statistical Methods

Teachers College, Columbia University

	MEN		WOMEN		BOTH	
SECTION	NUMBER	PERCENT	NUMBER	PERCENT	NUMBER	PERCENT
I	28	68.3%	13	31.7%	41	100.0%
II	24	88.9	3	11.1	27	100.0
III	18	64.3	10	35.7	28	100.0
Not identified	1	50.0	1	50.0	2	100.0
All sections	71	72.4%	27	27.6%	98	100.0%

The use of more than one kind of unit (numbers and percents) makes it imperative that the unit be named at the head of each column. To aid the eye in distinguishing which figures represent numbers and which represent percents, it is customary to place the percent sign (%) after the first figure in a column of percents and after the summary, if there is one. A similar convention applies to the use of the dollar sign.

The introduction of percents requires identification of the base on which each percent is computed. The entries of 100.0 percent in the final column make it clear that each number in that column is the base of the percents in its row. Thus, $\frac{28}{41} = 68.3$ percent, $\frac{13}{41} = 31.7$ percent, $\frac{41}{41} = 100.0$ percent, and we have as a check 68.3 percent $+$ 31.7 percent $= 100.0$ percent. Some authors record percents without indicating the numbers on which the percents are based. This practice is unfortunate because it leaves the reader uncertain as to what significance to attach to the percents. Clearly 50 percent means something quite different when it is based on 200 cases than when it is based on 2 cases.

In the final row of Table 3-3, the entries in the columns headed "Number" were obtained by adding the other entries in those columns, but the entries in the columns headed "Percent" were not obtained by adding other entries but by computing a new percent. Thus, $\frac{71}{98} = 72.4$ percent and $\frac{27}{98} = 27.6$ percent. Although the summaries for the columns headed "Number" could properly be termed totals, those for the columns headed "Percent" cannot be so designated. To provide a heading appropriate for all entries the summary row has been labeled "All sections."

EXERCISE 3-1

1. From the data of Table 3-2, construct a table with percents based on the totals of the sexes instead of on the totals of the classes, as in Table 3-3.

2. From the same data, construct a table with column heads as follows:

	NUMBER			PERCENT		
SECTION	MEN	WOMEN	BOTH	MEN	WOMEN	BOTH

A Table Containing Two Sets of Data.   Table 3-4 is more complex than the preceding tables since it not only classifies the subjects by section and by sex but also presents two sets of data on the same subjects—one set based on scores in a midterm test on the content of the course, and one on scores in a prognostic test in arithmetic given before the start of the course. The two sets of data are presented in a single table because the second set helps to explain the first.

Column headings not only indicate the two separate parts of the table, they also manage to name the tabular entries in the table's field. The stub is divided into four distinct blocks each of which designates a complete table, so that actually eight related subtables are presented together in such a way as to reveal new meaning.

Another feature of this table is that for the first time in this chapter use is made of continuous variables. These are the scores on the two tests.

Table 3-4   Number of Students in Each Section Above or
Below the Middle of the Entire Group on Each
of Two Tests

*A First Course in Statistical Methods, Teachers College,
Columbia University*

		NUMBER OF STUDENTS SCORING					
		ON MIDTERM TEST			ON ARITHMETIC TEST		
SECTION*	SEX	BELOW 53	53 OR HIGHER	TOTAL	BELOW 33	33 OR HIGHER	TOTAL
I	Men	11	17	28	7	21	28
	Women	5	8	13	3	10	13
	Both	16	25	41	10	31	41
II	Men	11	13	24	14	10	24
	Women	3		3	2	1	3
	Both	14	13	27	16	11	27
III	Men	12	6	18	13	5	18
	Women	6	4	10	8	2	10
	Both	18	10	28	21	7	28
	Men	34	36	70	34	36	70
All	Women	14	12	26	13	13	26
sections	Both	48	48	96	47	49	96

* Two cases with section unidentified have been omitted.

The choice of the two breaking points at 53 and 33, respectively (noted in the column heads), may at first glance seem arbitrary. However, the entries in the final row of the table show that these scores divide the entire group into equal subgroups for the midterm test and into nearly equal subgroups for the arithmetic test. A procedure for finding such dividing scores will be described in Chapter 5.

The frequencies in Table 3-4 were obtained by a simple sorting procedure which could be easily carried out by anyone concerned with similar data. For each individual, a small slip of paper was prepared and on that slip was recorded the individual's code number, section, sex, score on midterm test, and score on arithmetic test. Not all these data were used in the first tabulation, but they were all recorded in order to be available for later tabulations. The slip used for Student 1 is indicated below.

1	III	M	33	58

The 96 slips of paper were filled out and checked for accuracy. The slips were then sorted by midterm test score and arranged in descending score order. From this sort, it was ascertained that 48 students scored 53 or above and 48 scored below 53.

The high (53 and above) and the low (below 53) subgroups were separated. Each subgroup was then sorted into six piles, by section and by sex within section. Obtaining the entries for the left-hand half of Table 3-4 was then merely a matter of counting the numbers of slips of paper in the piles. The right-hand half of the table was constructed similarly, using the same 96 slips of paper.

The analysis of the frequency distribution forms a major part of this book, and in effect the procedure just described was an elementary study of a simple frequency distribution.

Although this chapter is concerned mainly with the structure of tables rather than with the content of any one of them, the content of Table 3-4 foreshadows so much of the reasoning in this book, and in the authors' more advanced book, *Statistical Inference*, that an analysis of this table seems desirable. An additional reason for the discussion which follows is that it gives an indication of how a table can be analyzed.

First, we may note that although the entire group is divided into equal subgroups by the score of 53 in the midterm test, the same is not true of the sections separately. In Section I three fifths of the students scored 53 or higher, but in Section II fewer than half, and in Section III only a third scored 53 or higher.

It is natural for the reader to wish to account for the differences among the sections. A first speculation might be that instruction was most effective in Section I and least in Section III. However, before accepting this explanation, it is necessary to look at the data on the prognostic test of arithmetic. Interestingly enough here also there are differences between the sections which parallel the differences observed in the other half of Table 3-4. In Section I three fourths were among the highest half of the entire group, but in Section II the corresponding fraction is two fifths, and in Section III it is only one fourth. It seems logical to assume that the observed differences in achievement of sections in the course may be partly or wholly accounted for by differences in mathematical ability of the students in these sections.

The relation between midterm score and arithmetic score for these 98 students and the extent to which the former score can be predicted from the latter are taken up in Chapters 12 and 13.

In practice, data like those in Table 3-4 would be treated by more powerful statistical methods than those exhibited in the table. However, a tabulation like that appearing in this table has various uses. It is helpful to the investigator in showing him direction for more extensive research. It is also helpful in displaying to readers who are not technically trained the relationships discovered by technical means.

Summary of the Structural Elements of a Table.   It seems desirable at this point to bring together in summary form a description of the structural elements of the table. These structural elements are illustrated in relation to Table 3-4 in Figures 3-2, 3-3, 3-4, and 3-5.

Table 3-4   Number of Students in Each Section Above or Below the Middle of the Entire Group on Each of Two Tests

A First Course in Statistical Methods, Teachers College, Columbia University

		NUMBER OF STUDENTS SCORING					
		ON MIDTERM TEST			ON ARITHMETIC TEST		
SECTION*	SEX	BELOW 53	53 OR HIGHER	TOTAL	BELOW 33	33 OR HIGHER	TOTAL
I	Men Women Both						
II	Men Women Both						
III	Men Women Both						
All sections	Men Women Both						

HEADING

STUB

Field

* FOOTNOTE → Two cases with section unidentified have been omitted.

Figure 3-2   Major Parts of a Table

Such elements are as follows:

1. The heading.   The portion of the table which includes the table number, the title, the subtitle, if any, the headnote, if any, the column headings, and column spanners.

(a) *Table number.*   An identifying numeral placed at the left of the title or above it. If there are several tables in the same study, the tables should be numbered in sequence, either through the entire study or through each chapter. Reference to a table should usually be by its number. In referring to a table in the text, it is better to say "In Table 15—" than to say "In the following table—" or "In the preceding table—." In preparing a manuscript the writer cannot be sure what will be the exact position of a table in the final printed version—whether it will actually follow or precede the reference—but its number is clear and unambiguous.

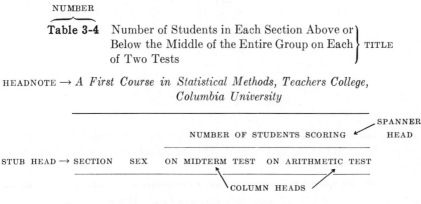

Figure 3-3   Parts of the Heading

(b) *Title.*   A clear description of the content of the table in as concise form as possible. It should help the reader understand quickly what the table is about. To begin the title with unnecessary words, such as "Table showing—" or "A comparison of the numbers of—" blunts its effect.

(c) *Subtitle.*   A second title used to supplement a briefer main title. Sometimes a table has a short display title and a longer, more explanatory, main title. Sometimes in a series of similar tables the titles might be so much alike as to be confusing. To distinguish them, each may be given a short main title calling attention to what is unique in the specific table and a subtitle which is similar for all tables in the series. The distinction between a subtitle and a headnote is not very important.

(d) *Headnote.*   A statement or phrase below the title which serves to qualify the title or to provide information relating to the table as a whole.

It may show units when all the units in the table are alike; it may show the source of the data. Not every table needs a headnote.

(e) *Column heading.* The descriptive title for all data in the column which it heads. Ideally, the column heading or the spanner heading should *name* the tabular entries in the column.

SECTION	SEX	
I	Men	← LINE CAPTION
	Women	
	Both	← SUBTOTAL
II	Men	
	Women	
	Both	
III	Men	
	Women	← Block
	Both	
All sections	Men	
	Women	
	Both	← Grand total

(SEX ← STUBHEAD)

Figure 3-4 Parts of the Stub

(f) *Stubhead.* The column head or caption of the stub, used to describe the stub listing as a whole. This is sometimes considered part of the stub.

(g) *Spanner head.* A classifying or descriptive caption spreading across and above two or more column headings, and applying equally to all columns thus covered.

2. The stub. A description of the row classification appearing in the left-hand column. The stubhead is sometimes considered as part of the stub even though it is physically placed in the heading.

(a) *Line caption.* A descriptive title for the basic data appearing on a line. As a line is also called a row, the line caption is also called row caption.

(b) *Subhead.* A title which serves as a heading for several line captions.

(c) *Caption for summary line.* A description of the group to which the summary applies. When the summary is merely the sum of several preceding entries, this caption is usually "Total" as in Tables 3-1 and 3-2. When, for a combined group, the summary gives a percent or an average or similar measure *not* obtained by adding related measures for the component groups, as in Table 3-3, the term "Total" does not apply, but instead some such phrase as "Combined group," "All cities," "All ages," or the like is used.

(d) *Block.* A segment of the stub consisting of several related line captions and subheads.

3. The field. The part of the table extending downward from the heading to the bottom rule of the table, and to the right of the stub. It contains the tabulated data of the table.

(a) *Cell.* The space allotted for an entry. It lies at the intersection of a column and a row.

(b) *Tabular entry.* The information appearing in a cell of the table.

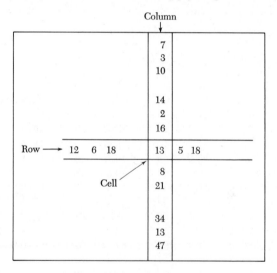

Figure 3-5   Parts of the Field

(c) *Line.* A horizontal array of cells with a common classification.

(d) *Column.* A vertical array of cells with a common classification.

4. The footnote. Information needed for interpreting a table which is placed in a note directly under the table. This may be information relating to the table as a whole, such as a statement concerning its source. A footnote may furnish an explanation of symbols used if their meaning is not obvious. When specific entries in a table are explained by footnotes, reference is usually made by means of such diacritical marks as *, **, ≠, or #, or by means of small letters a, b, c, ... Numbers are not used for such reference because they might be mistaken for exponents. The tabular footnote does not become a part of the general system of footnotes used in the book or monograph. A footnote to a table is not placed at the bottom of the page but directly beneath its table.

Tables with Averages as Entries. Tables 3-5 and 3-6 are presented as examples of tables in which the entries are neither frequencies nor percents, but averages. Tables in which the tabular entries are standard deviations, percentiles, correlation coefficients, or other summary measures occur in many scientific journals and also in later chapters of this text.

The figures in Table 3-5 are predictions, based on evidence available in the year indicated, as to what would be the average number of years

Table 3-5   Expectation of Life at Birth in the United States

(in Years)

	WHITE			NONWHITE			ALL RACES		
YEAR	MALE	FEMALE	BOTH SEXES	MALE	FEMALE	BOTH SEXES	MALE	FEMALE	BOTH SEXES
1900	46.6	48.7	47.6	32.5	33.5	33.0	46.3	48.3	47.3
1910	48.6	52.0	50.3	33.8	37.5	35.6	48.4	51.8	50.0
1920	54.4	55.6	54.9	45.5	45.2	45.3	53.6	54.6	54.1
1930	59.7	63.5	61.4	47.3	49.2	48.1	58.1	61.6	59.7
1940	62.1	66.6	64.2	51.5	54.9	53.1	60.8	65.2	62.9
1950	66.5	72.2	69.1	59.1	62.9	60.8	65.6	71.1	68.2
1951	66.5	72.4	69.3	59.2	63.4	61.2	65.6	71.4	68.4
1952	66.6	72.6	69.5	59.1	63.8	61.4	65.8	71.6	68.6
1953	66.8	73.0	69.7	59.7	64.5	62.0	66.0	72.0	68.8
1954	67.5	73.7	70.5	61.1	65.9	63.4	66.7	72.8	69.6
1955	67.4	73.7	70.5	61.4	66.1	63.7	66.7	72.8	69.6
1956	67.5	73.9	70.5	61.3	66.1	63.6	66.7	72.9	69.7
1957	67.2	73.7	70.3	60.7	65.5	63.0	66.4	72.7	69.5
1958	67.4	73.9	70.5	61.0	65.8	63.4	66.6	72.9	69.6
1959	67.5	74.2	70.7	61.3	66.5	63.9	66.8	73.2	69.9
1960	67.4	74.1	70.6	61.1	66.3	63.6	66.6	73.1	69.7
1961	67.8	74.5	71.0	61.9	67.0	64.4	67.0	73.6	70.2
1962	67.6	74.4	70.9	61.5	66.8	64.1	66.8	73.4	70.0
1963	67.5	74.4	70.8	60.9	66.5	63.6	66.6	73.4	69.9
1964	67.7	74.6	71.0	61.1	67.2	64.1	66.9	73.7	70.2
1965	67.6	74.7	71.0	61.1	67.4	64.1	66.8	73.7	70.2

SOURCE: This table is taken from page 94 of the 1967 *Life Insurance Fact Book*, Institute of Life Insurance. The Institute obtained the data from the National Center for Health Statistics, U.S. Department of Health, Education, and Welfare.

Table 3-6   Expectation of Life at Various Ages
in the United States 1965

(in Years)

AGE	WHITE		NONWHITE		ALL RACES
	MALE	FEMALE	MALE	FEMALE	
0	67.6	74.7	61.1	67.4	70.2
20	50.2	56.6	45.1	50.8	52.7
40	31.7	37.5	28.3	32.8	34.1
45	27.3	32.9	24.5	28.8	29.7
50	23.2	28.5	21.0	25.0	25.5
55	19.4	24.2	17.9	21.4	21.6
60	16.0	20.1	15.1	18.2	17.9
65	12.9	16.3	12.6	15.5	14.6
70	10.3	12.8	11.2	13.5	11.7

SOURCE: This table is taken from page 94 of the 1967 *Life Insurance Fact Book*, Institute of Life Insurance. The Institute obtained the data from the National Center for Health Statistics, U.S. Department of Health, Education, and Welfare.

in the life span for persons in the category indicated. Thus in 1900, it was predicted that for all white males born that year in the United States the average age at death would be 46.6 years.

The figures in Table 3-6 are predictions, based on evidence available in 1967, as to how many more years, on the average, persons of a given age might be expected to live. Thus it is predicted that for all white males aged 50 the average number of years before death would be 23.2 and, therefore, that for those men who have already survived to age 50 the average age at death would be 73.2.

EXERCISE 3-2

1. Locate and name the structural elements in Tables 3-5 and 3-6.

2. From Table 3-5 answer the following questions and indicate on what your answers are based:
   (a) By how many years did the average life span increase for white males during the period 1900–1965. For nonwhite males? For white females? For nonwhite females? Has the increase been greater for males or for females? For whites or nonwhites?
   (b) Was there any decade during this period in which the life expectation was greater for males than for females?

(c)   Has the discrepancy between the average life span of males and that of females increased or decreased during this period? Does your answer hold good for both whites and nonwhites?

(d)   Has the discrepancy between the average life span of whites and nonwhites increased or decreased during this period? Does your answer apply both to males and to females?

(e)   Would it be correct to use the word *Total* as heading for the summary columns here called *Both Sexes?* As a spanner heading instead of the phrase *All Races?*

(f)   Construct a new table in which the classifications by sex and by color are placed in the stub. Let the column headings be 1910, 1920, 1930, 1940, 1950, and 1960. Over these columns place a spanner head reading "Increase in Life Expectation During the Decade Ending". From the given data make the necessary computations to obtain the tabular entries for this new table.

   (1)   In which decade occurred the greatest increase in life expectation for white men? White women? Nonwhite men? Nonwhite women?

   (2)   For all whites, which decade showed the greatest increase? For all nonwhites?

3.   Find data in Table 3-6 from which to answer the following questions:

(a)   What is the prediction as to the average age at death of all babies born in the United States in 1965? Of all persons who are 20 years old in 1965? Of all persons who are 70 years old in 1965?

(b)   Is there any age at which the life expectation of women is less than that of men?

(c)   Is there any age at which the life expectation of white men is less than that of nonwhite men? Any age at which the life expectation of white women is less than that of nonwhite women?

4.   On a small slip of paper write for each student listed in Appendix Table VIII his number, his score in arithmetic, and his score on the midterm test. Sort these slips of paper into four classes to provide entries for the following tabulation:

	Arithmetic Scores	
	Below 33	33 and higher
Midterm test scores — 53 and higher		
Below 53		

Construct a formal table using the tabulation just completed. Name the structural parts of the table. Discuss the significance of the contents of the table.

4

DATA ON A SCALED
VARIABLE: TABULATING
AND GRAPHING

This chapter and the three which follow it will be devoted to problems
of summarizing and analyzing data on a scaled variable. The most elementary of those procedures, tabulating and graphing, are the subjects
of this chapter.

Several of the frequency distributions in Chapter 3 relate to data
classified in unordered categories. Thus in Table 3-1, students are classified in three sections. True, these sections are labeled I, II, and III, but
the numerals are merely names, and the sections might as well have been
called A, B, and C. In other tables in Chapter 3, students were classified
according to sex as well as according to class section; these are unordered
classes.

For a scaled variable, the values have a necessary order and numerical
relation to each other.

Frequency Distribution on a Scaled Variable. A tabulation of data
on the scaled variable age is shown in Table 4-1. The original records in
the decennial census were for age at the last birthday, but here, for convenience of presentation, the original data have been grouped in larger
intervals. Each interval except the highest is an interval of 5 years. No
upper limit is stated for the highest interval, and so it is called an *open*
interval. Every individual—or subject—in the study is allocated to one
of these intervals which we call *class intervals*. Thus approximately
180,676,000 persons enumerated in the 1960 census were classified in the
age groups named. These classes are called *mutually exclusive* because
it is impossible for any one subject to belong to more than one of them.

A set of mutually exclusive classes and the number of individuals belonging in each class constitute what is called a *frequency distribution*. The number of individuals in any one class is called a *frequency* and in this text will usually be indicated by the letter f unless it is named in words, as it is in Table 4-1. Table 4-1 presents five different frequency distributions all having the same set of classes.

Table 4-1   Population of the United States in 5 Census Years, Classified by Age

| Age in Years* | Number of Persons† Enumerated in the Census | | | | |
	1960	1950	1940	1930	1920
Total, all ages	179,324	150,697	131,671	122,775	105,711
Under 5	20,322	16,164	10,542	11,444	11,573
5–9	18,659	13,200	10,685	12,608	11,398
10–14	16,816	11,119	11,746	12,005	10,641
15–19	13,287	10,617	12,334	11,552	9,431
20–24	10,803	11,482	11,588	10,870	9,277
25–29	10,870	12,242	11,097	9,834	9,086
30–34	11,952	11,517	10,242	9,120	8,071
35–39	12,508	11,246	9,545	9,209	7,775
40–44	11,567	10,204	8,788	7,990	6,346
45–49	10,929	9,070	8,255	7,042	5,764
50–54	9,696	8,272	7,257	5,976	4,735
55–59	8,596	7,235	5,844	4,646	3,549
60–64	7,112	6,059	4,728	3,751	2,983
65–69	6,187	5,003	3,807	2,771	2,068
70–74	4,661	3,412	2,570	1,950	1,395
75 and over	5,359	3,855	2,643	1,913	1,470
Not reported	—	—	—	94	149

SOURCE: United States Census Report for 1960.

* As of last birthday at time of census enumeration.

† To the nearest thousand. For example, the total for all ages in 1950 was 150,697,361, which is approximately 150,697 thousands. Annexing 000 to each figure will give a good approximation to the original value.

In order to understand something of the contribution to statistical thought which a frequency distribution can make, it may be worthwhile to take a closer look at the entries in Table 4-1. Look first at the figures for 1920. Letting the eye run down that column, it will be noticed that the entry for any 5-year period selected is smaller than the entry for the preceding 5-year period. In other words, the number of persons living at a given age decreases as age increases. This phenomenon is a natural consequence of mortality. Of course, the frequency for the open period "75 and over" is larger than the frequency for the previous period, but this is understandable since "75 and over" covers more than a 5-year period.

None of the four later censuses shows this same steady decrease down the entire column. In 1930 there are fewer children under 5 than in the age group 5–9, obviously reflecting a lowered birth rate in the late 1920s, but after that youngest group, the number of persons living decreases as age increases.

The 1940 figures have been affected also by a second decrease in birth rate during the depression in the 1930s, and so we have the remarkable phenomenon of the number of living persons actually increasing over the first four age groups. The sharp rise in birth rate at the close of the Second World War is reflected in the 1950 figures in the large frequencies for the two youngest age groups. The three groups between ages 10 and 25 are still showing the effect of the lower birth rate in the earlier decades.

Reading across the table, comparing frequencies in the same row, is also instructive. Although the total population of the country increased every decade, the number of children under 5 decreased from 1920 to 1940 not only in proportion to the population but in actual numbers. The entries in rows near the bottom of the table show striking evidence of our aging population. The number of persons aged 75 and over has increased much faster than the total population. The total population increased about 70% from 1920 to 1960: $179,324/105,711 = 1.70$ while the number in this oldest age group increased about 3.65%: $5,359/1,470 = 3.65$.

Construction of a Frequency Distribution.   The frequency distributions in Table 4-1 were presented to us in completed form, and our role in regard to them has been more or less passive. But suppose we had the active task of constructing such a distribution, what problems would we face? It has already been said that a frequency distribution is the result of assigning each individual in the group studied to one of a set of mutually exclusive classes. Then, obviously, the problems are the selection of classes and the allocating of individuals to them.

The matter of allocating individuals to classes has already been discussed in Chapter 3 and need not be treated further here. For the census data we have been discussing, a card was punched for each individual and the cards were sorted and tabulated electrically according to the age intervals or classes shown in Table 4-1. The number of cards in each age interval is the frequency in the interval.

We shall give some attention to the manner in which the intervals of a frequency distribution are set up. To accomplish this, it will be necessary to examine closely the nature of a scaled variable and the way in which it can be divided up to provide appropriate classes for a frequency distribution.

Symbols for Variables and Scores. In this book the term *variable* will be used to mean the set of all possible outcomes of an experiment. If the experiment is giving a test to a class, then the set of all possible scores obtained on that test is a variable. If the experiment is the inclusion on the test form of an item calling for the sex of the respondent, then the two responses "male" and "female" constitute a variable.

Variables will be named by capital letters, usually but not necessarily, X or Y. Suppose that the possible scores on a test run from 0 to 100; then the set of all these numbers might be denoted by capital X and a specific score by lower-case x. Thus a lower-case letter will represent a specific number, and the corresponding capital letter will represent a collection of numbers.

A subscript may be used to associate a score with an individual. Thus x_1, x_2, and x_3 might denote the particular scores made by individuals 1, 2, and 3 on variable X.

Sometimes it is convenient to use the same capital letter with different subscripts to name different variables. Thus X_1 might denote all possible scores on one variable, and X_2 all possible scores on another. Then x_1 would be an individual score on the first variable and x_2 an individual score on the second. In this case if it seems desirable to associate a score with the individual making it, a second subscript will be required: x_{14} might designate the score made by individual 4 on variable X_1 and x_{38}, the score made by individual 8 on variable X_3.

If several individuals have the same score, say x, the number of those individuals will be denoted as f or f_x. The total number of individuals participating in an experiment will be denoted by n. Thus n is the total of all the f's.

Discrete and Continuous Variables. Some scaled variables—such as the number of children in a family, the number of books loaned by a library in a week, the number of automobiles sold by a dealer in a specified period—can take values only at certain discrete points on a scale. Such variables have already been termed *discrete variables*.

Other scaled variables—such as age, weight, or temperature—are termed *continuous* because they can have values at every point on a continuous scale.

Obviously, not all the infinitely many actual values of a continuous variable can be reported. One of the first tasks confronting a person who is to gather data on such a variable is to decide in advance what values he will report. Thus in measuring weight he might decide to report no fractions of a pound, so that a child who weighed a little less than 52.5 pounds would be reported as 52. In this case 50, 51, 52, 53, 54, and so forth would be reportable values, whereas 52.5 would, by prior decision, not be a reportable value. However, if it were agreed to report to the nearest half pound, such values as 51.5, 52.0, 52.5, and 53.5 would be reportable values, whereas 51.25 would not be.

For the analytic procedures which follow, it will be convenient to treat all scaled variables as continuous. Consequently, a rather detailed discussion of the scale of a continuous variable will be presented.

The Scale of a Continuous Variable.   In order to graph the frequency distribution of a continuous variable or to make and report computations with scaled data, it is essential to understand the representation of scores on the scale of such a variable.

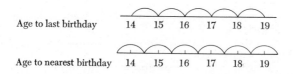

For the census data shown in Table 4-1, age was reported *to the last birthday.* Thus some of the persons reported as 16 years old have barely passed their sixteenth birthday, and others are on the verge of their seventeenth. Age is a continuous variable, but age reported by the census enumerator is a discrete variable, and only integers are reported. If an age scale is marked off in units of 1 year, for the census data or any data recorded as "age in years at last birthday," the single number 16 represents all ages in the interval from 16 to 17 with midpoint at 16.5. If age is reported "to the nearest birthday," the number 16 represents the interval from 15.5 to 16.5 with midpoint at 16.0. Many people would find great difficulty in deciding which is their nearest birthday, and therefore many studies of age call for the last birthday. Because both methods of reporting are in use, there is considerable ambiguity in age data unless an author tells exactly what a unit means. A column head should never be simply "age" or even "age in years," unless somewhere in title or subhead or footnote there is an explanation of where the year begins.

A year is sometimes indicated by its end point, as when a small child says he is "going on four" or an old man says, "I will be 95 at my next birthday." Because we name centuries by their end points, the century in which we are now living is called the twentieth. If a questionnaire item reads, "How many years have you been in your present position? Include the present year," each answer would represent a year interval named by its end point. Thus an interval named 6 would extend from 5 to 6, with midpoint at 5.5. The foregoing discussion is intended to emphasize the importance of thinking carefully about the meaning of scale units and giving the reader adequate information for interpreting them correctly.

For most scaled variables other than age, it has become customary to name a unit by the value of its midpoint. The width of the unit in which measurement is made indicates the degree of precision of the measurement—the narrower the unit, the more precise the measurement. Thus in measuring lines to the nearest inch, the line OA will be reported as 3 inches long; in measuring to the nearest half inch, as 3; in measuring to the nearest quarter, as $2\frac{3}{4}$.

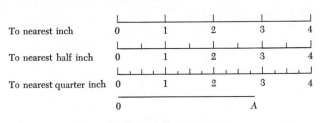

Scale: ½ inch = 1 inch

In measuring to the nearest inch, the number 3 represents a unit 1 inch wide extending from $2\frac{1}{2}$ to $3\frac{1}{2}$, with midpoint at 3.

In measuring to the nearest half inch, the number 3 represents a unit one-half inch wide, extending from $2\frac{3}{4}$ to $3\frac{1}{4}$ with midpoint at 3.

In measuring to the nearest quarter inch, the number 3 represents a unit one quarter of an inch wide, extending from $2\frac{7}{8}$ to $3\frac{1}{8}$, with midpoint at 3.

EXERCISE 4-1

1. Suppose that in measuring a continuous variable X the scores used in reporting measurements are as given below, each number being considered as representing the midpoint of a unit on the scale, exactly what portion of the scale is covered by the unit labeled 7? How wide is the unit?

(a)   5, 6, 7, 8, 9, 10, 11, 12.
(b)   3, 5, 7, 9, 11, 13, 15, 17, 19.
(c)   5.5, 6.0, 6.5, 7.0, 7.5, 8.0.
(d)   4, 7, 10, 13, 16, 19, 22.
(e)   6.8, 6.9, 7.0, 7.1, 7.2, 7.3.
(f) $5\frac{3}{4}$, 6, $6\frac{1}{4}$, $6\frac{1}{2}$, $6\frac{3}{4}$, 7, $7\frac{1}{4}$, $7\frac{1}{2}$.

2. Suppose a continuous variable X is measured with the degree of precision
shown below. Write five values of x representing five successive scale units,
and let 20 be the second of the five numbers.
(a)   0.5   Answer: 19.5, 20.0, 20.5, 21.0, 21.5.
(b)   0.1   .   20.0   .   .   . .
(c)   3.0   .   20.0   .   .   . .
(d)   0.01   .   20.0   .   .   . .

Grouping Scores into Wider Intervals.   In setting up Table 4-1, ages
originally recorded in units of 1 year have been combined into 5-year
intervals. Thus the ages originally recorded as 5, 6, 7, 8, or 9 have been
combined into an interval labeled 5–9. From the original meaning of the
year unit, it is clear that the interval 5–9 includes ages of children who
have passed their fifth birthday but not their tenth, and so on a linear
scale it represents an interval of width 5 extending from 5 to 10 with
midpoint at $7\frac{1}{2}$.

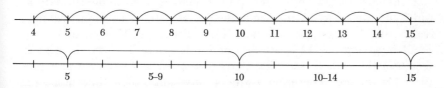

For almost any other type of scaled data, an interval 5–9 would repre-
sent an interval of width 5 extending from $4\frac{1}{2}$ to $9\frac{1}{2}$ with midpoint at 7.

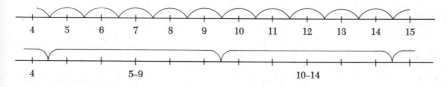

Examine the preceding sketch in which a scale unit is named by its
midpoint, and look at the interval which includes the five units, 10, 11,
12, 13, and 14; it has been named 10–14, and these numbers are called its
score limits. However, the interval actually extends from point 9.5 to
point 14.5, and the interval limits 9.5–14.5 are called *true limits* or *real*

limits. The midpoint of this interval is at 12, which is halfway between the real limits, $\frac{1}{2}(9.5 + 14.5) = 12$; and it is also halfway between the score limits, $\frac{1}{2}(10 + 14) = 12$. This midpoint is the *class value* or the *class index* or the *class mark.* The *size of the class interval,* or the *width of the interval,* is the difference between the real limits, $14.5 - 9.5 = 5.0$, and shows the *degree of precision* of the measurement. It is commonly denoted by the letter i (for interval). The class interval is also called the *step interval* or the *class sort.* Study the relation of these various terms as illustrated for several successive intervals:

Scores Included in the Interval	Real Limits	Score Limits	Class Index	i
20, 21, 22, 23, 24	19.5–24.5	20–24	22	5
15, 16, 17, 18, 19	14.5–19.5	15–19	17	5
10, 11, 12, 13, 14	9.5–14.5	10–14	12	5
5, 6, 7, 8, 9	4.5–9.5	5–9	7	5

Some authors prefer to name an interval by its real limits. However, that practice seems to invite mistakes in tabulation, as the person tabulating might place the number 14 in the interval marked 14.5–19.5 instead of in the one marked 9.5–14.5 where it belongs. This book will follow the practice of naming an interval by its score limits.

A careless mistake often made by beginners is to underestimate the width of interval by taking the difference between score limits instead of real limits. For the data we have been discussing, this mistake would result in saying $i = 14 - 10 = 4$, or $i = 9 - 5 = 4$. Look again at the sketch of the scale to see what parts of that scale $9 - 5 = 4$ and $14 - 10 = 4$ cover and to see exactly why i is not 4 but 5. The width of interval is, of course, also the difference between successive class indices which are 7, 12, 17, and 22.

EXERCISE 4-2

1. Draw a scale and mark on it the score points 6.0, 6.5, 7.0, 7.5, 8.0, 8.5, 9.0, and 9.5.
    (a) Draw a brace to represent the interval which includes scores 6.5, 7.0, and 7.5.
    (b) Fill in the blanks below:

Scores in Interval	Real Limits	Score Limits	Class Index	i
9.5, 10.0, 10.5	____	____	____	____
8.0, 8.5, 9.0	____	____	____	____
6.5, 7.0, 7.5	____	____	____	____

2. If each score is an integer and real limits are as indicated, fill in the blanks:

(a)
Score Limits	Scores in Interval	Real Limits	Class Index	i
17–19	____	____	____	____
14–16	____	____	____	____

(b)
Score Limits	Real Limits	Class Index	i
30–39	____	____	____
20–29	____	____	____
10–19	____	____	____

(c)
Score Limits	Real Limits	Class Index	i
70–76	____	____	____
63–69	____	____	____
56–62	____	____	____

3. If the score limits of two consecutive intervals are 29–31 and 32–34, what are the score limits of the next two higher intervals? What is the value of i?

Choice of Class Interval. The choice of interval depends partly on the unit in which the original observations were made and partly on the purpose for which the distribution is to be used. Inasmuch as the census data on age were recorded by the census enumerators in units of 1 year, no subsequent tabulation could possibly be made with intervals of, say, 1 month or 3 months or 18 months. It behooves any research worker to consider before he gathers data the use to which the data will be put and the degree of precision which will be desirable in later tabulations.

If considerable precision is required in analysis or computation, nothing is gained by reclassifying subjects into intervals larger than the units in which observations were first recorded. If great precision is not required, the use of larger intervals may simplify computation and clarify analysis. The presentation of a frequency distribution in a report is more effective if it can be so planned that the number of intervals is not too great and not too small. When the number of cases is small and the number of intervals is large, the distribution is likely to be irregular or ragged. By grouping cases in larger intervals or by obtaining a larger number of cases, the form of the distribution may be presented more vividly. However, the intervals can be made so wide that much of the available information is lost. If the number of intervals were reduced to the absurd extreme of having only one interval, we would lose all information about the form of the frequency distribution. In general, for purposes of effective presentation of data, it is a good plan to use between 10 and 20 intervals.

Combination of an odd number of units into an interval has the slight advantage of making the midpoint of the interval fall on one of the scores, whereas combining an even number makes the midpoint fall halfway between two scores. However, an interval of 10 units is generally preferred to one of 9 or 11 partly because tabulation by tens seems easy and natural. Intervals of 3, 5, or 10 units are most often used in reports.

The Cumulative Distribution. An important aid to interpreting a frequency distribution is the cumulative distribution which will now be illustrated from the 1960 data of Table 4-1. In Table 4-2 the column

Table 4-2 Cumulative Distribution of the Population of the United States in 1960, Classified by Age

Age in years*	Frequency†	Cumulative Frequency	Cumulative Percent
Under 5	20,322	20,322	11.3
5–9	18,659	38,981	21.7
10–14	16,816	55,797	31.1
15–19	13,287	69,084	38.5
20–24	10,803	79,887	44.5
25–29	10,870	90,757	50.6
30–34	11,952	102,709	57.3
35–39	12,508	115,217	64.3
40–44	11,567	126,784	70.7
45–49	10,929	137,713	76.8
50–54	9,696	147,409	82.2
55–59	8,596	156,005	87.0
60–64	7,112	163,117	91.0
65–69	6,187	169,304	94.4
70–74	4,661	173,965	97.0
75 and over	5,359	179,324	100.0
Total	179,324		

SOURCE: United States Census Report for 1960.

* To the last birthday.

† Rounded to the nearest 1000.

labeled *frequency* is identical with the corresponding column in Table 4-1 except for the position of the row for all ages. An entry in the frequency column shows the number of persons in the United States of the ages indicated in the given interval; an entry in the cumulative frequency shows the number of persons of those ages or younger. The entries in the cumulative-frequency column are obtained by the successive addition of the entries in the frequency column, beginning with the frequency in the lowest interval on the scale of age. Thus the cumulative frequency in the interval 5–9 is 20,322 + 18,659 = 38,981 and means that the number of persons who are 9 or younger—that is, persons who have not passed their tenth birthday—is approximately 38,981, to the nearest 1000. The next cumulative frequency is 38,981 + 16,816 = 55,797, and so on.

Entries in the cumulative percent column are obtained by expressing the corresponding cumulative frequencies as percents of the total. Thus for the interval 20–24, 79,887 ÷ 179,324 = .445, indicating that 44.5% of the population were not over 24 years of age. Obviously the last entry in this column must be 100%.

The cumulative distributions of Table 4-3 facilitate comparison of the 1920 and 1960 populations, but because the totals are different such comparisons can be most easily and meaningfully made after cumulative

Table 4-3 Cumulative Distribution of the Population of the United States in 1920 and in 1960, Classified by Age

Age in Years*	1920		1960	
	ƒ†	Percent	ƒ†	Percent
Under 5	11,573	11.0	20,322	11.3
5–9	22,971	21.8	38,981	21.7
10–14	33,612	31.8	55,797	31.1
15–19	43,043	40.8	69,084	38.5
20–24	52,320	49.6	79,887	44.5
25–29	61,406	58.2	90,757	50.6
30–34	69,477	65.8	102,709	57.3
35–39	77,252	73.2	115,217	64.3
40–44	83,598	79.2	126,784	70.7
45–49	89,362	84.7	137,713	76.8
50–54	94,097	89.1	147,409	82.2
55–59	97,646	92.5	156,005	87.0
60–64	100,629	95.3	163,117	91.0
65–69	102,697	97.3	169,304	94.4
70–74	104,092	98.6	173,965	97.0
75 and over	105,562	100.0	179,324	100.0

SOURCE: United States Census Report for 1960.

* To the last birthday.

† Rounded to the nearest 1000.

frequencies have been reduced to percents. Let your eye travel between the two cumulative percent columns, comparing the two figures for the same interval. After the first row, the percent for 1960 is smaller than for 1920, showing that for any age we choose the percent of persons younger than that age was greater in 1920 than in 1960.

The figures in Table 4-2 have been cumulated in the direction of highest age beginning with the interval for the lowest age. It may be equally interesting to begin with the interval in which age is highest. Thus for 1960 in the two intervals of highest age, the sum of the frequencies (in thousands) is 5,359 + 4,661 = 10,020; in the three intervals of highest age it is 10,020 + 6,187 = 16,207, and so forth. From such a cumulation we could learn that in 1920 the percent of the population 60 years or older was 7.5 and in 1960 that percent had risen to 13.

Graphs of the Frequency Distribution. Although there are many different ways of picturing a frequency distribution, certain devices contribute especially to the general understanding of the statistical measures; these will be discussed in later chapters. These devices are the histogram, the frequency polygon, the cumulative-frequency graph, and the cumulative-percent graph.

It will be convenient in discussing frequency graphs to use the conventional mathematical terms *ordinate* and *abscissa*. When a point P is located in reference to two axes at right angles to each other, the vertical distance from the horizontal axis to the point is called the *ordinate* of

that point, and the horizontal distance from the vertical axis to the point is called its *abscissa*. Together the ordinate and abscissa are called the *coordinates* of the point. The term abscissa originated in medieval surveying manuals written in Latin, in which that portion of a horizontal distance cut off by the shadow of a vertical staff was called the *pars abscissa* or the "part cut off."

1. *The Histogram.* Table 4-4 has been prepared from the data presented in Appendix Table IX. To prepare a histogram from the fre-

quencies in Table 4-4, lay off the scale of scores on a horizontal axis, the scale increasing from left to right. This scale must be marked in such a way that the division points between intervals can be identified. With infinitely many points on the scale to select from, a few must be chosen

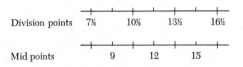

Division points   7½   10½   13½   16½

Mid points   9   12   15

to mark for identification. The most common methods of marking the scale are to show the values of the division points between intervals or the values of the midpoints of intervals as in the sketch above.

Table 4-4   Scores on Modern School Achievement
Test of Reading Speed Made by 109
Fourth-Grade Children

Score	f	Cumulative Frequency	Cumulative Percent
50–52	2	109	100%
47–49	2	107	98
44–46	7	105	96
41–43	7	98	90
38–40	6	91	83
35–37	4	85	78
32–34	5	81	74
29–31	14	76	70
26–28	18	62	57
23–25	16	44	40
20–22	7	28	26
17–19	5	21	19
14–16	7	16	15
11–13	3	9	8
8–10	5	6	6
5–7	1	1	1
Total	109		

It is most important to label the scale of scores properly. At right angles to the scale of scores, erect a scale of frequencies. *The zero point must be shown on the frequency scale* but not necessarily, or even usually, on the scale of scores. Now represent each of the 109 individuals in Table

4-4 by a small rectangle with horizontal length equal to one class interval on the scale of scores and vertical height equal to one unit on the scale of frequencies. This rectangle is an area unit which represents one frequency unit. In Figure 4-1 the 109 rectangles are shown but ordinarily

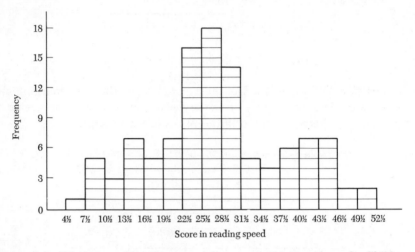

Figure 4-1   Histogram: Distribution of Scores of 109 Fourth-grade Children of City X on Modern School Achievement Test of Reading Speed (Based on frequency distribution of Table 4-4.)

only the column outlines need be drawn. Often even the vertical lines between columns are omitted.

In Figure 4-1 the frequency in an interval appears to be satisfactorily represented either by the area of a bar or by its height. However, in later sections we shall encounter situations (the frequency polygon in this chapter and the normal distribution in Chapter 14) where frequency cannot be represented by a vertical line but only by an area. It is wise to form the habit of thinking of frequency as represented by area.

2. *The Frequency Polygon.*   If we should try to compare two distributions by presenting two histograms on the same grid, the result would probably be a confusing mass of vertical lines difficult to identify. The picture is much easier to interpret if frequency polygons are used instead of histograms.

For a frequency polygon, the horizontal and vertical scales are laid off exactly as for a histogram. For each interval, a point is located directly above the middle of the interval so that its vertical distance represents the frequency in that interval measured on the frequency scale. In other words, the abscissa of the point is the score at the middle of an interval and its ordinate is the frequency in that interval. Note that if an interval has zero frequency, the point representing that frequency will lie on the horizontal axis, and also that there is always such a point at each end of the distribution. The points thus located are connected by straight lines to form a frequency polygon.

The relation between the histogram and the frequency polygon can be seen in Figure 4-2 where they have been graphed on the same axes.

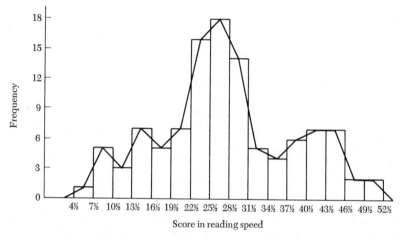

Figure 4-2   Frequency Polygon Superimposed on Histogram (From the data of Table 4-4. Compare with Figure 4-1.)

Examination of this figure shows that the areas of the two figures are exactly equal, because every triangle which is in the histogram but not in the frequency polygon is exactly matched by one which is in the polygon but not in the histogram. The fact that both figures represent frequency and that they have equal areas but not uniformly equal ordinates suggests that frequency is more satisfactorily represented by area than by ordinate in these two figures. Because the ordinate of a frequency polygon is related to frequency but does not actually measure frequency, the ordinate is often called the *frequency density*, and the vertical scale is called the *density scale*.

3.   *The Cumulative Graph.*   To draw a cumulative chart, the horizontal scale is laid off in score intervals as before, and a vertical scale is drawn to show cumulative frequency (or cumulative percent). Points are plotted very much as for the frequency polygon except that (a) it is cumulative frequency or percent instead of interval frequency that is plotted on the vertical scale, and (b) the points are not placed above the middle of an interval but above its upper real limit. The third and fourth columns of Table 4-4 show cumulative-frequency and cumulative-percent distributions, and Figure 4-3 shows the graph of the cumulative-percent distribution. The graph of a cumulative curve is usually a reflex curve,

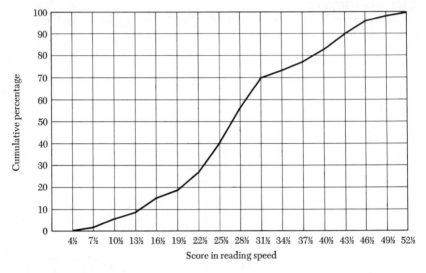

Figure 4-3 Cumulative-percent Curve: Percent of Children with Reading Speed Score Less than a Given Score (From the data of Table 4-3.)

with the center of curvature lying to the right in one portion of the curve and to the left in another portion. In architecture, an ogive arch (also called ogee) is one which has a reflex curve. Because cumulative-frequency or percent graphs are usually reflex, Francis Galton named them *ogives*, and the name is still used for this type of presentation even when the graph is not reflex.

EXERCISE 4-3

1. Make a graph for the cumulative-percent columns of Table 4-3, placing the data for 1920 and 1960 on the same axes. You will notice that these cumulative graphs do not have the usual ogive form.

2. Construct a table showing the frequency distributions of the artificial language scores for the three sections from the data of Appendix Table VIII. First, select an appropriate interval. Since the highest score is 60 and the lowest is 14, an interval of 3 units would give 16 or 17 intervals, and one of 5 would give 10 or 11 depending on where the interval starts.

3. From the table constructed in Question 2, construct three frequency polygons on the same base line. It might be noted in passing, how confusing it would be to try to present three histograms on the same base line.

4. From the table constructed in Question 2, construct and graph the three cumulative-frequency distributions.

5

PERCENTILES

Here and in Chapter 6 we shall consider methods of describing and comparing distributions in terms of certain typical scores and of describing the relative positions of individuals in those distributions. The methods considered in this chapter are based on the *order*, or the *ranking*, of scores in the distribution of scores.

Some very simple data will illustrate the basic ideas. A count has been made of the number of times a monkey changes his activity during a 20-minute period, with results as shown in Table 5-1 for three monkeys

Table 5-1 The Number of Times Each of Three Monkeys Changed His Activity During a 20-minute Period, Recorded for Each of Five Periods

| Monkey | *Number of Changes of Activity in Period* | | | | | Median | Range |
	1	*2*	*3*	*4*	*5*		
A	95	72	80	66	103	80	103 − 66 = 37
B	77	56	0	61	101	61	101 − 0 = 101
C	125	13	29	107	98	98	125 − 13 = 112

each observed over five periods. It would be convenient to have a single number to characterize each monkey, and this might be obtained by arranging the numbers for each monkey in ascending order and taking the middle one. Thus for A, the rearrangement is 66–72–80–95–103. The middle number, 80, describes the characteristic pattern of activity for monkey A and is called his *median* number. We can easily see that for these five periods C had the largest median, 98, and B the smallest, 61. We might also be interested in the consistency of the monkeys from period to period, as indicated by the range or the difference between their largest and smallest numbers. This range was only 37 for monkey A, indicating that he did not fluctuate much from one period to another and so the median gives a fairly good idea of his score in any selected period. For B and C, however, the numbers differ widely from one period to another, the ranges being, respectively, 101 and 112. If a sixth observation was to be made on each monkey, we could make a much better guess as to how A would come out than as to B or C. Thus each monkey has been characterized in terms of a typical score, the median, and the spread or variability of his scores, as indicated by the range of his scores.

If the number of records had been even, say, six instead of five, how would we define the median? Let us suppose that A's record in a sixth period was 69, so that his rearranged scores are 66–69–72–80–95–103. There is now no midscore, and any value between 72 and 80 will have half the values smaller and half larger than itself. Actually, any number between 72 and 80 will satisfy the definition of a median, as given below. However, in such a case it is customary to take an average and call $\frac{1}{2}(72 + 80) = 76$ the median.

The Median.   The three sections of a class in elementary statistical method, raw data for which are presented in Appendix, Table VIII have already received some attention. Table 3-4 on page 26 shows in a general way that Section I had the most and Section III the fewest members making high scores in the midterm test and also in the arithmetic test. Now we shall make a more explicit comparison of the three sections in terms of their midterm test scores.

From Table 3-4, we note that exactly half of the combined group of 96 students had scores of 53 or more, and half had scores of less than 53. The lower limit of the interval in which 53 stands—that is, 52.5—is, therefore, a point on the scale of scores having unique interest. It serves to describe this distribution by the fact that half the cases fall below and half fall above it. The score corresponding to this point is called *the median*, or the *fiftieth percentile*.

The median of a distribution is a value such that half of the scores in the distribution are larger and half are smaller than that value.

The medians of the three sections cannot be obtained from Table 3-4. To obtain them we must have the frequency distribution of each section, as shown in Table 5-2 without grouping, or in Table 5-3 with scores grouped in intervals of 5. We shall first compute the section medians from the ungrouped scores of Table 5-2 and then make more general computations from the grouped scores of Table 5-3.

Table 5-2   Distribution by Sections of Midterm Scores from
Appendix Table VIII

Ungrouped Frequencies

| | NUMBER OF CASES | | | | | NUMBER OF CASES | | | |
SCORE	*I*	*II*	*III*	TOTAL	SCORE	*I*	*II*	*III*	TOTAL
71	1	1		2	48				
70					47	4	1	4	9
69					46	1	1		2
68	1			1	45				
67					44	1	2	1	4
66					43	1	2		3
65	1			1	42				
64	2	1		3	41	1			1
63					40		1	1	2
62	5	3	1	9	39				
61	4	2	2	8	38			1	1
60					37			3	3
59	2	2		4	36				
58	5	1	3	9	35				
57					34				
56	1		2	3	33				
55	1	1	1	3	32				
54					31				
53	2	2	1	5	30	1	1		2
52	3	4	1	8	29				
51					28				
50	2	2	1	5	27			1	1
49	2		4	6	26				
					25			1	1
					Total	41	27	28	96

Let us first examine the distribution for Section III. Since $n = 28$ and $\frac{1}{2}n = 14$, the median is a score such that 14 scores are smaller and 14 larger. Counting up from the lower end of the distribution, we discover that there are 12 cases below the point 48.5 and 16 cases below 49.5. The median, therefore, is between 48.5 and 49.5. We may think of the 4 cases for which $x = 49$ as having scores not precisely identical but distributed evenly throughout the score interval. It turns out very neatly that 2 of

Table 5-3 Grouped Frequency Distribution by Sections of Midterm Scores from Table 5-2

| Score | Number of Cases | | | |
	I	II	III	Total
70–74	1	1		2
65–69	2			2
60–64	11	6	3	20
55–59	9	4	6	19
50–54	7	8	3	18
45–49	7	2	8	17
40–44	3	5	2	10
35–39			4	4
30–34	1	1		2
25–29			2	2
Total	41	27	28	96

the 4 may be considered as below and 2 above the midpoint, which is 49.0. Thus 49.0 is a point on the scale of scores such that $12 + 2 = 14$ scores are below it and $2 + 12 = 14$ are above it. The median of Section III is 49.

For Section I, $\frac{1}{2}n = \frac{41}{2} = 20.5$. Counting up from the lower end, we find that 20 cases have scores of 56 or less, and 25 have scores of 58 or less. The point we seek is in the interval from 57.5 to 58.5, where there are 5 cases. If the 5 cases are evenly distributed over that interval, we may picture them as in the following sketch.

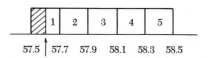

There are 20 cases below the beginning point of this interval; we are looking for a point below which 20.5 cases fall. Thus we want to divide the interval by a point such that the $20.5 - 20.0 = 0.5$ case may be represented as below (that is, to the left of) that point and the other $5 - 0.5 = 4.5$ cases above (that is, to the right). The point marked with an arrow in the adjacent sketch meets these requirements, and its score value is readily seen to be 57.6.

In similar fashion, the median of Section II is computed as 52.375 or 52.4. For a small group it is usually not important to carry the computation of the median to more than the first decimal place.

Go back now to Table 5-2, and place an asterisk in each column to show the location of the median for the section to which the column belongs. Clearly the larger the median of a group, the higher is the general placement along the scale of scores. The medians show a difference between Sections I and II which was not brought out by their ranges, inasmuch as in both sections the lowest score was 30 and the highest 71.

Computation of the Median from a Grouped Frequency Distribution. In many situations the original data are no longer available, and the statistician has to work from grouped frequencies. To illustrate this general procedure, Table 5-3 has been constructed by grouping the frequencies of Table 5-2 in intervals of 5. We shall now carry out in detail the computations for Section I and leave those for the other sections as an exercise for the reader.

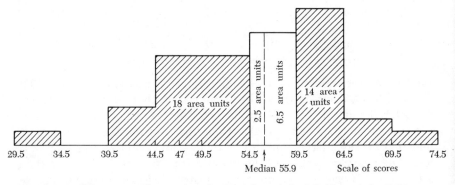

Figure 5-1   Histogram Illustrating the Position of the Median of the Midterm Scores of Section I (Data from Table 5-3.)

Reference to the histogram in Figure 5-1 will make the general process clear. As before, we are looking for a point on the scale of scores below which 20.5 cases will fall. In the histogram, frequency is represented by area, so the 18 cases below 54.5 are represented by 18 area units. The point sought must fall somewhere in the interval marked in Table 5-3 as 55–59, which is actually the interval from 54.5 to 59.5. This interval includes 9 cases, of which $20.5 - 18 = 2.5$ are to be to the left of the median. From the sketch below we see that $AB/AC = 2.5/9$ and, therefore, $AB = (2.5/9)$ (AC). Since $AC = 59.5 - 54.5 = 5$, or the width of the interval, the length of AB is $(2.5/9)$ $(5) = 1.4$. Thus the value attaching to the point B, which value is the median, is $54.5 + 1.4 = 55.9$. This result is not precisely identical with that obtained previously

Table 5-4   Procedure for Computing the Median

Step	General Statement	Application to Table 5-3
1	Find 50% of number of cases $= \frac{1}{2}n$	$\frac{1}{2}(41) = 20.5$
2	Find the interval which contains the median	Counting to 20.5 cases either from the top or the bottom brings us into the interval with limits 54.5 to 59.5
3	Find the number of cases below the true lower limit of the interval located in Step 2	Number of cases below 54.5 is 18
4	Subtract the number obtained in Step 3 from the number obtained in Step 1	$20.5 - 18 = 2.5$
5	Find the frequency in the interval which contains the median	$f = 9$
6	Find the ratio of the number in Step 4 to the number in Step 5	$\dfrac{2.5}{9}$
7	Find the width of the step interval	$i = 5$
8	Compute the median as follows: Lower limit of interval which contains the median plus the product of the ratio in Step 6 by the width of interval in Step 7	$\text{Median} = 54.5 + \dfrac{2.5}{9}\,5 = 55.9$

from the ungrouped scores, because we have here assumed that the 9 scores in the interval 54.5–59.5 are distributed evenly throughout the interval and such is not actually the case.

The procedure for computing a median may be schematically expressed as follows:

$$\text{Median} = \begin{bmatrix} \text{Lower limit} \\ \text{of interval} \\ \text{which contains} \\ \text{the median} \end{bmatrix} + \begin{bmatrix} \text{Width} \\ \text{of} \\ \text{interval} \end{bmatrix} \begin{bmatrix} \dfrac{\frac{1}{2}n - \begin{bmatrix} \text{Frequency below lower} \\ \text{limit of interval which} \\ \text{contains the median} \end{bmatrix}}{\begin{array}{c} \text{Frequency in interval} \\ \text{which contains the median} \end{array}} \end{bmatrix}$$

This procedure is displayed step by step in Table 5-4. It is an example of the general process of interpolation discussed in Chapter 14 of Walker, *Mathematics Essential for Elementary Statistics.*

Try your skill on the other distributions, computing their medians from the grouped frequency distribution of Table 5-3. The results should be as follows:

	I	*II*	*III*	*Total Group*
Median from ungrouped frequencies	57.6	52.4	49.0	52.5
Median from grouped frequencies	55.9	52.9	48.2	53.1

It is interesting to note that sometimes the grouping has increased, sometimes decreased, the median. Such changes due to grouping are called "grouping errors."

Other Percentiles.   The three sections have now been compared with respect to their central positions as measured by the median, but this tells only part of the story. The median alone gives no indication of the clustering or scatter of scores about this central value, and no basis for comparing distributions at points other than the central values.

Because 50 percent of the scores in a distribution are smaller than the median, the latter score is called the *fiftieth percentile*. By selecting other percents we can compute other percentiles. The only difference in the procedure is at the beginning. For example, to obtain, let us say, the 81st percentile, we start by finding, not $\frac{1}{2}n$, but 81 percent of n; to obtain the 13th percentile, we start by finding 13 percent of n. From that stage on, the procedure you have already learned for computing a median is followed exactly.

In this text we shall use the symbol $x_{.50}$ to denote the 50th percentile which is also the median, $x_{.81}$ to denote the 81st percentile, and so forth. There is no generally accepted symbol to represent a percentile, but these symbols are convenient and simple.

The 81st percentile of a distribution is defined as a *score such that 81 percent of the cases in that distribution have smaller scores.* In some situations this definition does not lead to a clear computational routine, and so it will be modified as follows: *The 81st percentile is the score corresponding to a point on the scale of scores such that 81 percent of the area of the histogram lies below that point.* These definitions may be adapted to any other percentile by substituting the appropriate number for 81.

We shall now show the computation of several selected percentiles for Section I. When several percentiles are to be computed for the same distribution, it is convenient to set up a cumulative frequency distribution, as in Table 5-5. To understand the following computations, you should

Table 5-5   Cumulative Frequencies for the Midterm Scores
of Section I, Taken from Table 5-3

Score	f	Cumulative f	Cumulative Percent
70–74	1	41	100.0
65–69	2	40	97.6
60–64	11	38	92.7
55–59	9	27	65.9
50–54	7	18	43.9
45–49	7	11	26.8
40–44	3	4	9.8
35–39		1	2.4
30–34	1	1	2.4
	41		

examine Table 5-5 to identify each number appearing in the computation. What is the 25th percentile?

$$25\% \text{ of } n = (0.25)(41) = 10.25$$

$$x_{.25} = 44.5 + \left(\frac{10.25 - 4}{7}\right)(5) = 49.0$$

$x_{.25}$ is also called the *first quartile* or the *lower quartile* and is sometimes denoted Q_1 or Q_L.

What is the 75th percentile?

$$75\% \text{ of } n = (0.75)(41) = 30.75$$

$$x_{.75} = 59.5 + \left(\frac{30.75 - 27}{11}\right)(5) = 61.2$$

$x_{.75}$ is also called the *third quartile* or the *upper quartile* and is sometimes denoted Q_3 or Q_U.

What is the 12th percentile?

$$12\% \text{ of } 41 = 4.92$$
$$x_{.12} = 44.5 + \left(\frac{4.92 - 4.0}{7}\right)(5) = 45.2$$

In each of the three illustrative computations note the fraction enclosed in parentheses. Its denominator is the frequency in the interval in which lies the particular score we are seeking. (Verify this from Table 5-5.) Its numerator is the additional frequency we need in order that the frequency below the score point sought will be the correct proportion of n. Thus the fraction is the ratio of one frequency to another frequency and is a *pure* number. That ratio is always a multiplier of the width of the step interval, which is 5 for the data we are using. The product of this ratio and the step interval gives us the distance which we must move into the interval along the scale of scores. Thus we have:

$$\begin{matrix} \text{Lower limit} \\ \text{of interval} \end{matrix} + \left(\text{Ratio}\right)\left(\begin{matrix} \text{Width of} \\ \text{interval} \end{matrix}\right) = \text{Percentile sought}$$

In addition to the terms *median, quartile,* and *percentile,* with which you are now familiar, you are likely to meet the term *decile.* The nine percentile values $x_{.10}, x_{.20}, x_{.30}, \ldots$ and $x_{.90}$ are also called the first, second, third, . . . , and ninth deciles. All these values belong to the same general family, and all of them may be called *quantiles.* Many attempts have been made to drop the first syllable of the term *percentile* on the ground that *centile* would be more consistent with *decile* and *quartile.* This change would be a very sensible one, but somehow it never catches on. Most people who have put any thought on the matter express approval of the shorter term but go on using the longer one. Not only were the terms median, quartile, percentile, and decile invented by Francis Galton before 1885 but also certain other related terms such as octile, dodecile, and permille, which have long since passed out of the statistician's vocabulary. The generic term *quantile* is of recent origin. The related terms "quartile" and "quarter" are often confused, as are also the terms "percentile" and "percent." The entries in the box, which is at the top of the following page, serve to distinguish the meaning of these terms.

Quartile versus Quarter

The upper *quartile* $x_{.75}$ is a score.
The upper *quarter* of the distribution is that fourth of the cases ($\frac{1}{4}n$) with scores above $x_{.75}$.

Percentile versus Percent

"Percentile" relates to a score.
"Percent" relates to a portion of frequency.

Computation of Percentiles for a Small Number of Scattered Scores.
Gaps in a small set of scattered scores sometimes produce a puzzling situation. For example, suppose a teacher has obtained scores for 20 students which when arranged in order are as follows:

6, 7, 9, 12, 12, 13, 15, 16, 16, 18, 19, 21, 22, 22, 24, 26, 27, 30, 32, 35

She wishes to know $x_{.75}$ so she computes 75 percent of 20 = 15 and looks for a score such that 15 cases have lower scores. Obviously, this condition is met by any value between 24.5 and 25.5. Some arbitrary decision must be made, and the one which seems most reasonable is to select a value 75 percent of the way from 24.5 to 25.5, namely, 25.25. In a similar manner, $x_{.15}$ would be computed as lying 15 percent of the way between 9.5 and 11.5, or as 9.8.

Variability. It is often as important to know something about the variability of a group as to know a measure of central position such as the median. Neither the median nor any single percentile can give any clue as to whether the scores in a distribution scatter widely or cluster closely about a typical value. Measures of variability are also called measures of *dispersion,* of *scatter,* of *spread,* or of *variation.*

The following examples show the importance of studying variability: The Chamber of Commerce of a community, which shall be nameless, advertised that the median annual temperature there was 70° (a true statement) and claimed that this implied a Bermuda-like climate. They failed to mention that sometimes the temperature rose to 110° in summer and fell to −20° in winter and that the number of days was small on which it actually hovered in the neighborhood of 70°. The median temperature would be considerably less important than the range for a person planning to move into the community.

Two classes in college physics contain the same number of students and have the same median score on a prognostic test, but the range of scores in one is three times as great as the range in the other. The problems of teaching these classes will be very different.

The people who live in village A and those who live in village B have almost the same median annual income. However, A is a very homogeneous community in which the difference between the highest income and the lowest income is small. In B, on the other hand, although there are a few middle income families, the population consists largely of two extreme groups, one composed of high-salaried business and professional people and the other composed of domestic servants and unskilled workers. To describe these communities, it is clear that information about median income needs to be accompanied by some measure of income dispersion.

The Range.   As a measure of the scattering, or dispersion, of the scores in a distribution, the range comes first to mind. The range is simply the difference between the highest score and the lowest. From Table 5-2, we see that the range for each section and for the total of all sections is

Section I	$71 - 30 = 41$
Section II	$71 - 30 = 41$
Section III	$62 - 25 = 37$
All sections	$71 - 25 = 46$

The range is meaningful but notoriously unstable, subject to the accidental presence or absence of extreme cases in the group studied. Note for yourself how the range for Sections I or II would have been affected if the person scoring 71, or the person scoring 30, or both, had dropped out of the course or changed to another section.

Interpercentile Range.   To obtain a measure of dispersion which is more reliable than the total range, it is common practice to report the difference between two symmetrically placed percentiles, such as $x_{.90} - x_{.10}$, which gives the range of scores of the middle 80 percent of the cases or $x_{.95} - x_{.05}$, which gives the range of the middle 90 percent. Most widely used of all such interpercentile ranges is the difference between the first and third quartiles, called the *interquartile range*. For Section I, $x_{.75} - x_{.25} = 61.2 - 49.0 = 12.2$. Interquartile ranges for the other sections will be found in Table 5-6, and it is interesting to see how similar the three sections are in regard to this measure of variability.

Relation of Quantiles to the Cumulative-percent Curve.   In Chapter 4, we became acquainted with the cumulative-frequency curve and the cumulative-percent curve. Now we shall put the latter to new uses, first

Table 5-6   Selected Quantiles Computed from the Data
of Table 5-3

	Value of the Quantile			
Quantile	I	II	III	All Sections
$x_{.95}$	66.9	64.2	62.2	64.3
$x_{.90}$	64.0	63.1	59.8	63.1
$x_{.80}$	62.1	60.8	57.3	60.7
$x_{.75}$	61.2	59.7	56.2	59.5
$x_{.70}$	60.3	58.1	55.0	58.2
$x_{.60}$	58.2	54.8	50.8	55.7
$x_{.50}$	55.9	52.9	48.2	53.1
$x_{.40}$	53.4	51.2	46.5	50.4
$x_{.30}$	50.4	49.6	44.8	47.7
$x_{.25}$	49.0	46.4	42.0	46.3
$x_{.20}$	47.5	43.9	39.0	44.9
$x_{.10}$	44.6	41.2	35.5	40.3
$x_{.05}$	41.2	39.8	28.0	35.5
Range of				
middle 50%	12.2	13.3	14.2	13.2
middle 60%	14.6	16.9	18.3	15.8
middle 80%	19.4	21.9	24.3	22.8
middle 90%	25.7	24.4	34.2	28.8
entire group*	45.0	45.0	40.0	50.0

* Upper limit of top interval minus lower limit of bottom interval.

for determining the quantiles graphically and later for comparing two or
more groups.

The relation between the first and last columns of Table 5-5 is displayed graphically in Figure 5-2. Such a graph, in which the scale of
scores is shown on one axis and the cumulative percent of the frequency
below a given score is shown on the other axis, is called an *ogive*. The
scale of scores may be placed on either axis. However, when two or more
groups are to be shown together (as in Figure 5-3) and the scale of scores
is placed on the horizontal axis, the group with higher scores is represented by the lower line; that effect tends to confuse a person not thoroughly familiar with graphs. Therefore, when showing such a graph to
persons to whom this type of presentation is new—as perhaps to a school
board or a parents' organization—less explanation will be required if the
scores are placed on the vertical rather than the horizontal axis.

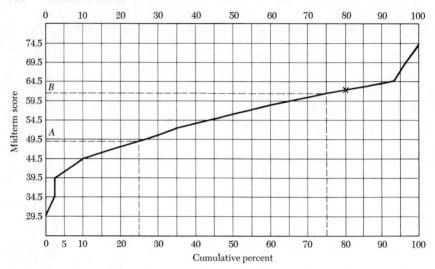

Figure 5-2 Cumulative-percent Curve for the Midterm Scores of Section I (Data from Table 5-5.)

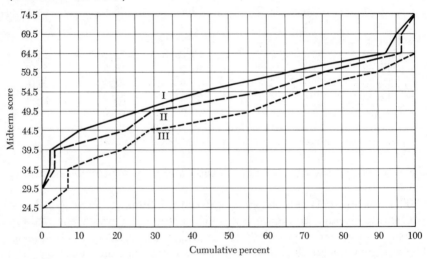

Figure 5-3 Cumulative-percent Curves for Midterm Scores for Three Sections (Data from Table 5-3.)

As this graph was constructed, the scale of scores is on the vertical axis; plotted points lie on horizontal lines through those score points which mark the boundaries between consecutive intervals. Now let us plot on the same grid the percentile data for Section I shown in Table 5-6. The subscript for one of the percentiles will indicate a point on the horizontal axis and the value of that percentile will indicate a point on the vertical axis. Consider, for example, $x_{.80} = 62.1$. The pair of values 80 and 62.1 is represented by a small cross which falls exactly on the ogive previously drawn. In a similar fashion, plot the values $x_{.10}$, $x_{.20}$, and $x_{.30}$ in relation to the respective percents 10, 20, and 30. Each of the plotted points should lie exactly on the ogive. In drawing the graph, points were located by finding the cumulative percent corresponding to given score points. Points are now being located by finding scores (that is, percentiles) corresponding to given cumulative percents. Points located in either way fall on the same line of relation.

Now use Figure 5-2 to determine graphically some of the other entries in Table 5-6. For example, to find $x_{.25}$ locate the 25-percent point on the base line, draw a vertical line from it to the curve, then a horizontal line which meets the scale of scores at the point A. The scale value of A is $x_{.25}$. Verify that the point B represents $x_{.75}$. Then the segment AB represents the range of the middle 50 percent or the interquartile range. Estimate the length of AB as accurately as possible and compare that estimate with the computed value shown in Table 5-6.

Use of Ogives for Comparing Groups. The ogive for each of the three sections is shown in Figure 5-3, drawn in the same way in which the ogive for Section I was drawn in Figure 5-2. It will be noticed that the percentile values for the three Sections found in any one row of Table 5-6 can be read in this figure on a vertical line through the appropriate cumulative-percent point. It will also be noticed that the graph immediately conveys the correct impression—that Section I has the highest scores and Section III the lowest. If the axes were reversed, this relation would not be so immediately obvious.

We have already seen in Table 5-6 on page 63 that the median of the combined group was 53.1. Find this point on the scale of scores, and draw through it a horizontal line cutting each of the three section curves. On the horizontal scale, read the values of the corresponding percents. These appear to warrant the statement, "The percent of cases with scores below the median of the combined group is 39 for Section I, 51 for Section II, and 65 for Section III." If computed from the ungrouped data of Table 5-2, the exact percents are $\frac{16}{41} = 39$ percent for Section I, $\frac{14}{27} = 51.8$ percent for Section II, and $\frac{18}{28} = 64.3$ percent for Section III.

EXERCISE 5-1

1. By computing from the data of Table 5-3, verify enough of the entries in Table 5-6 to assure yourself that you have no difficulty in computing percentiles.

2. The combined group has a greater total range than any of the sections.
   (a) Could it possibly have a smaller range than one of the sections?
   (b) In terms of the other four measures of spread, does the total group appear to be more variable than any one of the three sections?

3. Could the values for the total group in Table 5-6 be found exactly by adding the corresponding values for the sections and dividing by 3? Are the values for the total group exactly equal to the median of the three corresponding values for the sections?

4. Which section appears least variable if you base your opinion on the total range? Which if you base it on the interquartile range? On the value of $x_{.80} - x_{.20}$? On the value of $x_{.90} - x_{.10}$? On the value of $x_{.95} - x_{.05}$?

5. What is the ninth decile for the total group? Is the ninth decile a score or a portion of the frequency? Then, would you say that a particular person has a score *above* the ninth decile or *in* the ninth decile?

6. In Section I, does the median lie nearer to $x_{.75}$ or to $x_{.25}$, or is it exactly halfway between them? In Section II? In Section III? In the combined sections?

7. In Figure 5-2, locate $x_{.90}$ and $x_{.10}$ on the vertical scale, and mark the line segment which represents the range of the middle 80 percent of the frequency.

8. Make a graph for either Section II or Section III or the combined group, similar in plan to Figure 5-2.

9. The medians of four groups are respectively: Group A, 38.6; Group B, 49.2; Group C, 24.7; Group D, 41.3. Arrange the four letters in the order in which the groups appear to stand on the scale of scores, placing first the group with highest standing.

10. The percents of individuals with scores below 85 in five groups are as follows: Group A, 23 percent; Group B, 36 percent; Group C, 18 percent; Group D, 76 percent; Group E, 53 percent. Arrange the five letters in the order in which the groups appear to stand on the scale of scores, placing first the group with highest standing.

11. If 28 percent of Group A have scores below the median of Group B, which group appears to stand higher on the scale of scores?

12. In the sketch below, one tenth of the area of the histogram lies to the left of the line BN and one tenth to the right of the line GI. The line EK divides the area in half. One fourth of the area is to the left of CM and one fourth to the right of FJ. The point D is at the middle of the interval having the greatest frequency. Answer each of the following questions by naming some part of the sketch. If the part is a line segment, name it by the letters on the ends, as "Line AB." If it is an area, you can name it by several letters, as "Area $BCMN$" or by such a phrase as "Area between BN and CM." If the

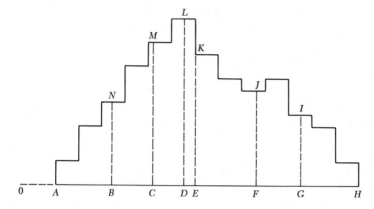

part is a point, name it by a single letter. You may assume that the zero point of the scale is at 0.

What represents:

(a)   The median?

(b)   The mode (= midpoint of interval having greatest frequency)?

(c) $x_{.75}$?

(d) $x_{.25}$?

(e)   The interquartile range?

(f)   The middle half of the cases?

(g)   The upper 10 percent of the cases?

(h)   The tenth percentile?

(i) $x_{.90} - x_{.10}$?

13.   From the sketch below, answer the following questions:

(a)   On which axis are percentile values measured?

(b)   What name would you give to the score of point A? Of B?

(c)   How many different names can you think of for point C?

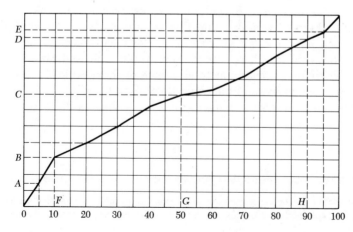

(d) What does the segment AE represent? The segment BD?

(e) What does the segment FH represent? The segment OG?

Comparisons among Individuals within a Group. Suppose a very important position is to be filled and the committee responsible for the appointment has narrowed its choice to five persons each of whom has different strengths and weaknesses. The committee finds itself unable to give *scores* to the candidates but after long deliberation places the names in an order of preference. In other words, it *ranks* the candidates.

There are many such situations in which it is not feasible to *measure* subjects but very feasible to *rank* them. For this purpose it is not necessary to know by how much one subject exceeds another but only to know that he does exceed him. There are other situations in which, although it is possible to obtain scaled measures, it is preferable to use ranks.[1] It is common practice to assign rank one to the subject standing highest in the characteristic under consideration and to assign succeeding ranks in descending order. That practice will be followed in this text. However, the reader should be warned that several recent writers on mathematical statistics reverse the order, using one for the lowest rank.

Changing Scores to Ranks. To transform scores to ranks presents no difficulty unless there are ties—that is, two or more individuals with the same score. For example, suppose we have six individuals with scores as indicated:

Subject:	A	B	C	D	E	F
Score:	12	9	8	11	9	10

We assign rank 1 to A, because he has the highest score, rank 2 to D and rank 3 to F. Now B and E are tied and must receive the same rank, but, if we give them both rank 4 and then give C rank 5, the group of individuals will have no one ranking sixth. If it had been possible to distinguish between B and E, one of them would have had rank 4 and one rank 5. Therefore, it is proper to give each of them rank $\frac{1}{2}(4 + 5) = 4.5$ and to give the individual following them rank 6.

When n individuals are ranked, the sum of their ranks is always

(5-1)
$$\frac{n(n + 1)}{2}$$

and this fact may be used as one check on the correctness of the ranks

---

[1] Most of the reasons why ranks would be used in preference to scaled scores cannot be clearly explained at this stage. The reader is referred to more advanced texts.

which have been assigned. Thus, for 6 individuals, the sum of the ranks should be $6(7)/2 = 21$. Add the ranks indicated below to see that their sum is 21.

Subject:	A	B	C	D	E	F
Score:	12	9	8	11	9	10
Rank	1	$4\frac{1}{2}$	6	2	$4\frac{1}{2}$	3

EXERCISE 5-2

For each of these sets of scores, change the scores to ranks, and check the sum of the ranks by formula (5-1):

1.

Subject:	A	B	C	D	E	F	G
Score:	25	21	18	24	20	21	21

2.

Subject:	A	B	C	D	E	F	G	H	I
Score:	6	9	13	6	14	8	12	5	2

3.

Subject:	A	B	C	D	E	F	G	H	I	J	K	L
Score:	12	8	7	10	6	7	9	7	13	8	14	5

4.

Subject:	A	B	C	D	E	F	G	H	I	J	K	L	M	N
Score:	5	4	7	9	8	5	3	5	9	12	8	2	13	6

Ambiguities in Ranks. Suppose your neighbor tells you that his son has written home that he stands tenth in his freshman class in college; should you congratulate or commiserate? The meaning of this rank of 10 depends on the size and the nature of the group with which he is being compared. The boy may be near the top of a class of several hundred students or near the bottom of a very small class. The college in which he is enrolled may have very strict or very loose admission requirements.

A measure of relative position must always derive part of its meaning from the nature of the group with which the individual is compared. A child of 10 may feel very tall and very mature in a group of younger children, but, if transferred to a group in which he is the youngest, he may feel very small and inexperienced. He has not changed, but the standard of comparison has changed. A measure of an individual's position relative to his group has obvious meaning for understanding persons and group behavior. Especially in the field of education and psychological testing scores are completely without meaning unless they can be referred to some standard, some bench mark. What does it mean to say merely that a student scored 29 on a test? Nothing at all. This may be a phenomenally high score or a phenomenally low one, or it may be the average for students of comparable age and training.

Now consider the ambiguity in ranks due to size of the group. Suppose John, whose height is 52 inches, ranks third as to height in a group of boys. Can we express his position in terms of the percent of boys who are shorter than he?

If there are 10 boys in the group, one might at first thought say that 7 boys, or 70 percent, are shorter and 2 boys, or 20 percent, are taller. This statement accounts for only 70 percent + 20 percent = 90 percent of the group, John himself constituting the other 10 percent.

If there are 50 boys in the group and John is third, one might on the same basis say that 47 boys, or 94 percent, of the group are shorter and 2 boys, or 4 percent, are taller. This statement accounts for 94 percent + 4 percent = 98 percent of the group, John himself constituting the other 2 percent.

These two statements are much more meaningful than the mere statement that John has rank 3. The statements can be further improved if we think of histograms for the two distributions of ranked scores. In the histogram for 10 scores, John's rank is at the seventy-fifth percentile of the distribution. In the histogram for 50 scores, John's score is at the ninety-fifth percentile. We can rephrase these statements conveniently by saying that John's score has a *percentile rank* of 75 in the first distribution and a *percentile rank* of 95 in the second.

Percentile Score and Percentile Rank.   If John is 52 inches tall and his height has third rank among 10 boys, the following statements may be made:

(a)   75 percent of the group have height less than John.
(b)   75 percent of the group have height less than 52 inches.
(c)   The 75th percentile of height for this group is 52 inches.
(d)   John's height is the 75th percentile.
(e)   John has percentile rank 75 in height.
(f)   The height 52 inches has percentile rank 75.

Study these statements carefully noting that they all express the same relationship but with different emphases, just as the two sentences "John is Mary's brother" and "Mary is John's sister" express the same relationship with different emphases. Note that statements (c) and (d) employ the term *percentile*, and statements (e) and (f) employ the new term *percentile rank*. Compare each of these with statements (a) and (b), noting that 52 inches, which is a score, is called the percentile, and 75 percent, which is a relative frequency, furnishes the percentile rank.

EXERCISE 5-3

Give two translations of each of these statements by filling blanks in the two sentences which follow the statement.

1. Exactly 57 percent of the teachers in a given school system receive salaries less than $7800.
   _____ is the _____ percentile of the distribution of salaries.
   _____ is the percentile rank of _____.

2. Of the deaths recorded from a given disease for a given year in a given community, 87 percent were of persons under 12 years old.
   _____ is the _____ percentile of the age of persons dying from this disease.
   _____ is the percentile rank of age _____.

3. Exactly one half of the scores in a given distribution are less than 72.0.
   The _____ percentile is _____.
   The percentile rank of _____ is _____.

4. Exactly 5 percent of all persons consulted had annual incomes larger than $10,850.
   The _____ percentile is _____.
   The percentile rank of _____ is _____.

Determining Percentile Rank from a Frequency Distribution and from a Graph.   What is the percentile rank in Section I of a student who scored 47? Since we have already available the ogive of Figure 5-2 on page 64, the easiest way to answer this question is to read the required percentile rank there. We locate 47 in the scale of scores, which is the vertical scale, and draw a horizontal line through that point to the curve. Then on the horizontal scale we read the corresponding cumulative percent. That cumulative percent is the percentile rank. In Section I the percentile rank of 47 is approximately 18. Note that in Figure 5-2 *percentiles (which are scores) are measured along the vertical scale, and percentile ranks (which are cumulative percents) are measured along the horizontal scale.*

The same result could, with a little more work, be obtained from Figure 5-1 on page 56 or Table 5-3 on page 55 if the ogive were not already drawn. Note the position of score point 47 on the scale of scores in the interval 45–49. We need to know what proportion of the frequency (that is, the area of the histogram) would lie to the left of this point. Either from the histogram or the table we can see that the number of cases below this point is $1 + 3 +$ part of the 7 cases in the interval 45–49. To decide how many of the 7 cases in the interval 45–49 should be considered as below 47, we may examine the sketch below. If x repre-

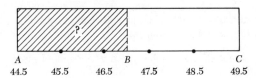

sents the number of cases in the given interval below the point B, then

$$\frac{x}{7} = \frac{AB}{AC} \quad \text{or} \quad \frac{x}{7} = \frac{47 - 44.5}{49.5 - 44.5} \quad \text{and} \quad x = 7\left(\frac{2.5}{5}\right) = 3.5$$

The number of cases below the point 47 is, therefore,

$$1 + 3 + \left(\frac{2.5}{5}\right)7 = 7.5$$

and the percent of cases below that point is $7.5/41 = 18.3$. The score 47, therefore, has percentile rank of 18 in Section I.

Another way to think of this computation is to note in Table 5-3 that 4 cases fall below the score point 44.5 and 11 below 49.5, and to interpolate to find the missing number:

Score Point	Number of Cases Below Score Point
49.5	11
47	?
44.5	4

What would the percentile rank of score 47 be in Section II?

$$1 + 5 + \left(\frac{2.5}{5}\right)2 = 7$$

and

$$\frac{7}{27} = 25.9 \text{ percent}$$

Its percentile rank would be 26.

What would its percentile rank be in Section III?

$$2 + 4 + 2 + \left(\frac{2.5}{5}\right) 8 = 12$$

$$\frac{12}{28} = 42.9 \text{ percent}$$

Its percentile rank would be 43.

Table 5-7 Procedure for Computing the Percentile Rank
of a Given Score

Step	*General Statement*	*Application to score 47 for Section I in Table 5-3 on page 55*
1	Find the interval in which the given score is located	Score 47 is in the interval with true limits 44.5 to 49.5
2	From the given score, subtract the lower limit of the interval in which it is located	$47 - 44.5 = 2.5$
3	Find the width of the interval	$i = 49.5 - 44.5 = 5.0$
4	Find the ratio of the number in Step 2 to the number in Step 3	$\dfrac{2.5}{5.0}$
5	Find the number of cases below the true lower limit of the interval located in Step 1	$1 + 3 = 4$
6	Find the frequency in the interval located in Step 1	$f = 7$
7	Compute the following: Value of Step 5 plus the product of the ratio in Step 4 by the frequency in Step 6	$4 + \left(\dfrac{2.5}{5.0}\right)(7) = 7.5$
8	Compute the percentile rank as the ratio of the value obtained in Step 7 to the total number of cases, expressed as a percent	$\dfrac{7.5}{41} = 18.3\%$

Note that the fraction enclosed in parentheses in each of these computations has for its denominator the width of the step interval and for its numerator the distance from the beginning of that interval up to the score point in question. The ratio of these two linear values is a *pure* number such that, when it is multiplied by the frequency in the interval, it gives us the portion of that frequency falling below the score point in which we are interested.

The calculation proceeds in two steps. First the frequency below the score point is computed thus:

$$\begin{matrix} \text{Frequency} \\ \text{below} \\ \text{ordinate} \\ \text{at score point} \end{matrix} = \begin{matrix} \text{Sum of} \\ \text{frequencies} \\ \text{in all lower} \\ \text{intervals} \end{matrix} + \left[\frac{\text{Score minus beginning point of interval}}{\text{Width of interval}} \right] \left[\begin{matrix} \text{Frequency} \\ \text{in} \\ \text{interval} \end{matrix} \right]$$

Then the percentile rank is computed as:

$$\text{Percentile rank} = 100 \left(\frac{\text{Frequency below score point}}{n} \right)$$

The procedure is set out step by step in Table 5-7.

EXERCISE 5-4

The following tabulation shows the percentile rank, for each section and for the combined group, of four scores. Verify these by computation based on Table 5-3 and by reference to Figure 5-3.

		Percentile Rank		
Score	I	II	III	Total Group
41	5	9	24	11
54	42	56	67	53
62	79	85	95	85
66	94	96	100	96

6

MEAN AND STANDARD DEVIATION

Chapter 5 was devoted to three problems: (*a*) how a frequency distribution can be characterized by a measure of position and a measure of variability; (*b*) how two or more distributions can be compared; and (*c*) how the relative position of individuals within a distribution can be described. This chapter will deal with the same three problems, the difference being that the measures used in Chapter 5 belong to the family of measures known as quantiles, all of which are based on some aspect of order or ranking, whereas in this chapter we shall use the mean as a measure of position and the standard deviation as a measure of variability. In Chapter 7, the two types of measures will be compared and criteria suggested for choice between them.

The Mean. Suppose five undernourished children are placed on an enriched diet, and during a given period of time make gains, measured in pounds, as follows:

John, 8; Sam, 2; Max, 3; Bob, 5; George, 12

What *average* gain might be reported? By the methods of Chapter 5, the median gain is immediately seen to be 5 lbs, and this is one type of average. Most people are also familiar with another type of average obtained by adding the scores of all the individuals and dividing by the number of individuals. This type of average is called the *arithmetic mean* (pronounced arithme'tic) or, usually, just *the mean*. The sum of the gains of these 5 boys is 30 and so the mean gain is $30/5 = 6$.

Symbols. To reduce the burden of writing long sentences, we shall employ a few convenient symbols. As before, we shall use the letter x to represent a score on variable X for any one of a set of n individuals. "The sum of" is customarily denoted by the capital Greek letter sigma, Σ, comparable to the capital S in our alphabet. Then Σx means "the sum of all measures represented by the letter x." Throughout this text n will be used to indicate the number of individuals under consideration. The mean will be denoted $\bar{x}$, and this symbol can be read either "mean of x" or "x bar" or "bar x." Note how vivid the definition of a mean becomes when translated into these symbols:

The mean	can be obtained by	finding the sum of all the scores	and dividing it by	the number of scores
$\bar{x}$	$=$	Σx	$\div$	n

(6-1)   or better $\bar{x} = \dfrac{\Sigma x}{n}$

This formula constitutes a definition of the mean.

EXERCISE 6-1

Find the mean and the median of each of these sets of scores:

1. 3, 2, 7, 12, 9, 6, 4, 5, 7, 2

2. 19, 12, 18, 8, 23, 13, 19

3. 415, 402, 401, 405, 409, 410, 416, 412, 414, 406

A Measure of Variability.   When the median is used as a measure of central position, variability can be measured by some interpercentile range such as $x_{.95} - x_{.05}$. However, when central position is measured by the mean, variability should be measured by a statistic based on deviations from the mean.

For each of the 5 boys mentioned at the beginning of this chapter, we might find the deviation of his score from the mean score, which is 6, $x - \bar{x} = x - 6$ and perform the following computations:

Boy	x	$x - \bar{x}$	$|x - \bar{x}|$	$(x - \bar{x})^2$
John	8	2	2	4
Sam	2	-4	4	16
Max	3	-3	3	9
Bob	5	-1	1	1
George	12	6	6	36
	30	0	16	66

The sum of the deviations from the mean is always identically zero no matter how great or how small the variability,

$$(6\text{-}2) \qquad\qquad \Sigma(x - \bar{x}) = 0$$

because positive and negative deviations from the mean exactly balance each other. Therefore, that sum provides no useful indication of variability.

We should like to have a statistic which will be zero only when there is no variability—that is, when all individuals have exactly the same score—and which becomes larger and larger as the spread among individuals is increased. To obtain such a statistic we have to get rid of the minus signs. Two methods of banishing them probably occur to you.

One method is simply to ignore the signs treating all deviations as absolute values. The customary symbol for the *absolute or unsigned value* of a number a is $|a|$. Thus $|-3| = |+3| = 3$. For the 5 individuals under consideration, the sum of the absolute values of the deviations from the mean is 16, and the mean of the unsigned deviations is $16/5 = 3.2$. This value is known as the *mean deviation from the mean*.

$$(6\text{-}3) \qquad \text{Mean deviation from the mean} = \frac{\Sigma|x - \bar{x}|}{n}$$

It tells us how much on the average the individuals in a group deviate from the mean. It is not easy to compute when the mean is not an integer or an easy fraction and when n is large. It has some other theoretical disadvantages and so is used only for very small samples, such as are sometimes drawn in industrial plants. We shall not consider it further in this text.

The other method of removing minus signs is to square each deviation and take the sum of the squared deviations. For our data, $\Sigma(x - \bar{x})^2 = 66$. However, this number reflects not only the variability of the group but also its size, and so we divide it by $n - 1$. The result is a measure of variability known as *the variance*.

$$(6\text{-}4) \qquad\qquad \text{Variance} = \frac{\Sigma(x - \bar{x})^2}{n - 1}$$

For our data, the variance is $\frac{66}{4} = 16.5$.

The computation of the variance for these five scores was particularly simple because the mean was an integer, and values of the deviations $x - \bar{x}$ could be expressed exactly. This pleasant situation will not usually obtain, and the method of computation we have used will involve large rounding errors unless deviations are carried to many decimal places. A formula equivalent to (6-4) but in most cases requiring less arithmetic work is

(6-5)
$$\text{Variance} = \frac{n\Sigma x^2 - (\Sigma x)^2}{n(n - 1)}$$

For our data, $\Sigma x^2 = 2^2 + 3^2 + 5^2 + 8^2 + 12^2 = 246$. Hence, the variance is $[5(246) - 30^2]/5(4) = 330/20 = 16.5$ as before.

Graphic Representation.   The variance is a measure of spread for which no graphic representation can be made. Do not try to form a visual image of it. Its square root, however, can be represented as a distance measured along the scale of scores. That square root is called the *standard deviation*. Later in this chapter, it will be used as a *standard* for measuring the *deviations* of individuals from the group mean, a use which appears to be responsible for its name. At present, however, we are considering its use as a measure of the variability of a group. The symbol[1] we shall use for the standard deviation is s.

(6-6)
$$s = \sqrt{\text{Variance}} = \sqrt{\frac{\Sigma(x - \bar{x})^2}{(n - 1)}}$$

In Figure 6-1 each of the five scores we have been discussing is represented by a point on the scale of scores. Each score has a frequency of 1,

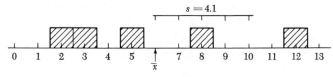

Gain in weight in pounds

Figure 6-1   Gain in Weight in Pounds Made by Five Boys. $\bar{x} = 6$ and $s = 4.1$

---

[1] When the term *standard deviation* was coined by Karl Pearson in 1894, he used the small Greek letter sigma (σ) and defined it as

$$\sigma = \sqrt{\frac{\Sigma(x - \bar{x})^2}{n}}$$

Today, some writers use n and some use $n - 1$ in the denominator. The use of n gives a biased estimate and the use of $n - 1$ an unbiased estimate of the variance in the population from which the sample was drawn. Today many writers prefer, so far as possible, to use English letters as symbols for statistics computed from a sample and the corresponding Greek letters for theoretical rather than observed values. Following this practice we shall use s to denote a standard deviation computed from observations, but in the chapter on the normal distribution we shall use σ for the standard deviation of that theoretical distribution. In Chapters 9 to 18, both symbols will be used.

and these frequencies are represented by the shaded areas. The mean is at the point 6. The standard deviation is the line segment 4.1 scale units long, which is marked s.

The standard deviation and the variance will be used in most of the statistical problems you are likely to find of interest. The feeling of strangeness which you now entertain in relation to them will soon wear off as you meet them in a variety of situations. The rest of this chapter and much of Chapter 7 will serve to give you experience with the standard deviation, but before you can use this statistic in an actual situation you need to be able to compute it from a frequency distribution.

Computation When There Are Duplicate Scores.   Now let us suppose that instead of 5 children there had been 25 and that several had the same score, as in Table 6-1. Here, there are 3 children with score 11, and

Table 6-1   Computation of the Mean and Standard Deviation of the Gain in Weight for 25 Children During a Specified Period (Artificial Data)*

x Gain in lb	f	fx	fx^2
12	1	12	144
11	3	33	363
10	4	40	400
9	1	9	81
8	5	40	320
7	2	14	98
6	4	24	144
5	3	15	75
4	2	8	32
	25	195	1657

$$\bar{x} = \frac{195}{25} = 7.8$$

$$\text{Variance} = \frac{25(1657) - 195^2}{25(24)} = \frac{3400}{600} = 5.67$$

$$s = \sqrt{5.67} = 2.4$$

* Note that the purpose of this table is not to display data but to illustrate a computational method, and for that reason a title has been used which would be inappropriate in a treatise emphasizing content.

so 11 must enter 3 times into Σx. An easy way to achieve this is to multiply 11 by 3 and similarly to multiply every other score by its frequency before adding. The resultant sum may be appropriately labeled Σfx. In the same manner, the 3 scores of 11 must contribute $11^2 + 11^2 + 11^2 = 3(121) = 363$ to Σx^2. An easy way to accomplish this is to set up a final column headed fx^2 and to obtain each entry in this column by multiplying an entry in the fx column by its corresponding x. Thus for $x = 11$, we have $11(33) = 363$ which is, of course, equal to $3(11^2)$.

Then for a distribution having duplication of scores, it is appropriate to insert an f in formulas (6-1) and (6-5), making them read

(6-7)
$$\bar{x} = \frac{\Sigma fx}{n}$$

(6-8)
$$\text{Variance} = \frac{n\Sigma fx^2 - (\Sigma fx)^2}{n(n-1)}$$

Even when the f is not written in the formula, it is always understood.

EXERCISE 6-2

1. Make a graph of the histogram for the distribution of Table 6-1. On the scale of scores mark the position of $\bar{x} = 7.8$, $\bar{x} - s = 5.4$, $\bar{x} - 2s = 3.0$, $\bar{x} + s = 10.2$, and $\bar{x} + 2s = 12.6$. Draw a line segment of length $s = 2.4$.

2. Compute the percentile rank of score point $\bar{x} + s = 10.2$.

3. Compute the percentile rank of score point $\bar{x} - s = 5.4$.

4. Compute $x_{.50}$. For these data it happens that $\bar{x}$ and $x_{.50}$ are identical.

5. For these data, the range is equal to how many standard deviations?

6. Compute the median, mean, variance, and standard deviation of each of the following sets of scores:
   (a) 7, 15, 10, 9, 4, 3, 7, 10, 10, 8, 16
   (b) 24, 27, 22, 28, 26
   (c) 1, 2, 5, 6, 10, 11, 12, 14, 15, 15
   (d) 1, 9, 0, 1, 5, 1, 0, 11
   (e) 2, −3, 7, −9, +8
   (f) 9, −7, 0, −4, −6, 3, −9

7. For each of the distributions in the preceding question, what is the ratio of the range to the standard deviation?

8. From the values of $\bar{x}$ and s obtained in Questions 6(d), 6(e), and 6(f), would you say it is possible for s to be larger than $\bar{x}$?

Computation from an Arbitrary Origin.   Suppose it is necessary to find the mean of the 5 numbers 27309, 27321, 27305, 27299, and 27296. No doubt many people would compute $\Sigma x = 136530$ and $\bar{x} = 136530/5 = 27306$. However, some of you who are more inventive may say to yourselves, "Now, obviously, the mean cannot be very far from 27300. I will just add together the amounts by which the various numbers differ from 27300 to see how much the mean differs from 27300. That gives me $9 + 21 + 5 - 1 - 4 = 30$. If all 5 cases produce an excess of 30, the mean excess must be 30/5 or 6. So I think the mean is $27300 + 6 = 27306$."

Now let us generalize this process, which is certainly much easier than adding the original numbers as required by the definition of formula (6-1). The number 27300 may be called an *arbitrary origin* and designated as a. Then,

The mean	is equal to	the arbitrary origin	increased by	the mean of the deviations from that origin
$\bar{x}$	$=$	a	$+$	$\dfrac{\Sigma(x - a)}{n}$

(6-9)
$$\bar{x} = a + \frac{\Sigma(x - a)}{n}$$

Again suppose you were asked to compute the standard deviation of these numbers. Because the mean has already been found to be 27306 we have

x	$x - \bar{x}$	$(x - \bar{x})^2$
27309	3	9
27321	15	225
27305	-1	1
27299	-7	49
27296	-10	100
	0	384

By formula (6-4), variance $= \frac{384}{4} = 96$, and $s = \sqrt{96} = 9.8$. If x had not been an integer, it would not be economical to use formula (6-4), and we might have computed $\Sigma x = 136{,}530$ and $\Sigma x^2 = 3{,}728{,}088{,}564$. Then by formula (6-5)

$$\text{Variance} = \frac{5(3{,}728{,}088{,}564) - (136{,}530)^2}{5(4)} = \frac{1920}{20} = 96$$

as before. This procedure is correct but unnecessarily laborious. To shorten the process we shall again take an arbitrary origin at the convenient point $a = 27300$. Then

$$\Sigma(x - a) = 30$$
$$\Sigma(x - a)^2 = 9^2 + 21^2 + 5^2 + (-1)^2 + (-4)^2 = 564$$

and variance $= [5(564) - 30^2]/5(4) = 1920/20 = 96$ as before.

The procedure is symbolized by the formula

$$(6\text{-}10) \qquad \text{Variance} = \frac{n\Sigma(x - a)^2 - [\Sigma(x - a)]^2}{n(n - 1)}$$

and this can be seen to differ from formula (6-5) only in that the x of (6-5) has become $(x - a)$ of (6-10).

The original values (here denoted x) measured from an origin at zero are called *raw scores* or *gross scores*. When each score is reduced by some arbitrary value, the results (here denoted $x - a$) are called *deviations from an origin at a*. When gross scores are large, computation by such gross score formulas as (6-1), (6-5), (6-7), and (6-8) may be very laborious by hand computation though quite convenient if a machine is available. An arbitrary origin can be used to reduce the size of the numbers involved and thus to reduce the labor of computation.

To assure yourself that the choice of a is really arbitrary, compute the mean and variance of the numbers 23, 29, 21, 35, 31, and 36, using first one origin and then another, as for example:

a	$\Sigma(x - a)$	$\Sigma(x - a)^2$	$\bar{x}$	*Variance*
0	175	5293	$\dfrac{175}{6} = 29.17$	$\dfrac{6(5293) - (175)^2}{30} = \dfrac{1133}{30}$
20	55	693	$20 + \dfrac{55}{6} = 29.17$	$\dfrac{6(693) - (55)^2}{30} = \dfrac{1133}{30}$
30	-5	193	$30 - \dfrac{5}{6} = 29.17$	$\dfrac{6(193) - (5)^2}{30} = \dfrac{1133}{30}$
32	-17	237	$32 - \dfrac{17}{6} = 29.17$	$\dfrac{6(237) - (17)^2}{30} = \dfrac{1133}{30}$
25.5	22	269.5	$25.5 + \dfrac{22}{6} = 29.17$	$\dfrac{6(269.5) - (22)^2}{30} = \dfrac{1133}{30}$

The outcome will be exactly the same no matter what value is given to a, so the computer has free choice. However, in order to keep down the labor of computation it is wise to let a be an integer (20 or 26 would be better than 25.5), which is easy to subtract (20 would be better than 19 or 26) and which is small enough so that all deviations are positive (20 would be a better choice than 26).

Computation When Scores Are Grouped in Intervals.   Until we could take up the situation in which scores are grouped in intervals, it has seemed wise to restrict attention to distributions with small n and narrow range. Mastering the type of problem discussed in this section, however, will make it possible to work with any distribution.

Even though you may be heartily tired of the midterm scores of the three sections of a statistics class, it will be instructive to go back to them again in order to see the relation of mean and standard deviation to the measures computed in Chapter 5.

In Table 6-2 three computations have been made from the same set of scores. Computation I makes use of gross scores, treating each score in

Table 6-2   Three Computations of the Mean and Standard
Deviation for the Midterm Scores of Section I
Taken from Table 5-3 on page 55

Score	f	x	fx	fx^2	$x-32$	$f(x-32)$	$f(x-32)^2$	x'	fx'	$f(x')^2$
		I				**II**			**III**	
70–74	1	72	72	5184	40	40	1600	8	8	64
65–69	2	67	134	8978	35	70	2450	7	14	98
60–64	11	62	682	42284	30	330	9900	6	66	396
55–59	9	57	513	29241	25	225	5625	5	45	225
50–54	7	52	364	18928	20	140	2800	4	28	112
45–49	7	47	329	15463	15	105	1575	3	21	63
40–44	3	42	126	5292	10	30	300	2	6	12
35–39		37			5			1		
30–34	1	32	32	1024	0	0	0	0	0	0
Sum 41			2252	126394		940	24250		188	970

Mean:

$$\frac{2252}{41} = 54.93$$

$$32 + \frac{940}{41} = 54.93$$

$$32 + \left(\frac{188}{41}\right)5 = 54.93$$

Variance:

$$\frac{41(126394) - (2252)^2}{41(40)}$$

$$= \frac{110650}{1640} = 67.47$$

$$\frac{41(24250) - (940)^2}{41(40)}$$

$$= \frac{110650}{1640} = 67.47$$

$$25\left(\frac{41(970) - (188)^2}{41(40)}\right)$$

$$= 25\left(\frac{4426}{1640}\right) = 67.47$$

Standard deviation:

$$\sqrt{67.47} = 8.21$$

$$\sqrt{67.47} = 8.21$$

$$\sqrt{67.47} = 8.21$$

the distribution as if it were situated at the midpoint of its interval and using formulas (6-7) and (6-8) to complete the computation. Do not feel obliged to carry out the arithmetic, but examine the table to make sure you know what has been done. The numbers are large enough to seem rather formidable in hand computation, and you will doubtless prefer Methods II and III.

Method II is exactly like Method I except that an arbitrary origin has been taken at the midpoint of the interval 30–34, and so each value of x has been reduced by 32, thus making the numbers considerably smaller. Note that the final values of mean and variance are identical with those of Method I.

Method III makes use of what are called *coded* scores. Note that each value in column x' is exactly one fifth as large as the corresponding value in column $x - 32$ of Method II, and so $\Sigma x'$ is one fifth as large as $\Sigma(x - 32)$ and must be multiplied by 5 to give the same result in computing $\bar{x}$. Now 5 is the step interval, and we shall designate the width of the interval by i. Then,

(6-11) $$ix' = x - a$$

An x' is called a *coded score*. Such coded scores are obtained by writing 0 in the interval where the arbitrary origin is placed and labeling the intervals above that origin successively 1, 2, 3 . . . and the intervals below it $-1, -2, -3$ . . . . Thus the values of x' provide a new scale in which each step is one interval wide. To change this scale back to the scale of scores, we must multiply by i or i^2 as indicated in formulas (6-12) and (6-13).

(6-12) $$\bar{x} = a + i\left(\frac{\Sigma fx'}{n}\right)$$

(6-13) $$\text{Variance} = i^2\left(\frac{n\Sigma f(x')^2 - (\Sigma fx')^2}{n(n-1)}\right)$$

Note that the only difference between formula (6-13) and the earlier (6-8) is that x' has taken the place of x, and the whole has been multiplied by the square of the step interval. Note that the arbitrary origin is always taken at the midpoint of an interval.

In Method II, each value of x has been reduced by 32 but is still measured in score units. Therefore, $a = 32$ and $i = 1$.

Other Formulas Related to the Variance.   The formulas given here for the variance, (6-5), (6-8), (6-10), and (6-13) have been selected out of many algebraically equivalent formulas because they present an economical computing routine. The student who reads widely in other books will probably discover some of the following formulas which are set down here only to reassure him that they are algebraically consistent with those presented in this chapter.

$$(6\text{-}14) \qquad \Sigma(x - \bar{x})^2 = \Sigma x^2 - \frac{(\Sigma x)^2}{n}$$

$$(6\text{-}15) \qquad \Sigma(x - \bar{x})^2 = i^2 \left[\Sigma f(x')^2 - \frac{(\Sigma f x')^2}{n} \right]$$

$$(6\text{-}16) \qquad s^2 = \frac{\Sigma x^2 - [(\Sigma x)^2/n]}{n - 1}$$

$$(6\text{-}17) \qquad s^2 = i^2 \left\{ \frac{\Sigma f(x')^2 - [(\Sigma f x')^2/n]}{n - 1} \right\}$$

As formulas for the standard deviation are obtained from those for the variance merely by taking the square root, they need not be listed here. As noted previously, many texts use n where this uses $n - 1$ as denominator for the variance.

Checking Results. Retracing the steps of a computation seldom catches mistakes. Check methods are needed to give a computer confidence in the results of his work, especially if he is computing without access to a machine.

If coded scores and an arbitrary origin are used, shifting the origin, recomputing $\Sigma x'$ and $\Sigma(x')^2$, and comparing these with the values under the first computation provides an easy check.

Raising the arbitrary origin by r intervals

(6-18)   changes $\Sigma x'$ to $\Sigma x' - nr$
(6-19)   and changes $\Sigma(x')^2$ to $\Sigma(x')^2 - 2r\Sigma x' + nr^2$

Lowering the origin by r intervals

(6-20)   changes $\Sigma x'$ to $\Sigma x' + nr$
(6-21)   and changes $\Sigma(x')^2$ to $\Sigma(x')^2 + 2r\Sigma x' + nr^2$

Now let us apply these principles to the computation by Method III in Table 6-2, where $i = 5$ and the origin was taken at the midpoint of the lowest interval in which any frequency occurs. This is a good place to take the origin because it keeps the values of x' small but positive. If we move it up, some values of x' will be negative, which you will find is a nuisance when a computing machine is used. In the columns $\Sigma f x'$ and $\Sigma f(x')^2$ below are given the values you would obtain if you made a specified shift (r) in the value of a. Verify these from the data, and then verify that they agree with the values in the check columns:

a	r	$\Sigma fx'$ from data	Check	$\Sigma f(x')^2$ from data	Check
37	1	147	$= 188 - 41(1)$	635	$= 970 - 2(1)(188) + 41(1)$
47	3	65	$= 188 - 41(3)$	211	$= 970 - 2(3)(188) + 41(9)$
22	-2	270	$= 188 + 41(2)$	1886	$= 970 + 2(2)(188) + 41(4)$

Relation of Standard Deviation to Range.   Sometimes a large error—such, for example, as may result from misplacing a decimal point—can be detected by comparing the computed standard deviation with the range. The ratio of the range to s is almost never smaller than 2 or larger than 6; in fact, it is not often smaller than 2.5 or larger than 5.5. For example, the range estimated from the data of Table 6-2 on page 83 is $74 - 30 = 44$ and $s = 8.21$, so the ratio of the range to the standard deviation is $44/8.2 = 5.4$. Suppose the decimal point had been misplaced in computing the variance, so the standard deviation was reported not as $\sqrt{67.47} = 8.2$ but as $\sqrt{6.747} = 2.6$. The presence of some serious error could have been surmised because Range/$s = 44/2.6 = 17$, which is much too large to be reasonable. If the decimal point had been misplaced in the other direction so that s appeared to be $\sqrt{674.7} = 26$, the ratio of range to s would be $\frac{44}{26} = 1.7$, which is small enough to rouse suspicion. The comparison of the standard deviation to the range cannot prove the correctness of the computation of the standard deviation but it can sometimes reveal an error.

Comparison of Groups.   The mean and the standard deviation have been developed to this point as summary measures useful in characterizing a distribution. The mean achieves this by identifying a point on the scale, usually near the center of the distribution; the standard deviation describes the spread of the distribution. One can also speak of this mean and standard deviation as characterizing the group on which the scores making up the distribution were computed. Thus we may speak of the mean intelligence quotient (IQ) of a class being 100 and its standard deviation being 10.

The mean and the standard deviation can also be used to compare groups. Suppose, for example, that the mean IQ of one eighth-grade class is 103 and standard deviation 4.1, and the mean IQ of a second eighth-grade class of about the same size is 102 and standard deviation is 9.3. The means are approximately the same, but the spread of the second class is so much greater it will present different and more difficult teaching problems.

Both variance and standard deviation are expressed in the units of the original distribution. The standard deviation of the weights of a group of 12-year-old children from underprivileged homes might be compared with the standard deviation of the weights of a group from privileged homes because both standard deviations are in terms of weight units, and the means are not very disparate. One could not, in the same way, ask whether a group of children is more variable in weight than in height, because one standard deviation is expressed in weight units and one in height units, and there is no basis for comparison. Nor would it be reasonable to say that adults are more variable in weight than babies merely because the standard deviation of adult weights is larger than that of infant weights. The means of the two groups would be so diverse that one would be inclined to inquire how large is the variability relative to the average. The ratio of the standard deviation to the mean, $s/\bar{x}$, is called the *coefficient of variation*.

Caution.   At this stage there is a great temptation for the beginning student to run ahead of his information. He has, let us say, found that a group of 150 girls had a higher mean score and a larger variability than a group of 120 boys on a test of art appreciation. That is an observation true for these particular cases. But the research worker is tempted to go further and to say not that "these girls had a higher mean than these boys and were more variable" but that "girls have a higher mean and are more variable than boys," making a generalization from the particular cases observed to boys and girls at large. Such a generalization is known as a statistical inference, being an inference from an observed sample to the unknown population from which that sample is drawn. The more extended analysis on which such inferences must be based will be introduced in Chapter 9 of this text. For the present, the student would be well advised not to generalize on his data and to report his conclusions in the past tense so that his readers will understand that they apply specifically to the groups studied.

EXERCISE 6-3

1.   The basic data for computing the mean, variance, and standard deviation for the midterm scores of each section and the combined sections are given on page 88. Complete the computations, and write your results on the blank lines.

	I	II	III	Combined Group
From the ungrouped data of Table 5-2 on page 54				
Σfx	2257	1425	1349	5031
Σfx^2	126,979	77,291	67,569	271,839
$\bar{x}$	—	—	—	—
s^2	—	—	—	—
s	—	—	—	—
$\Sigma f(x - 50)$	207	75	−51	231
$\Sigma f(x - 50)^2$	3779	2291	2669	8739
$\bar{x}$	—	—	—	—
s^2	—	—	—	—
s	—	—	—	—
From grouped scores of Table 5-3 with $a = 52$ and $i = 5$				
$\Sigma fx'$	24	4	−22	6
$\Sigma f(x')^2$	122	82	120	324
$\bar{x}$	—	—	—	—
s^2	—	—	—	—
s	—	—	—	—

2. You will notice that there is very close agreement between the results obtained from the grouped and the ungrouped data. Why is the agreement not perfect?

3. Do the three sections stand in the same order with respect to the mean as they do with respect to the median?

4. For each section find the value of $\bar{x} - x_{.50}$.

   Section   I   _____
   Section   II   _____
   Section III   _____
   Combined group _____

   Does there seem to be fairly close agreement between the mean and the median? Are they identical?

5. In terms of the standard deviation, place the three groups in order of variability, from most variable to least variable. Do the same in terms of the interquartile range. Is the order the same?

6. Compute the ratio of the range to the standard deviation for each section and for the combined group, working (a) with the ungrouped data and (b) with the grouped data.

7. Suppose you were reviewing the work of someone else and you found the statement that $\Sigma x = 920$, $n = 30$, $\bar{x} = 3.06$; it would be clear that a mistake had been made because these three numbers are inconsistent. If you found

the statement that $\bar{x} = 23.4$ and $x_{.50} = 16.9$, you would not be able to say that the results were correct, but you would certainly have no evidence that they were not. For each of the following sets of data in which there is an inconsistency indicating that a *mistake* has certainly or probably been made, write M on the blank line. If there is no evidence of inconsistency and the data are quite *possibly* correct, write P on the line.

_____ (a) $x_{.20} = 32; x_{.50} = 61; x_{.40} = 63.$
_____ (b) $\bar{x} = 1; s = 7.2.$
_____ (c) $n = 30;$ if $a = 6, \Sigma x' = 23;$ if $a = 5, \Sigma x' = 18.$
_____ (d) $n = 37; \bar{x} = 12; \Sigma(x - 12) = 26.$
_____ (e) $n = 60; \bar{x} = 7; \Sigma x = 420.$
_____ (f) The extreme scores in a distribution are 3 and 69; $\bar{x} = 31, s = 7.4.$
_____ (g) The extreme scores in a distribution are 5 and 36; $\bar{x} = 29; s = 7.9.$
_____ (h) $n = 60; \Sigma(x - 8) = 33; \Sigma(x - 9) = -27.$
_____ (i) $n = 80; \Sigma(x - 12) = -15; \Sigma(x - 11) = -95.$
_____ (j) $n = 100; \bar{x} = 31; \Sigma(x - 31)^2 = 502.$
_____ (k) $n = 100; \bar{x} = 31; \Sigma(x - 31)^2 = 502; \Sigma(x - 35)^2 = 421.$
_____ (l) $n = 30; \Sigma(x - 25) = 120; \bar{x} = 25 + \frac{120}{30} = 29.$
_____ (m) $x_{.25} = 52; x_{.50} = 60; x_{.75} = 73.$

Mean and Standard Deviation of Combined Group from Those of Its Subgroups. Sometimes the values of n, $\bar{x}$, and s are available for several subgroups, and there is need to obtain the corresponding values for the combined group. This situation might arise when one is working over published data without access to the original scores. Even if original scores are available, going back to them and reconstructing a frequency distribution for the combined group may be unnecessarily laborious. Formulas (6-22) to (6-25) show an easy numerical procedure by which the statistics of the combined group (indicated by the subscript c) may be obtained from those of the various subgroups. The letter k is used to indicate the number of subgroups.

(6-22)
$$n_c = n_1 + n_2 + \cdots + n_k$$

(6-23)
$$\bar{x}_c = \frac{1}{n_c}(n_1\bar{x}_1 + n_2\bar{x}_2 + \cdots + n_k\bar{x}_k)$$

(6-24) $d_i = \bar{x}_i - \bar{x}_c$, or in words, d_i represents the amount by which the mean of the ith subgroup differs from the mean of the combined group

(6-25) $(n_c - 1)s_c{}^2 = (n_1 - 1)s_1{}^2 + (n_2 - 1)s_2{}^2 + \cdots + (n_k - 1)s_k{}^2$
$$+ n_1d_1{}^2 + n_2d_2{}^2 + \cdots + n_kd_k{}^2$$

The use of these formulas will now be illustrated by application to the data of Table (5-2) on page 54. There, of course, the original scores are available, and direct computation from them produces the values $\bar{x}_c =$ 52.406 and $s_c = \sqrt{86.18} = 9.28$. We shall now apply the formulas just as if all the information at hand were what is provided in Table 6-3 in the columns headed n, $\bar{x}$, and s.

Table 6-3 Routine for Obtaining $\bar{x}_c$ and s_c for a Composite Group

Subgroup	n	$\bar{x}$	s	$n\bar{x}$	d	$(n-1)s^2$	nd^2
I	41	55.05	8.27	2257	2.64	2736	286
II	27	52.78	8.95	1425	.37	2083	4
III	28	48.18	9.77	1349	−4.23	2577	501
	96			5031		7396	791

$$\bar{x}_c = \frac{5031}{96} = 52.406$$

$$95\,s_c^2 = 7396 + 791 = 8187$$

$$s_c = \sqrt{(8187/95)} = \sqrt{86.18} = 9.28$$

The sum of the column n gives us 96 as the value of n_c according to formula (6-22).

The sum of the column $n\bar{x}$ divided by 96 gives us $\bar{x}_c = 52.406$ according to formula (6-23).

The values in column d are obtained by subtracting $\bar{x}_c$ from each value of $\bar{x}$.

The values in columns $(n-1)s^2$ and nd^2 are filled in. The sums of these columns are added to provide the value on the right of the equality sign in formula (6-25). This number is 8187.

The computation of s_c is completed by finding $s_c^2 = 8187/95 = 86.18$ and $s_c = \sqrt{86.18} = 9.28$.

By direct computation from raw data, we had obtained $\bar{x}_c = 52.406$ and $s_c = 9.28$, and by application of the formulas we have the same values.

No similar procedure is possible for obtaining percentiles or percentile ranks of a combined group without recourse to the frequency distribution.

Standard Scores. If the deviation of a score from its mean $x - \bar{x}$ is divided by the standard deviation s the result

$$z = \frac{x - \bar{x}}{s}$$

is called a standard score. Such conversion of scores to standard scores is a device which facilitates some important comparisons described in the following paragraphs.

The usefulness of the transformation to standard scores is due to the fact that when all the scores in any distribution are thus standardized, their sum is always zero,

$$\Sigma z = \Sigma \frac{x - \bar{x}}{s} = 0$$

their mean is always zero,

$$\bar{z} = \frac{\Sigma z}{n} = 0$$

and their standard deviation is always 1

$$s_z = \sqrt{\frac{\Sigma(z - \bar{z})^2}{n - 1}} = \sqrt{\frac{\Sigma z^2}{n - 1}} = \sqrt{\frac{1}{s^2} \cdot \frac{\Sigma(x - \bar{x})^2}{n - 1}} = 1$$

One use of standard scores is to compare the standing of a single individual in relation to his group when he has been measured on several variables. For example, in Appendix Table VIII we note that the student with code number 3 has scores 38, 38, and 39, respectively, on the three tests of reading, artificial language and arithmetic. Can this be interpreted to mean that his standing in the group is about the same on the three tests?

By the methods of Chapter 5 we could compute his percentile rank in each test, and we would thus discover that in reading he had percentile rank 62; in artificial language, 19; in arithmetic, 86. Obviously, the raw scores alone do not properly reflect his relative positions on the three tests. To compute a percentile rank it is necessary to have the entire distribution of scores. Moreover percentile ranks can be used appropriately only when central position is measured by the median. If the mean is being used as measure of central position, relative standing should be indicated in terms of the mean and standard deviation, and this can be done without reference to the distribution of original scores.

Comparing the scores of student 3 with the group means shown in Table 6-4, we see that he stood 2.3 points above the mean in reading, 8.4 below the mean in artificial language, and 7.6 above in arithmetic. But can these differences be considered comparable when scores on the three tests were not equally variable? To allow for differences in the standard deviations of the three tests, his score on each test may be converted to a standard score, as shown in Table 6-4. The three values are then directly comparable.

Table 6-4   Computation of Standard Scores on Three Tests
for Student 3 of Appendix Table VIII

Test	x_3 Score of Student 3	$\bar{x}$	$x_3 - \bar{x}$	s	$\dfrac{x_3 - \bar{x}}{s}$	$10\left(\dfrac{x_3 - \bar{x}}{s}\right) + 50$
Reading	38	35.7	2.3	7.7	.30	53
Artificial language	38	46.4	−8.4	11.0	−.76	42
Arithmetic	39	31.4	7.6	7.3	1.04	60

To avoid the danger of overlooking a decimal point, it has generally been found convenient to multiply such standardized deviations by some number large enough to give a result from which we feel justified in dropping the final digit. A convenient number to use for this purpose is 10. To eliminate the negative signs, we add some number large enough to make all results positive. The number 50 is ordinarily, but not necessarily, used for this purpose. To avoid confusion we shall refer to the transformed scores thus produced as *modified standard scores*. The modified standard scores for student 3 in the final column of Table 6-4 indicate, as did his percentile ranks, that his standing was highest in arithmetic and lowest in artificial language.

It should be noted that when scores are transformed to

$$10\,\frac{x - \bar{x}}{s} + 50$$

their mean becomes 50 and standard deviation 10. In general, when they are transformed to

$$a\,\frac{x - \bar{x}}{s} + b$$

their mean is changed to b and their standard deviation to a. This relation is underscored by the discussion on the effect of increasing every score in a distribution by the same amount or multiplying every score by the same amount on page 104.

When scores on several tests are to be averaged and it is desired to give them all the same weight in the average, they should be put in standard form before being averaged. If raw scores are averaged, the test with the largest standard deviation will exercise the greatest weight in the composite.

In the previous example, the three standard deviations (7.7, 11.0, and 7.3) were not very diverse, and so the advantage of transforming scores to standard form before averaging—if scores were to be averaged—is not very great. Suppose, however, that a teacher wants to average scores on three class tests in which the means and standard deviations are

$$\bar{x}_1 = 9 \quad s_1 = 3$$
$$\bar{x}_2 = 41 \quad s_2 = 7$$
$$\bar{x}_3 = 63 \quad s_3 = 2$$

Consider the four students listed in Table 6-5, and note that if raw scores are averaged, the students all seem to have the same standing but that if scores are transformed to standard form before being averaged, it is clear that student D excels the others by a considerable amount. The individual scores and the standard deviations here were chosen deliberately to produce an effect somewhat more striking that you are likely to encounter in practice.

Table 6-5   Raw Scores and Standard Scores for Four Students
on Three Tests (Artificial Data)

Test	$\bar{x}$	s	Raw Score for Student				Modified Standard Score for Student			
			A	B	C	D	A	B	C	D
I	9	3	15	7	2	15	70	43	27	70
II	41	7	41	50	57	33	50	63	73	39
III	63	2	61	60	58	69	40	35	25	80
Sum			117	117	117	117	160	141	125	189
Mean			39	39	39	39	53	47	42	63

EXERCISE 6-4

1. For the midterm grades listed in Appendix Table VIII, the mean and standard deviation were approximately $\bar{x} = 52.5$ and $s = 9.3$. By inspection of the table, count to see how many students had scores in the range indicated, and then change these numbers to percents.

Range in Symbol Form	Range in Scores	Number of Students	Percent of Students
Below $\bar{x} - s$	Below 43.2	—	—
Between $\bar{x} - s$ and $\bar{x}$	Between 43.2 and 52.5	—	—
Between $\bar{x}$ and $\bar{x} + s$	Between 52.5 and 61.8	—	—
Above $\bar{x} + s$	Above 61.8	—	—
Below $\bar{x} - 2s$	Below 33.9	—	—
Between $\bar{x} - 2s$ and $\bar{x}$	Between 33.9 and 52.5	—	—
Between $\bar{x}$ and $\bar{x} + 2s$	Between 52.5 and 71.1	—	—
Above $\bar{x} + 2s$	Above 71.1	—	—

2. Suppose marks on four class tests with means and standard deviations as shown are to be averaged. Convert the marks of the pupils named to standard scores and average those standard scores.

Test	$\bar{x}$	s	Mark for Jones	Mark for Smith	Mark for Brown
A	80.3	5.6	89	82	75
B	25.4	8.4	27	24	28
C	37.3	12.1	38	46	40
D	65.2	4.3	72	68	63

3. Below are given John's score, the mean, and the standard deviation on each of three tests. On which did he stand highest in relation to the group for which $\bar{x}$ and s were computed? On which lowest?

Test	$\bar{x}$	s	John's Score
English usage	73.2	12.6	72
Algebra	58.6	4.2	61
U.S. History	41.3	6.8	50

7

THE FREQUENCY
DISTRIBUTION:
A SUMMARY

In a sense most of the material presented up to this point has dealt with the frequency distribution—tabulation, graphing, scaled variables, measures of central position, and measures of variability. Certainly all of Chapters 4, 5, and 6 and considerable portions of 3 dealt with aspects of the frequency distribution. Nor will the rest of the book turn away from this extremely important topic. Before taking up problems of correlation and regression which involve two variables, it will be wise to summarize, amplify, and consolidate what has been said about the frequency distribution of a single variable. The discussion in the present chapter relates chiefly to the distribution of a scaled variable.

Summary Measures. In Chapters 5 and 6 we have learned to describe frequency distributions by certain summary measures such as the mean, median, various percentiles, the difference between two percentiles, and the standard deviation. Other summary measures will be described in this chapter.

With so many measures at hand, the reader may well wonder what selection of them he should make for adequate description of a frequency distribution. Some guides will be presented in the following paragraphs. It may be pointed out at once that, if a distribution is fairly symmetrical (a matter which is discussed later in this chapter) and has its largest frequencies near the middle of the distribution, nearly as much information can be conveyed by giving a measure of central position and a measure of variability as by presenting the entire distribution. The question still remains as to which measure of central position and which measure of variability are to be chosen.

Measures of Position.   The mean, median, and the percentiles all refer to points or positions on the scale of the frequency distribution, so that they may all be called measures of position or location. The mean and median in particular commonly occupy a central position in the distribution so that they are called measures of central position. These two measures are also called averages because they convey the meaning usually associated with the concept of the average of a group of numbers.

Another summary measure, which is usually associated with the mean and median as both a measure of central position and as an average, is the mode. It should be noted, however, that the mode does not necessarily have to be either central or even unique and that it can be computed when there is no scale. It is sometimes defined as that value which occurs most frequently in a distribution, just as the style of dress or hat or coat which occurs most frequently in a given season is called the mode. A more satisfactory definition would be to say that the mode is a value at which the distribution attains a peak. In a grouped frequency distribution, the mode is considered to be at the midpoint of the interval in which such a peak occurs. If two or more adjacent intervals have the same frequency and it is larger than the frequency in the intervals on either side, the mode is considered to be at the midpoint of the entire range covered by the intervals of high frequency. In a smooth frequency curve, the mode is the abscissa corresponding to a peak on the curve. Two kinds of mode should be distinguished. The *mathematical mode* is a concept of considerable interest in theoretical work and is far more useful than the *crude mode*. However, the mathematical mode is difficult to obtain; it can, in fact, be obtained only if the equation of a theoretical frequency curve is known, so that its maximum point can be found. This is almost always a task calling for some mathematical erudition. Therefore, the mathematical mode is not discussed in elementary treatises, and the term "mode" is ordinarily applied to what is more properly called the crude mode.

The crude mode is the measure described in the first four sentences of the preceding paragraph.   It is extremely easy to obtain, requiring no computation but merely inspection of the distribution.

Two other averages may be mentioned in passing. Problems concerning time, rate, and distance—or time, rate, and work—often require the use of what is called the *harmonic mean*. Problems that relate to a variable which is changing over a period of time at a fairly constant rate (not constant amount) often require the use of the *geometric mean*. Rate of growth of an organism or a population is such a variable. Further description of these means or presentation of their formulas is outside the scope of this book. These means are seldom needed, but when they are needed there is no alternative. Any person who has a problem in averaging rates would do well to consult a text in which the geometric and harmonic means are discussed.

The Choice of a Measure of Central Position.   Let us begin by looking at certain situations in which there really is no choice because not all three measures—mean, median, and mode—can be computed:

1.   If the variable is expressed in unordered categories only, the mode is the only average which can be found. Suppose a panel of judges is deciding which of four exhibits shall receive a prize and they vote as follows:

Exhibit	A	B	C	D
Number of votes:	2	13	1	4

It is quite correct to say that B is the mode of this distribution. There is no mean and no median.

2.   If the variable is not scaled but consists of a set of ordered categories, there is no mean but a rough median can usually be found, and a mode.

3.   If the variable is scaled but has one or both ends open, there is a mean, but it cannot be computed. The median and the mode can usually be obtained. An example of an open-ended distribution would be one in which the highest interval is designated as "75 years and over," or "more than $50,000," or the like; or the lowest as "less than 5 years," or the like.

Certain qualities which are desirable in an average are possessed by more than one measure of central position and so do not furnish criteria for choice between them:

1.   An average should be easy to comprehend and interpret. The mean, median, and mode meet this requirement.

2.   It is desirable that an average be easy to compute. None of the three is difficult to compute. However, it might be noted that to obtain either the mode or the median it is necessary first to organize the scores into a frequency distribution (which is easy enough to do on an electric sorter even if the number of cases is very large), but this step is not necessary for obtaining the mean.

3.   It is desirable that the average should not be greatly affected by the process of grouping scores into intervals. Such grouping has very little effect on either mean or median unless the number of cases is quite small. However, it often affects the mode greatly. For an illustration, look at Table 5-2 on page 54. Here the frequencies are so scattered that the modes are not clearly defined. Group I has two modes at 58 and 62, and the distribution for the combined sections has three modes at 47, 58, and 62. When the frequencies are grouped in intervals of 5, the position of the mode changes. In fact, as different beginning points are chosen for the intervals, the mode shifts. To illustrate the instability of the mode as scores are grouped into intervals, the modes for the data of Table 5-2 on page 54 are shown here under various conditions of grouping:

	I	II	III	All Sections
Ungrouped frequencies	58, 62	52	47, 49	47, 58, 62
Highest interval 67–71	59	54	49	59
Highest interval 68–72	60	60	50, 60	60
Highest interval 69–73	61	51	51	51
Highest interval 70–74	62	52	47	62
Highest interval 71–75	63	53	48	48

4. It is desirable that the average fall in the middle of the distribution. The mean and median meet this condition, but the mode sometimes stands at the very end of the distribution.

Now having disposed of the less weighty issues we come to some really important considerations which relate to the choice among mean, median, and mode:

1. It is very important that an average have *sampling stability*. In general, the mean varies less from one sample to another than the median does, and this gives great advantage to the mean. Table 9-2 on page 147 presents the results of drawing 20 small samples from the same population and computing for each both the mean and the median. A glance at this table will reveal that the mean showed less variability from sample to sample than did the median.

However, in a distribution in which frequencies pile up at the ends instead of in the middle, the mean may vary more from sample to sample than the median. The presence of such extreme scores, even if they occur at both ends of the distribution, suggests the possibility that in another sample a few very high scores might occur without compensating low scores, or vice versa.

2. It is helpful to have an average obtained in such a way that it may be expressed by a formula based on the values of *all* the scores in a distribution. Neither the mode nor the median possesses this characteristic. Consider, for example, the eleven values in row *A*. If the values below 15 are made much smaller, as in row *B*, the mean is lowered but neither the median nor the mode reflects that change. If the scores above 15 are made larger, the mean is increased but not the median or the mode.

		$\bar{x}$	$\bar{x}_{.50}$	Mode
A.	9, 10, 13, 14, 15, 15, 15, 16, 18, 19, 21	15	15	15
B.	2, 3, 6, 8, 15, 15, 15, 16, 17, 17, 18	12	15	15
C.	9, 10, 13, 14, 15, 15, 15, 19, 26, 35, 38	19	15	15

There are at least two distinct advantages which the mean possesses because it is an algebraic function of all the scores in the distribution. The first advantage is that it has been possible to develop a great deal more mathematical theory about the mean than about either of the other two averages. A second advantage is seen when several separate groups are to be combined into one single group. To obtain either the median or the mode of the combined group, it is necessary to construct its frequency distribution. The mean of the combined group can be obtained from the means and the n's of the component groups.

3.   When extreme scores occur at one end of a frequency distribution, careful thinking about the meaning of the data is necessary in order not to use a measure which will misrepresent central position. Quoting only the mean might give a fantastically wrong impression. A few atypical cases with extremely high or a few with extremely low scores may affect the mean unduly. For such data both median and mode are of genuine interest and should be reported.

If a teacher wishes to compare the average of her class on a standardized test with published norms for the grade, it is better to have both norm and average expressed as medians because one or two atypical scores might influence the mean too greatly.

Describing a Distribution in Terms of Its Variability.   Several different measures of variability have been described—the total range of scores, some interpercentile ranges, and the standard deviation. Each has its special usefulness. In general, the standard deviation and the mean belong to the same family of measures and are used together; an interpercentile range and the median belong together; the total range may be used with either average. The total range, however, is sometimes inflated by the presence of even one extreme case so that it grossly exaggerates the spread of the whole. If even 2 percent of the cases are eliminated at each end of the distribution, the reduced range $x_{.98} - x_{.02}$ may give a fairer representation of the scattering in the group.

The relation between various measures of dispersion for four distributions of different form is seen in Figure 7-1. The dotted ordinates are erected at the decile points. The brace below each figure extends from $x_{.25}$ to $x_{.75}$ with its cusp at the median. The mean is indicated by a large V, the points one standard deviation from the mean by smaller v's.

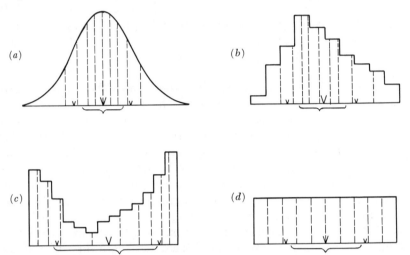

Figure 7-1 Position of Decile Points in Frequency Distributions of Different Forms
    Dotted Lines are Erected at the Decile Points. Brace Extends from $x_{.25}$ to $x_{.75}$ with Center at Median.
    Large Arrow Indicates Mean. Small Arrows Indicate Points at Distance of One Standard Deviation on Either Side of Mean.

Now take a card or piece of crisp paper, and mark on its edge a distance equal to the standard deviation of Figure 7-1(a). Lay this distance off along the base line to estimate the length of the range in standard deviation units. Do the same for each of the four figures. You will find that the ratio of the range to s is about $5\frac{1}{2}$ for (a), about $4\frac{1}{2}$ for (b), a little less than 3 for (c), and about $3\frac{1}{2}$ for (d). In general, you are unlikely to encounter a distribution in which the range is more than $6s$ or less than $3s$. In Figure 7-1(a), the interquartile range is about $\frac{4}{3}s$; in (b), about $\frac{3}{2}s$; in (c), slightly more than $2s$; and in (d), slightly less than $2s$. Obviously, the relation of these measures of variability depends on the form of the distribution.

Describing a Distribution in Terms of Symmetry or Skewness. If the graph of a distribution can be folded along an ordinate at the median in such a way that the two halves of the figure coincide perfectly, that distribution is called *symmetrical*. In such a distribution the mean and median are identical. If there is a single mode, it is also identical with the mean and median. Of course, absolute symmetry is practically never encountered in real data, but the concept provides a convenient method of describing distributions in a general sort of way.

When $\bar{x} - x_{.50}$ is positive, a distribution is said to be *positively skewed* or to have *positive skewness*.

When $\bar{x} - x_{.50}$ is negative, a distribution is said to be *negatively skewed* or to have *negative skewness*.

The terms "skewed to the right" and "skewed to the left" are sometimes met but they are ambiguous and should be avoided. Writers who use these terms are not agreed on their meaning, some applying one term and some the other to the same type of distribution.

Various statistics are available for measuring skewness, but are not used often enough to justify describing them here.

Describing a Distribution in Terms of Its Peakedness.   Figure 7-2 shows six histograms all of which are symmetrical, yet the six are obviously dissimilar. Their dissimilarity derives chiefly from the amount of

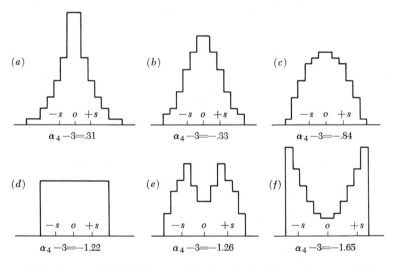

Figure 7-2   Symmetrical Histograms Illustrating Different Degrees of Kurtosis

peakedness they possess. The technical term for peakedness is *kurtosis* or "arching." A moderately peaked distribution such as Figures 7-1(a) and 7-3(a) is called *mesokurtic*. A distribution which goes up to a high, thin peak and has long tails, such as Figures 7-2(a) and 7-3(b) is called *lepto-*

kurtic. A squatty distribution, such as Figures 7-2(*c*) and 7-2(*d*), or one which sinks down in the middle, as Figures 7-2(*e*) and 7-2(*f*) is called *platykurtic.*

Measures of kurtosis exist but are not needed often enough to warrant describing them here. Suffice it to say that the value α_4 printed below the sketches in Figure 7-2 is such a measure; and that $\alpha_4 - 3$ is zero for a mesokurtic or moderately arched distribution; $\alpha_4 - 3$ is positive for a leptokurtic distribution and negative for a platykurtic distribution.

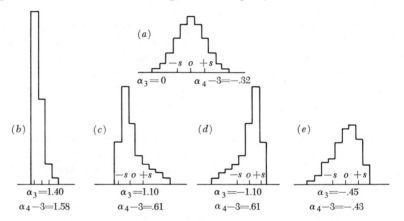

Figure 7-3   Histograms Illustrating Different Degrees of Skewness and Kurtosis

Figure 7-3 shows five distributions which differ both as to symmetry and as to skewness. The figure α_3 below them is a measure of skewness such that α_3 is positive for a distribution which is positively skewed; is negative for one which is negatively skewed; and is zero for one which is symmetrical.

U-shaped, I-shaped, J-shaped Distributions.   These terms are vivid but not explicit. A distribution such as Figure 7-2(*f*) or Figure 7-1(*c*) with a mode at each end of the distribution, is called U-shaped because it resembles the letter U.

A distribution which has a mode in the middle, no matter whether it is symmetrical or not, has some resemblance to a manuscript "i" and is called I-shaped. Sketches in Figures 7-1(*a*) and (*b*) and sketches in Figures 7-2(*a*), (*b*), and (*c*) might be called I-shaped.

A skewed distribution with mode at one end only—at either end—is termed J-shaped. Examples are Figures 7-3(*b*), (*c*), and (*d*).

Bimodal Distributions.   Distributions having two pronounced modes are unusual and, perhaps for that very reason, interesting. A classic example described in many texts and very real to the imagination of all persons who have anxiously scanned the skies in an attempt to predict the weather is that of the degree of cloudiness observed at Greenwich during the month of July in the years 1890 to 1904.[1] Cloudiness was measured on an 11-point scale with 58 percent of the cases in the two extreme categories.

Handedness is another familiar trait which is clearly bimodal. Durost[2] computed for 1300 children a handedness ratio which was negative for the left-handed, zero for the ambidextrous, and positive for the right-handed. He found a strong mode at +40, a smaller one at −35, and very few cases with scores near 0.

The length of stay of a patient in some hospitals has shown a bimodal distribution, with a large number of cases remaining only a few days and another large number remaining for a long period.

Some distributions which at first appear to be bimodal are actually composite distributions produced by combining two groups with unequal means or by eliminating cases near the middle of the distribution. Thus if we combined a group of 15-year-old and a group of 10-year-old boys and measured them on any variable related to age—such as height, weight, strength of pull, or pitch of voice—the resulting measures would have a distribution with two humps, but its apparent bimodality would be artificially produced and not genuine.

The Normal Curve.   The "normal" curve is a theoretical distribution of great importance which will be treated more fully in Chapter 8. It is a smooth curve, not a histogram; it is symmetrical, and moderately peaked. It has an unlimited range. Mean, median, and mode are identical. The semi-interquartile range is about two thirds of the standard deviation. In this particular curve, the standard deviation is the abscissa of the point of inflection, as illustrated in Figure 7-4.

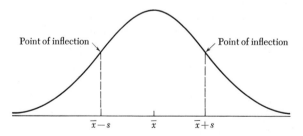

Figure 7-4   Standard Deviation in a Normal Curve

[1] Gertrude E. Pearse, *Biometrika*, **20A** (1928), p. 336.
[2] W. N. Durost, *Genetic Psychology Monographs*, **16** (October, 1934), pp. 225–335.

The Effect of Increasing Every Score in a Distribution by the Same Amount or of Multiplying Every Score by the Same Amount. Below is a list of nine scores and certain statistics computed from them.

	Scores	Mean	Median	Mode	Range	s^2	s
x	1 2 3 4 7 9 9 9 10	6	7	9	$10 - 1 = 9$	12.25	3.5
$x + 4$	5 6 7 8 11 13 13 13 14	10	11	13	$14 - 5 = 9$	12.25	3.5
$2x$	2 4 6 8 14 18 18 18 20	12	14	18	$20 - 2 = 18$	49	7
$3x$	3 6 9 12 21 27 27 27 30	18	21	27	$30 - 3 = 27$	110.25	10.5

In the second line, $x + 4$, each of the original scores has been increased by 4. Examine the statistics computed for this row, compare them with the corresponding statistics in the first row, and formulate a general statement about the effect of adding a constant amount to every score in a distribution or subtracting a constant amount from every score.

In the third line, $2x$, each of the original scores has been doubled and in the fourth line, $3x$, tripled. Examine the statistics for these lines, compare them with the corresponding statistics in the first line, and formulate a general statement about the effect of multiplying (or dividing) every score in a distribution by the same amount.

> *Increasing (or decreasing) every score in a distribution by a given number will have no effect upon any measure of variability (range, interpercentile range, variance, or standard deviation) but will increase (or decrease) by that number every measure of position (mean, median, mode, or any percentile).*
>
> *Multiplying (or dividing) every score in a distribution by a given positive number will have the effect of multiplying (or dividing) by that number every measure of variability except the variance and every measure of position. The variance will be multiplied by the square of the given number.*
>
> *Multiplying (or dividing) every score by a given negative number will have the effect of multiplying (or dividing) the mean, median, and mode by that number; the standard deviation by the absolute value of that number; and the variance by the square of that number. The percentiles other than $x_{.50}$ will be multiplied by that number and inverted so that if $-a$ is the multiplier, $-ax_{.30}$ will be $x_{.70}$.*

These rules may be applied to obtain the mean, variance and standard deviation of standard scores, as in the tabulation at the top of the following page.

Score	Mean	Variance	Standard Deviation
x	$\bar{x}$	s^2	s
$x - \bar{x}$	$\bar{x} - \bar{x} = 0$	s^2	s
$\dfrac{x - \bar{x}}{s}$	$\dfrac{0}{s} = 0$	$\dfrac{s^2}{s^2} = 1$	$\dfrac{s}{s} = 1$
$10\,\dfrac{x - \bar{x}}{s}$	$10(0) = 0$	$10^2(1) = 100$	$10(1) = 10$
$10\,\dfrac{x - \bar{x}}{s} + 50$	$0 + 50 = 50$	100	10

Geometric Representation of Various Statistics of a Frequency Distribution.   Visualizing the position of a statistic on a histogram is an aid to understanding it. You will find it worth while to examine Figure 7-5 with some care. Remember that the scale of scores is the horizontal

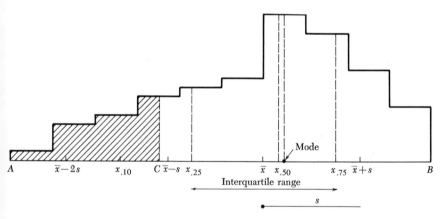

Figure 7-5   Geometric Representation of Various Statistics (Shaded Area Is Percentile Rank of Score C)

axis, and therefore the three averages, which are scores, are found on this horizontal scale. The median $x_{.50}$ is almost always found between the mean $\bar{x}$ and the mode. Note the position of these three values. The vertical line which divides the area into two equal parts serves to locate $x_{.50}$ on the base line. Its length does not represent any statistic. The vertical line at the midpoint of the interval with the largest frequencies serves to locate the mode on the base line, but the length of that vertical line is not the mode.

A measure of variability is represented by a line segment which is a portion of the horizontal axis. Note the segments representing the standard deviation s and the interquartile range. The segment AB represents the range.

Percentile rank of a score is represented by area to the left of the point which represents that score. Thus the shaded area represents the percentile rank of score C.

In Figure 7-5, $\bar{x}$ lies to the left of $x_{.50}$, indicating that the mean is smaller than the median, or $\bar{x} - x_{.50}$ is negative. Therefore, this distribution is *negatively skewed*.

8

THE NORMAL
DISTRIBUTION

I know of scarcely anything so apt to impress the imagination as the wonderful form of cosmic order expressed by the "Law of Frequency of Error." The law would have been personified by the Greeks and deified, if they had known it. It reigns with serenity and in complete self-effacement amidst the wildest confusion. The huger the mob and the greater the apparent anarchy, the more perfect is its sway. It is the supreme law of Unreason. Whenever a large sample of chaotic elements are taken in hand and marshalled in the order of their magnitude, an unsuspected and most beautiful form of regularity proves to have been latent all along.

Sir Francis Galton

Almost every person trained in any of the sciences knows and uses the term "normal curve," but many use it incorrectly or at least without understanding. Social scientists sometimes assert that when enough cases are observed, every human trait is normally distributed. This is not so. The curve is esthetically beautiful and mathematically interesting, and many people assume that mathematicians have proved it to have a universal cogency. Teachers sometimes talk foolishly about "grading on the curve" and deceive themselves into thinking they are being commendably scientific. However, in spite of popularly held misconceptions and all too frequent misuse, the normal distribution is a powerful statistical concept with extremely valuable practical applications.

The Normal Curve. Suppose you have measured the heights of 50 adult men and from those measures have constructed a histogram with step intervals of 1 inch. Try to imagine in a general way how that histogram might look. It would be somewhat irregular because the number of cases is small. Almost certainly the frequencies would pile up in the neighborhood of the mean and fall off at both ends of the distribution.

Suppose you increase the number of cases from 50 to 200, and construct another histogram. How would this differ from the first one? The range would be about the same, or possibly a little larger, the irregularity in the first distribution would tend to decrease, and the histogram would become more symmetrical.

Now suppose the number of cases is increased to 10,000, and the interval is made much narrower, say 0.1 of an inch. There will now be a great many narrow rectangles instead of a few broad ones so the steps will be smaller. Because the number of cases has been greatly increased, the distribution will be more regular. The piling up of frequencies around the mean is now very obvious, and we say the *density* is greater in that part of the distribution.

Imagine that the rectangles in this last histogram are outlined in heavy ink and that the graph is hung on a wall at some distance from you. The frequency distribution is still represented by a series of rectangles, but from a distance the tops of those rectangles may look to you like a smooth curve. Furthermore for the variable named—that is, the heights of adult men—the curve which this distribution suggests would strongly resemble the ideal mathematical curve known as *the normal curve*.

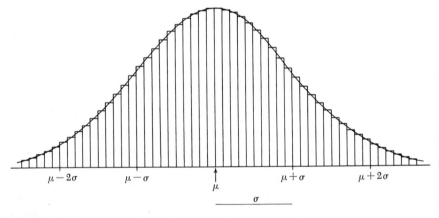

Figure 8-1 Histogram Approximating a Normal Curve, with a Normal Curve Superimposed

Figure 8-1 shows just such a histogram with many narrow step intervals and shows a normal curve superimposed on it. This smooth curve is both interesting and important, and in this chapter we shall discuss its characteristics and present some of its applications.

If, for any continuous variable, a histogram with narrow step intervals is drawn to show the frequency distribution for a very large number of cases, the tops of the rectangles will convey the general impression of a smooth curve. However, that curve will not necessarily resemble the normal curve you have seen in Figure 8-1. Suppose, for example, that the variable is the annual dollar income of all families in some fairly large city. Toward the lower end of the income scale, the density will be much greater than toward the upper end, and the distribution will be considerably skewed, with its mean larger than its median and its median larger than its mode. The smooth curve suggested by the tops of the rectangles in this histogram would be similarly skewed and would not bear much resemblance to the normal curve.

The Normal Distribution. The normal curve is a mathematical construct based on an equation which was first derived as a mathematical exercise without any reference to concrete data. It is an astonishing fact that many physical variables have frequency distributions which bear a striking similarity to this mathematical curve.

The theoretical normal curve never actually touches the base line, and its range is unlimited, although it comes closer and closer to the base line as you go farther out in either direction. For any physical variable, such as height, the range must of necessity be limited.

The area of all the rectangles which form a histogram represents the total frequency in the distribution. Similarly, the area under a normal curve represents the total frequency of a *normally distributed* variable.

In a histogram, the frequency in any class interval is represented by

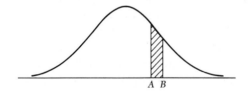

the area of a rectangle which has that interval as its base. In a smooth curve, if one ordinate is drawn at any point A on the base line and another drawn at a second point B, the area of the segment between those ordinates represents the frequency of values in the interval AB.

In graphing the distribution of a physical variable, it is impossible to make the intervals small enough for the distribution actually to become a smooth curve. Thus the graph of a set of *observed* data will always be a step curve, or histogram. However, if the normal curve appears to approximate closely the frequency distribution of a variable in a set of data, we are inclined to generalize and to say that the variable is *normally distributed*, or *has a normal distribution*.

Mean and Standard Deviation of a Normal Distribution. On the histogram of the heights of 10,000 adult men discussed earlier in this chapter, the mean height would be represented by a point on the base line in the manner made familiar in earlier chapters, and the standard deviation would be represented by the length of a horizontal line segment.

The normal distribution also has a mean which is usually called μ (*mu*, the small Greek letter corresponding to the English *m*) and has a standard deviation which is usually called σ (*sigma*, the small Greek letter corresponding to the English *s*). These could be obtained approximately by computation in which the frequencies are represented by segments of area, but accurate solutions are obtained by the methods of integral calculus.

Thus for the normal curve of Figure 8-1, the mean is located at the point marked μ; the standard deviation is the length marked σ, the point one standard deviation above the mean is marked $\mu + \sigma$, and so forth.

Principal Uses of the Normal Distribution. Though the content of such applications varies enormously, the principal applications of the normal distribution fall into the three general categories which follow:

1. Its most important and most interesting application is in connection with generalization from a limited number of measures on observed individuals to similar measures of a larger number of individuals that have not been observed. Much of the information basic to modern science, industry, education, and government depends upon investigations of this sort. As this use is discussed at some length in later chapters, it will not be elaborated on here.

2. Certain types of physical measurements have distributions which so closely approximate the normal that its use greatly facilitates their analysis. (*a*) The physical sciences furnish numerous distributions of this nature. The first tabulation of the normal distribution appeared in a book on the refraction of light near the horizon, published shortly before 1800. Another early work used it in relation to the determination of the position of the polar star. (*b*) The distributions of some measurements on human beings have a striking resemblance to the theoretical normal distribution; other measurements do not suggest normality. For example, in a large, unselected group of adults, homogeneous as to racial origin and living under fairly similar conditions, the distribution of height closely approximates the normal form. The distribution would probably not be normal if the group included both sexes or was made up of children of varying age or included persons born in diverse parts of the globe and reared under different climatic and dietary conditions. The distribution of weight, on the other hand, is likely to be very far from normal because weight is not so directly dependent on heredity as height is, being partly controlled by individual preference. It takes little thought to recognize that many measures influenced by social customs have distributions very far from the normal form. Consider, for example, the distributions, for adults in the United States, of dollar income, number of years of education, average caloric intake per day, number of days illness in the past year. About the middle of the nineteenth century, the great Belgian statistician, Adolphe Quetelet, made studies of the heights of army recruits and decided that they followed the "normal law." In that period there were almost no other large-scale compilations of data available, but Quetelet was convinced, by analogy to height, that when such became available they too would "obey the normal law," and his ideas were widely disseminated. This historical accident may be the origin of the current superstition that all variable traits are normally distributed. There are statistical tests for normality of distribution but these are outside the scope of this book.

3.   For some variables the true form of distribution is unknown, but it is convenient to scale the variable in such a way as to produce a normal distribution of scores. Consider intelligence, for example. It seems reasonable to assume that an unselected group of children of a given age would show a normal distribution of intelligence test scores, and most of the intelligence tests are scaled in such a way as to produce that result. In some situations only categorical ratings are available—as, for example, when a teacher assigns grades of A, B, C, D, and F. As nothing is known about the form of the underlying distribution, it often seems not unreasonable to transform these categorical ratings to scores on the assumption that the underlying trait has a normal distribution. The procedure of "normalizing" will be taken up at the end of this chapter.

Drawing a Normal Curve.   In order to make use of the normal distribution, one needs to be able to read values from such tables as Appendix Tables I and II and to have a clear understanding of what the tabular entries mean. An intuitive feeling for the shape of the curve is a real help. One of the best ways to develop such understanding is to make a drawing of the curve and to examine its properties.

Look once more at Figure 8-1. If x represents the score value of any point on the horizontal scale, then $(x - \bar{x})/s$ would be the standard score of that point expressed in terms of the mean and standard deviation of a set of observed scores. In relation to a normal distribution, such a standard score would be denoted

(8-1)
$$z = \frac{x - \mu}{\sigma}$$

and called a *standard normal deviate*. We shall scale the horizontal axis of the curve we are about to draw in terms of z. On Figure 8-1 it is apparent that the normal curve comes very close to the base line at 2.5 standard deviations on each side of the mean, that is at $z = (x - \mu)/\sigma = 2.5$. Therefore, it would seem satisfactory to let our scale go from $z = -3$ to $z = 3$.

As a first step, lay off on a piece of graph paper a horizontal scale, and mark on it points which you label $-3, -2, -1, 0, 1, 2,$ and 3. It will be convenient to have the interval between two successive points subdivided into 5 or 10 small units in order that you can easily plot the values listed in Table 8-1. At the point marked 0, erect a perpendicular axis, and mark on it the points .05, .10, .15, .20, .25, .30, .35, and .40 as in Figure 8-2. You may use a scale different from that in this figure but do not make your drawing too small.

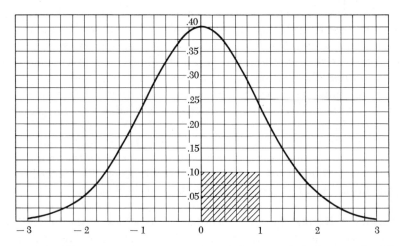

Figure 8-2 The Normal Curve (Area of Shaded Rectangle Is One Tenth of Total Area under the Curve)

For each pair of entries in Table 8-1 plot a point in the usual manner. Then plot a second point for which the value of z is negative, thus producing a symmetrical pattern of dots as in Figure 8-2. Draw a smooth curve through the points you have plotted. If you find you cannot draw the

Table 8-1   Abscissas (z) and Ordinates (y) of the Standard Normal Curve

z	y	z	y
0	.399	1.6	.111
.2	.391	1.8	.079
.4	.368	2.0	.054
.6	.333	2.2	.036
.8	.290	2.4	.022
1.0	.242	2.6	.014
1.2	.194	2.8	.008
1.4	.150	3.0	.004

curve *smoothly* and touch all the points marked, you may know you have estimated some values incorrectly and you should re-examine them.

Characteristics of the Normal Distribution.   Certain general characteristics of the normal distribution will now be stated and you should try to visualize them in relation to the figure you have just drawn.

1.   The curve is symmetrical. If the figure were folded along its vertical axis, the two halves would coincide.
2.   The mean, median, and mode of the normal distribution are identical.
3.   The range is unlimited, infinite, in both directions, but as the distance from μ increases the curve approaches the horizontal axis more and more closely. Theoretically, it never actually reaches the horizontal axis.
4.   The curve is smooth in contrast to the histograms or step curves we have previously drawn from observed data.
5.   The normal distribution has a standard deviation which we shall call σ. In the picture you have drawn, $\sigma = 1$. In other words, the horizontal scale has been laid off in standard deviation units, so that the horizontal variable is actually in standardized form. Then a score on the horizontal variable is $(x - \mu)/\sigma$.

The Family of Normal Curves.   We have said that the distribution of heights of a population of adult persons of one sex closely approximates the normal distribution. If we consider men and women as separate populations, the distribution of the heights of each population will also closely approximate the normal. However, the means of these two distributions of heights will surely differ, and there may also be a difference between the two standard deviations.

We see in this pair of normal distributions an example of the fact that there is not *a* normal curve but a family of normal curves. A member of this family is fully defined when its mean μ, and its standard deviation σ are known. Several members of the normal family are illustrated in Figures 8-3, 8-4, and 8-5. The curves in Figure 8-3 all have the same μ and σ but are drawn to different scales so that their areas are different.

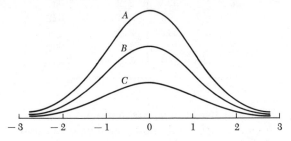

Figure 8-3 Three Normal Curves All Having the Same Mean and Standard Deviation but Different Areas

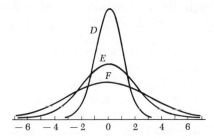

Figure 8-4 Three Normal Curves All Having the Same Mean and Area but Different Standard Deviations

For Curve $D, \sigma = 1$
For Curve $E, \sigma = 2$
For Curve $F, \sigma = 3$

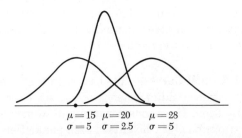

Figure 8-5 Three Normal Curves with Same Area but Different Values of μ and σ

The curves in Figure 8-4 all have the same μ but differ in their values of σ. The curves in Figure 8-5 differ in their values of both μ and σ.

Parameters.   The values μ and σ, which distinguish one member of the family of normal curves from other members of this family, are called *parameters*. This term is borrowed from mathematics where it is used generally to distinguish one member of a family of curves from other members of the same family. We shall find in Chapters 9 and 10 that the search for information about parameters plays a central role in statistical investigations.

Standard Normal Distribution.   When a normally distributed variable X is expressed in standard score form $z = (x - \mu)/\sigma$, the mean of these standard scores is 0, and the standard deviation is 1. The area can also be considered to be unity; then we have a kind of standard curve convenient to tabulate, to which all normal curves can be reduced by changing scale or position. Such a curve—with zero mean, unit standard deviation, and unit area—is the form to which Appendix Tables I and II relate. This form is called *standard normal* or *unit normal*. The equation for the general normal curve is

$$(8\text{-}2) \qquad y = \frac{1}{\sigma \sqrt{2\pi}} e^{-\frac{(x-\mu)^2}{2\sigma^2}}$$

In standard form, this equation becomes

$$(8\text{-}3) \qquad y = \frac{1}{\sqrt{2\pi}} e^{-\frac{z^2}{2}}$$

Formulas (8-2) and (8-3) have been presented for the information of those readers who are interested. These formulas will not be used in any subsequent portions of this book. The normal distribution has been so thoroughly tabulated that few persons ever need to make use of its formula.

Area Relationships under a Standard Normal Curve.   Ordinates are seldom needed for anything except for drawing the curve, and although there are some practical problems which call for knowledge of an ordinate, you will not meet any such in this book. The area under the curve between two values of z, on the other hand, is of great importance in a variety of problems, because such an area represents relative frequency. Before you can solve any problems of the kind presented in the last part of this chapter and in Chapters 9 and 10, you must be able to obtain such proportions of area from Appendix Table I. In order that you may clearly understand the entries in this table, we shall first estimate certain areas from the graph you have drawn and then check these crude results against the tabulated values.

To obtain an estimate of the fraction of area between $z = 0$ and any other value of z, you may count the number of small squares in that area—estimating as well as you can the parts of squares cut by the curve—then divide this number by the total number of squares in the entire area under the curve. The latter can be estimated fairly well by sheer counting but that is laborious, and the area in the tails is hard to estimate. There is a much easier way. Draw a small rectangle with length equal to σ and height equal to the interval from 0 to .1 on the vertical axis. (See the shaded rectangle in Figure 8-2.) Count the number of small squares in this rectangle, multiply that number by 10, and you will have the number of small squares in the entire area under your curve.

In the following examples, areas obtained by counting squares on the graph which you have constructed will be compared with areas read from Appendix Table I. Since the areas in Table I are based on the standard normal distribution which has total area unity, the areas in that table are in effect fractions of the total area. To get comparable readings from your graph you will need to divide the area in the specified segment by the total area.

The column headed *area* in Table I gives the areas between a specified value of z and $z = 0$. Such an area is shaded in the sketch below.

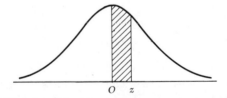

By counting squares on your graph, find the percent of area between $z = 0$ and $z = .2$. This should be approximately 8. Now turn to Table I, find the value of .2 in the column headed z, and read the corresponding entry in the column headed "area." It is .0793 or approximately 8 percent.

Again, from your graph, find the fraction of area between $z = 0$ and $z = 1.4$. Check the outcome by finding the value 1.4 in the z column of Table I and noting that the corresponding area is given as .4192. As the curve is symmetrical, the area between $z = -1.4$ and $z = 1.4$ would be .4192 + .4192 = .8384 or .84.

Find the area to the left of $z = 1.8$. The area between $z = 0$ and $z = 1.8$ is .4641, and you know that half of the total area is to the left of $z = 0$.

Therefore, the area to the left of $z = 1.8$ is $.5000 + .4641 = .9641$ or $.96$. What area is to the left of $z = -1.8$? Reference to your graph will convince you that this is $.5000 - .4641 = .0359$ or $.04$.

What area lies outside the values $z = -1.8$ and $z = 1.8$? This is the area to the left of -1.8 and the area to the right of 1.8, or $.0359 + .0359 = .0718$ or $.07$. It can, of course, also be computed as $1.0000 - 2(.4641) = 1.0000 - .9282 = .0718$.

EXERCISE 8-1

(Some students will need to go through only a few of the exercises in each group in order to achieve mastery. Others will need much practice. It is suggested that you check your results with the answer key and stop work in each group as soon as you find you are making no errors.)

1. First, by counting squares on your graph and then by reading from Appendix Table I, find the area between the mean and a point for which z has the value indicated.
   (a) .60   (c) 1.60   (e) -2.00 (g) 1.00
   (b) .40   (d) $-.10$ (f) 1.80   (h) .70

2. What is the area between the indicated symmetrically placed values of z? Read answers from the table.
   (a) $-.45$ and $.45$ (c) -1.00 and 1.00 (e) -2.57 and 2.57
   (b) $-.67$ and $.67$ (d) -1.96 and 1.96 (f) -1.03 and 1.03

3. What is the area to the left of the indicated value of z? Read answers from the table.
   (a) .15   (c) 1.90   (e) -1.85 (g) .67
   (b) -1.32 (d) 1.67   (f) -1.65 (h) $-.67$

4. What is the area outside the indicated values of z? Read answers from the table.
   (a) -1.96 and 1.96 (c) -2.32 and 2.32 (e) $-.75$ and $.75$
   (b) -1.64 and 1.64 (d) -2.58 and 2.58 (f) -2.81 and 2.81

Percentiles of the Normal Distribution.   Suppose you have read from Appendix Table I that 73 percent of the area under the standard normal curve lies below—that is, to the left of—$z = .61$. The same relationship may be stated in either of the following ways:

(a) $z = .61$ is the 73rd percentile of a standard normal distribution
(b) $z_{.73} = .61$

In earlier chapters the letter x with appropriate subscript was used to designate a percentile. As we are now using the letter z to denote a standardized deviate of the normal distribution, we shall use z with appropriate subscript to denote a percentile of that distribution. Thus $z_{.50}$ is the median (and also the mean because the curve is symmetrical). Whenever z has a subscript larger than .50, it represents a point to the right of the median, and such points have positive values. Whenever z has a subscript smaller than .50, it represents a point to the left of the median, and such points have negative values.

What is the value of $z_{.90}$? The subscript tells us that 90 percent of the area lies to the left of the desired point and, therefore, we know that 90 percent $-$ 50 percent $=$ 40 percent lies between that point and 0. In Table I we look along the area column to find the entry which is closest to .40. This is the entry .3997 corresponding to $z = 1.28$. We conclude that $z_{.90} = 1.28$. This solution required some maneuvering, but Appendix Table II would have given us immediately the statement that $z_{.90} = 1.282$.

What is the value of $z_{.10}$? Because the subscript is less than .50 we know the value must be negative. Since 10 percent of the area is to the left of $z_{.10}$, the area between $z_{.10}$ and the mean must be $.50 - .10 = .40$. As before, in Table I we find the number nearest .40 in the area column and read the corresponding value in the z column. As before, it is 1.28, but now we must attach a minus sign. Thus $z_{.10} = -1.28$. Now look in Table II. Here you read that $z_{.10} = -1.282$.

Since $z_{.10} = -1.282$ and $z_{.90} = 1.282$, the following relations hold:

$$z_{.10} = -z_{.90}$$
$$z_{.90} = -z_{.10}$$
$$z_{.10} + z_{.90} = 0$$

These statements might be generalized by using some letter, say, a, to represent one of the subscripts:

$$z_a = -z_{1-a}$$
$$z_{1-a} = -z_a$$
$$z_a + z_{1-a} = 0$$

If a normal distribution has $\mu = 72$ and $\sigma = 13$, what is its 40th percentile? From Table II we read that the 40th percentile of the standard normal curve is $-.253$. This tells us that $(x_{.40} - \mu)/\sigma = -.253$.

If $\mu = 72$ and $\sigma = 13$, then

$$\frac{x_{.40} - 72}{13} = -.253$$

Therefore,

$$x_{.40} - 72 = -.253(13) = -3.289$$
$$x_{.40} = 72 - 3.289$$
$$x_{.40} = 68.7$$

EXERCISE 8-2

In each of the groups of questions 1 through 11 answer only as many questions as you need to achieve mastery. Be sure to go on to Group 9, as the questions in that group provide important background for Chapter 9.

1. What area is to the left of each value?
   (a) $z_{.07}$ (b) $z_{.96}$ (c) $z_{.55}$ (d) $z_{.31}$ (e) $z_{.45}$

2. What area is to the right of each value?
   (a) $z_{.99}$ (b) $z_{.05}$ (c) $z_{.40}$ (d) $z_{.995}$ (e) $z_{.94}$

3. Which of the values named in groups 1 and 2 are negative?

4. What area lies between the stated values?
   (a) $z_{.01}$ and $z_{.99}$ (c) $z_{.25}$ and $z_{.75}$ (e) $z_{.025}$ and $z_{.975}$
   (b) $z_{.005}$ and $z_{.995}$ (d) $z_{.02}$ and $z_{.08}$ (f) $z_{.60}$ and $z_{.71}$

5. By reference to Appendix Table II, verify the following statements:
   (a) $z_{.01} = -2.326$ (e) $z_{.75} = .6745$ (i) $z_{.999} = 3.090$
   (b) $z_{.25} = -.6745$ (f) $z_{.80} = .842$ (j) $z_{.005} = -2.576$
   (c) $z_{.12} = -1.175$ (g) $z_{.20} = -.842$ (k) $z_{.995} = 2.576$
   (d) $z_{.70} = .524$ (h) $z_{.94} = 1.555$ (l) $z_{.975} = 1.960$

6. What area lies outside the stated values? (This area is the sum of the area to the left of the smaller value and the area to the right of the larger.)
   (a) $z_{.01}$ and $z_{.99}$ (c) $z_{.025}$ and $z_{.975}$ (e) $z_{.05}$ and $z_{.95}$
   (b) $z_{.005}$ and $z_{.995}$ (d) $z_{.02}$ and $z_{.98}$ (f) $z_{.10}$ and $z_{.90}$

7. Between which percentile values is the middle 30 percent of the area included? Solution: If 30 percent of the area is in a strip symmetrically

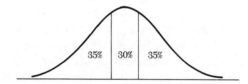

placed around the mean, one half of 30 percent = 15 percent must lie on each side of the mean. Therefore, the area below the lower edge of that strip must be $.50 - .15 = .35$ and below the upper edge, $.50 + .15 = .65$. The strip therefore, extends from $z_{.35} = -.385$ to $z_{.65} = .385$. These values were obtained by reading $z_{.34}$ and $z_{.36}$ from Appendix Table II and interpolating between them.

8. State the percentile values which would bound a middle strip of area symmetrically placed around the mean if that area is as follows:
   (a) .99   (c) .50   (e) .90   (g) .25
   (b) .98   (d) .95   (f) .80   (h) .999

9. In a normal distribution with $\mu = 64$ and $\sigma = 8$,
  (a)   What is the value of the 10th percentile?
  (b)   What is the value of the 70th percentile?
  (c)   What proportion of cases would have scores larger than 70?
  (d)   What proportion of cases would have scores less than 50?
  (e)   Between what scores would the middle 50 percent of the cases lie?
  (f)   Between what scores would the middle 95 percent of the cases lie?
  (g)   Beyond what score would the highest 5 percent lie?
  (h)   Outside what two scores, symmetrically placed in regard to the mean, would the most extreme 1 percent lie?

10. Answer the eight questions asked in Group 9 if $\mu = 91$ and $\sigma = 15$.

11. Answer the eight questions asked in Group 9 if $\mu = 59$ and $\sigma = 12$.

Probable Error.   The values $z_{.25}$ and $z_{.75}$ are the first and third quartiles of the standard normal distribution, and one half the area under the normal curve lies between them. From Appendix Table II it can be seen that $z_{.25} = -.6745$ and $z_{.75} = .6745$. The value $.6745\sigma$ is called *the probable error*.[1] The term is rather unfortunate because it is somewhat misleading but it has been in use since 1815 and seems to survive all efforts to do away with it. Neither the term nor the idea will be used again in this book. It is mentioned here only because you are likely to meet it in your reading elsewhere.

Normal Probability.   In certain very interesting and very important types of problems, some of which are treated more fully in subsequent chapters, the task is to assess the likelihood that a single individual, chosen "at random" from a group of individuals, will come from a particular category of that group. In such problems the *proportional frequency in each category is called the probability of that category and the proportional frequency distribution is called a probability distribution.*

There are many forms of probability distribution; not by any means all probability distributions are normal. A probability distribution may have a continuous scale or may consist of discrete classes. It may consist of just two classes. For example, it is well known that about 51 percent of newborn babies are boys and about 49 percent girls. The two classes,

---

[1] The term was introduced in 1815 by the great German astronomer and physicist, Friedrich Wilhelm Bessel, in a paper on the position of the pole star, in the Berlin Astronomical Yearbook. (See Walker, *Studies in the History of Statistical Method*, p. 24.) He used the term *der wahrscheinliche Fehler*. Almost at once the expression appeared in scientific journals all over Europe, as *error probabalis* in Latin, as *l'erreur probable* in French, as *l'errore probabile* in Italian, as *the probable error* in English. The roots of this term are deep and tenacious.

boys and girls, and the relative frequencies .51 and .49 constitute a probability distribution. One may say that there is probability .51 that the first baby born in Chicago in the year 2000 will be a boy. In general, a probability distribution consists of a set of classes and the relative frequencies related to them. Many extremely useful probability distributions are described in more advanced texts. At this point our emphasis will not be on the topic of probability in general but on the use of the normal distribution as a probability distribution in preparation for a better understanding of the important ideas taken up in later chapters.

First let us consider a rather artificial situation. Suppose that for each of 50,000 school children an intelligence quotient (IQ) is available and that the distribution of these is approximately normal. If each child's IQ is recorded on a small ticket, all the tickets are thoroughly mixed in a huge lottery, and one ticket is drawn at random,[2] what is the probability that the IQ recorded on that ticket will be at least two standard deviations above the mean of the entire group of 50,000? Because the distribution has been described as normal, this question can be answered by merely finding the area under a normal curve beyond $z = 2$. Appendix Table I shows this area to be .023. What is the probability that a ticket drawn at random will contain an IQ very close to the mean—say, one lying within 0.1 of a standard deviation from the mean? Answer $2(.0398) = .08$.

Now let us discard as irrelevant the fiction of writing the IQ's on tickets and physically drawing one by means of a lottery but retain the idea of a very large normally distributed population of IQ's. Selecting at random an individual from such a population is a matter of some difficulty because people of widely different intelligence levels do not tend to frequent the same places. So let us again place a question with respect to an unborn infant. What is the probability that the first child born in Chicago in the year 2000 will have an intelligence quotient at least 3 standard deviations below the general mean? In other words, what is the area under the normal curve below $z = -3$? Table I shows this to be .0013.

Classifying in Groups with Equal Range.   Suppose a research worker is carrying out a study for which he needs a measure of job satisfaction (or interest in teaching, or conservatism-liberalism, or any other attitude difficult to measure objectively). This research worker has drawn up a list of 150 statements of opinion ranging from extreme satisfaction to extreme dissatisfaction and wants to place these on a scale. Ranking the 150 statements looks like a formidable undertaking but it does not seem too difficult to place them in ordered categories.

[2] Some discussion of what drawing "at random" means will be found in Chapter 9. For the present, the term will be left admittedly vague.

The number of categories is an arbitrary choice. Let us suppose the investigator decides to use 9 categories. Suppose also that he makes the assumption that his set of statements represents a normal distribution of job satisfaction and that he wishes to have each of the 9 categories cover the same range on the scale of the distribution. Strictly speaking, this is impossible because the range of a normal distribution is infinite and $\frac{1}{9}$ of an infinite range would also be infinite. However, if he lets each category have a range of 0.5σ, the 9 categories will cover a range of 4.5σ, or the range from -2.25σ to 2.25σ.

Reference to the table of the normal distribution shows that only a little more than 1 percent of the area under the curve lies above 2.25σ or below -2.25σ. Consequently, the investigator may decide to make the end intervals open, so that all cases above 1.75σ constitute one category and all cases below -1.75 constitute another category. The following tabulation shows the 9 categories formed by such a subdivision of the scale, the corresponding percents of area under the curve, the number of statements in each category in accordance with the indicated percents, and code numbers ranging from 1 to 9 which are to be the scale values of the categories.

Category		Percent	Number	Code
Above	1.75	4.01 or 4	6	9
1.25 to	1.75	6.55 or 7	10	8
.75 to	1.25	12.10 or 12	18	7
.25 to	.75	17.47 or 17	26	6
−.25 to	.25	19.74 or 20	30	5
−.75 to	−.25	17.47 or 17	26	4
−1.25 to	−.75	12.10 or 12	18	3
−1.75 to	−1.25	6.55 or 7	10	2
Below	−1.75	4.01 or 4	6	1
		100.00	150	

The number of items to be placed in each of the 9 categories is shown in the accompanying sketch as an area under the normal curve between the points indicated.

The statements must now be distributed in accordance with the frequencies shown in the tabulation. Each statement is written on a separate

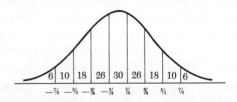

slip of paper, and these slips are sorted into 9 piles of the size specified. The 6 statements (no more and no less) which are judged as expressing the greatest degree of job satisfaction (or interest in teaching, or whatever variable is under consideration) are placed at the extreme right and coded 9, and the 6 which are judged to express the greatest dissatisfaction are placed at the extreme left and coded 1. The other slips are assigned to intermediate piles in such a way as to form a set of ordered classes. The code numbers from 1 to 9 which are assigned to these classes are usually treated as constituting a scale.

This is also the general line of procedure for what is called the "Q-sort" used in many psychological studies. Each of the items to be sorted is written on a separate card, and the judges work independently. If they are told how many cards to place in each pile, every judge will achieve the same mean and standard deviation. If they are left free to decide how many cards to place in each pile, there may be considerable discrepancy from judge to judge as to mean and as to standard deviation of the distribution of cards. The purpose of having more than one judge to do the sorting is twofold: (a) to ascertain how well different judges can agree on the placement of items and (b) to improve measurement by eliminating those items on which a prescribed degree of agreement was not reached. Thus the investigator in this problem might intend to reduce his original 150 items to 50 by first throwing out each one which was so ambiguous that the judges did not place it in the same category and then by random selection (discussed in the next chapter) eliminating enough others so that the frequencies in the categories were one third as large as' for the original 150 items, namely 2, 3, 6, 9, 10, 9, 6, 3, and 2.

EXERCISE 8-3

Verify enough of the following statements to be sure you understand them.

1. If individuals are classified in 5 categories each with range of 1σ, the division points will fall at -1.5, $-.5$, $+.5$, and 1.5, and the relative frequencies in each category will be .0668, .2417, .3830, .2417, and .0668, or roughly .07, .24, .38, .24, and .07.

2. If individuals are classified in 5 categories each with range of 1.2σ, the division points will fall at -1.8, $-.6$, .6, and 1.8, and the relative frequencies in the categories will be .0359, .2384, .4514, .2384, and .0359.

3. If 6 categories are used with range of 1σ each, the division points will be at -2, -1, 0, 1, and 2, and the relative frequencies will be roughly .02, .14, .34, .34, .14, and .02.

4. If 11 categories are used with range of $.4\sigma$, the division points will be -1.8, -1.4, -1.0, $-.6$, $-.2$, .2, .6, 1.0, 1.4, and 1.8, and the relative frequencies will be .036, .045, .078, .116, .146, .159, .146, .116, .078, .045, and .036.

5. If 11 categories are used with range of $.5\sigma$ each, the division points will be -2.25, -1.75, -1.25, $-.75$, $-.25$, .25, .75, 1.25, 1.75, and 2.25, and the relative frequencies will be approximately .01, .03, .07, .12, .17, .20, .17, .12, .07, .03, and .01.

Transforming Ordered Categories into Scaled Scores. Let us suppose that a wealthy citizen has given a sum of money to a high school for scholarships, with the stipulation that the recipients shall be chosen on the basis of average class marks. This presents a problem because the school's records are in the form of letter grades, and now it must average these letter grades in some way which will appear fair and objective to all concerned. One member of the selection committee proposes to use the arbitrary code: A = 10, B = 9, C = 8, D = 7, E = 6, and F = 5. Another member points out that reducing each code number by 5 so that A = 5, B = 4, C = 3, D = 2, E = 1, and F = 0 would simply reduce each pupil's average by 5, thus leaving pupils in the same relative order but would somewhat lessen the work of obtaining averages. The committee is about to adopt this latter code when someone calls attention to

Table 8-2   The Number and Percent of Students to Whom Each of Three Teachers Has Assigned the Specified Letter Mark

Mark	Number of Marks Given by			Percent of Marks Given by		
	Smith	Jones	Brown	Smith	Jones	Brown
A	24	36	106	9.6	12.0	26.5
B	130	78	152	52.0	26.0	38.0
C	68	95	96	27.2	31.7	24.0
D	12	55	36	4.8	18.3	9.0
E	8	12	6	3.2	4.0	1.5
F	8	24	4	3.2	8.0	1.0
Total	250	300	400	100.0	100.0	100.0

the marks given over a period of several terms by three teachers in the same department, teaching students who are presumably similar in ability. These three teachers, Smith, Jones, and Brown, have assigned marks as indicated in Table 8-2. One member of the committee objects to the code on the ground that the letters do not seem to have exactly the same meaning as used by these three teachers, that an A given by Brown does not present as strong evidence of high achievement as one given by Smith. Another objection he makes to the code is that it is based on the assumption that the 6 letters represent 6 evenly spaced points on a scale, that A is assumed to be as much better than B as B is better than C, and so forth. This he says is unrealistic. This committee member proposes an alternate plan of scaling, based on the assumption that there was really a normal distribution of performance for the students taught by each teacher. He admits that this assumption may not be true—that it can neither be established nor refuted because there is no evidence. He thinks the assumption of normality is at least as acceptable as the assumption that increments between consecutive letters represent equal scale intervals. In order to place the categorical ratings (the letter grades) on a scale, it is necessary either to make an assumption concerning the scale interval (that is, assign an arbitrary code) or to make an assumption concerning the form of distribution and to derive the scale equivalents from that distribution.

The committee member who argued in favor of making the assumption of normality carried out the appropriate computations and presented the following transformation code for the marks of Smith, Jones, and Brown:

Mark	Smith	Jones	Brown
A	67	66	61
B	54	57	51
C	43	49	43
D	36	42	35
E	33	37	29
F	29	32	24

He admitted, however, that the application of his method might be difficult to explain to the public, who were much interested in the allotment of the awards and that it would require a considerable amount of work to make such a transformation for the grades of every teacher in the school and then to average the transformed scores of individual pupils.

The method used in making the transformation may be illustrated for Jones' set of 300 marks. A normal distribution is assumed to be divided

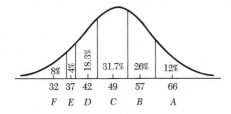

into segments with areas corresponding to the percents of frequency shown in Table 8-2. For each segment of area there must be found a value of the normal deviate z which will serve as typical value for that category. Either the median value of the category or the mean value can be used, but the median is easier to obtain, and so we shall use it. Consider category A. The median of that category is a point which has one half of the area of the segment on either side and, therefore, has .06 of the total area of the curve above it (to the right) and .08 + .04 + .183 + .317 + .26 + .06 = .94 of the total area below it. Appendix Table II gives 1.555 as the value of $z_{.94}$. If we let $T = 10z + 50$, the transformed value for grade A will be 15.55 + 50 = 66.

Now consider the median of category B. The area below it will be .08 + .04 + .183 + .317 + .13 = .75 and the area above it .13 + .12 = .25. The value of $z_{.75}$ is .6745 and $T = 6.7 + 50 = 57$. In similar manner a value of T is found for the median of each category.

Table 8-3 Computations Used in Transforming a Set of Ordered Categories to Scaled Scores (T) on the Assumption of Normality of Distribution

Category	Frequency	Percent in Category	Percent below Median of Category	z	T $= 10z + 50$
A	36	12.0	94.0	1.56	66
B	78	26.0	75.0	.67	57
C	95	31.67	46.17	−.10	49
D	55	18.33	21.17	−.80	42
E	12	4.0	10.0	−1.28	37
F	24	8.0	4.0	−1.75	32
	300	100.0			

The computations leading to the transformation are presented in Table 8-3. Here the first two columns provide the data. The third column is obtained by dividing each frequency by the sum of all the frequencies (300 in this case). The percent below the median of the category for category C is 8.0 + 4.0 + 18.33 + (31.67/2) = 46.17 and similarly for the other categories. The z values are read from either Table I or II and need not be recorded with great precision because after they are multiplied by 10 they will be rounded to the nearest integer. The entries in the final column are obtained by multiplying z by 10, adding 50, and rounding off the final digit.

Normalizing a Distribution of Scores. The procedure described in the preceding paragraph for transforming ordered categories into scores can also be used to convert a non-normal distribution of scores into a normal distribution. Each score interval defines a category, and the cumulative frequency distribution is treated in the manner illustrated in Table 8-4.

Table 8-4   Computation for Normalizing a Distribution of Scores

x	f	f to Cumulative Midpoint	Cumulative Percent	z	T
45	22	139	92.7	1.45	65
44	20	118	78.7	.80	58
43	24	96	64.0	.36	54
42	20	74	49.3	−.02	50
41	18	55	36.7	−.34	47
40	16	38	25.3	−.67	43
39	11	24.5	16.3	−.98	40
38	7	15.5	10.3	−1.26	37
37	4	10.0	6.7	−1.50	35
36	4	6.0	4.0	−1.75	32
35	3	2.5	1.7	−2.12	29
34	1	0.5	0.3	−2.75	22
	150				

Such normalized scores were named T-scores by William A. McCall in honor of E. L. Thorndike and Louis Terman, pioneers in the measurement movement.

These transformed T-scores must not be confused with the modified standard scores, discussed on page 92 of Chapter 6. There are three important distinctions: (a) Modified standard scores can be obtained only for a scaled variable, not from one expressed in ordered categories, whereas T-scores can be obtained for either kind of variable; (b) to obtain modified standard scores one first computes the mean and standard deviation and then computes $50 + 10\ (x - \bar{x})/s$ whereas to obtain T-scores one first computes percentile ranks (the percent below the midpoint of the category), reads the corresponding values of z from a table of normal probability, and then $T = 50 + 10z$, only the final step of the two procedures being the same, and (c) T-scores have a normal distribution regardless of the form of the original distribution, but modified standard z scores do not have a normal distribution unless the original variable had a normal distribution.

EXERCISE 8-4

1. Verify the T-score values given for the grades of Smith and Brown on page 125.

2. Suppose a supervisor has placed a group of 200 teachers in 6 qualitative categories as follows:

Excellent in every way	20
Superior in most traits	30
Above average	64
Fair	48
Poor but not subject to dismissal	30
Very poor, subject to dismissal	8

What transformed scores correspond to these classes?

3. For the data of Table 8-4 compute the modified standard scores $50 + 10(x - \bar{x})/s$, and compare them with the T-scores shown in the table.

Profile Chart. When an individual is to be evaluated relative to a group on several traits at once, it is customary to plot his scores on a graph known as a *profile chart*.

A profile chart of Pupil 1 of Appendix Table IX appears in Figure 8-6. The bars represent the percentile ranks of this individual on the five variables named at the left of the graph. To read the bars it is necessary to refer to the horizontal scale at the bottom of the graph. The values of this scale are percentile ranks referred to the standard normal distribution. A bar originates at the 50th percentile rank and extends to the scale value representing the individual's percentile rank on a particular test. Thus we see from the bars that Pupil 1 has percentile rank 26 in IQ, percentile rank 88 on arithmetic computation, and so on.

In addition to its value in showing the relationship of an individual to the group, the profile chart is useful in comparing individuals. Figure 8-7 shows profiles of Pupils 1, 2, and 3 from Table IX set side by side for ready comparison. It is immediately apparent from the figure that Pupil 3 has the highest IQ and the highest scores on the verbal tests of reading speed and language but is lowest in computation and history-civics.

A profile chart like that shown in Figure 8-6 can be made most simply by plotting percentile ranks on specially printed paper, which can be purchased as "arithmetic probability paper." If such paper is not available, the percentile ranks can be converted to T-scores, and these can then be plotted on an arithmetic scale.

The use of bars is a departure from the common practice of drawing line segments to connect the plotted percentile ranks. That common practice makes a profile chart look like a line graph and emphasizes especially the relationship of a variable to its immediate neighbors. In a profile chart the order in which the variables are listed is usually arbitrary, so that the appearance of a line graph of a profile might be greatly changed if the order of listing were changed. The use of bars rather than connecting lines does away with the false impression of a trend and simplifies both the evaluation of the individual on each variable and comparison of his standing on any one variable with his standing on all other variables.

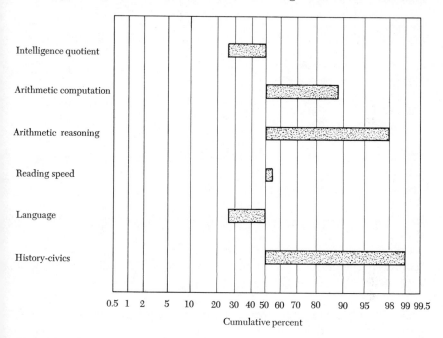

Figure 8-6 Profile on Normal Probability Scale for Pupil 1 of Appendix Table IX

A profile chart is sometimes made up by plotting percentile ranks directly on an arithmetic scale. This usage is unsatisfactory because it makes the distances on the scale for a difference between two percentiles near the middle the same as the distance for a like difference near the ends of the distribution. Most distributions ordinarily considered are less dense at the ends of the distribution than at the middle.

Another common practice is to make a profile chart by plotting the standard scores described in Chapter 6. The usefulness of this practice is limited when the shape of distribution differs from trait to trait. Consider, for example, two traits, one of which is measured by an easy test and the other by a difficult test. On the easy test, the top score might be only one and a half standard deviations above the mean, whereas on the difficult test the top score might be three standard deviations above the

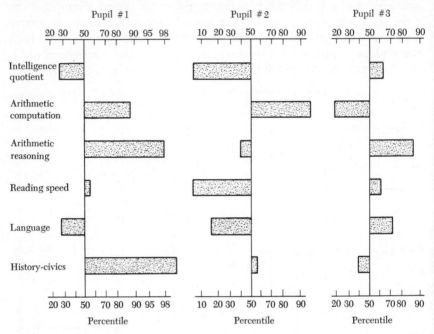

Figure 8-7   Profiles of Pupils with Code Numbers 1, 2, and 3 in Appendix Table IX

mean. Clearly a standard score of one and one half means something quite different on the two tests.

Transforming Ranks to Normal Deviates.   It often seems reasonable to assume that the underlying variable represented by the ranks is normally distributed. Appendix Table X may be used to transform a set of ranks into a set of scores from a normal distribution in which $\mu = 50$ and $\sigma = 10$, for samples in which $5 \leq n \leq 30$. To find the normal equivalent of a rank R when $n > 30$, first find the proportion

$$(8\text{-}4) \qquad\qquad p = 1 - \frac{R - 0.5}{n}$$

then read in Table II the normal deviate z_p in a unit normal curve corresponding to the proportion p. The corresponding normal value when $\mu = 50$ and $\sigma = 10$ is $10z_p + 50$.

Discovery of the Normal Curve.[3] The curve which we now call "normal" is the curve of normal probability or curve of error which attracted the interest of most of the great mathematicians and astronomers of the first half of the nineteenth century. More than two centuries ago, Abraham De Moivre (1667–1754), a refugee mathematician living in London, making his living partly by solving problems for wealthy gamblers, recognized that the random variation in the number of heads appearing on throws of n coins corresponds to the terms of the binomial expansion of $(.5 + .5)^n$ and that as n becomes larger this distribution approaches a definite form. In 1733 De Moivre derived the equation for this curve and presented it privately to some friends. To him it was only a mathematical exercise, utterly unconnected with any sort of application to empirical data.

At that time there were no collections of empirical data at hand for study. As yet no one had made any measurements of any large number of individuals. A Swiss mathematician named Jacques (or James) Bernoulli (1654–1705, eldest of the very remarkable family of mathematicians by that name) had suggested that the theory of probability might have useful applications in economic and moral affairs, but he, himself, was too near death to investigate the applications, and, moreover, he had no numerical data which he could have used for that purpose. The idea must have seemed fantastic to his contemporaries. In 1713, his great book on probability, *Ars Conjectandi*,[4] was published posthumously under the editorship of his nephew Nicolas, who seems to have tried in vain to induce other mathematicians to develop the argument further. Nearly a century passed before any scientific worker began to gather large masses of concrete data and to study the properties of distributions. The application of the normal curve in studies of concrete data begins with the work of the great mathematical astronomers who lived at the beginning of the nineteenth century, chiefly Laplace (1749–1827) in France and Gauss (1777–1855) in Germany, each of whom derived the law independently and presumably without any knowledge of De Moivre's derivation.

[3] For a more extended treatment, see H. M. Walker, *Studies in the History of Statistical Method*, Baltimore, The Williams & Wilkins Co., 1929, Chapter II; "Bicentenary of the normal curve," *Journal of the American Statistical Association*, **29** (1934), 72–75; "Abraham De Moivre," *Scripta Mathematica*, **2** (1934), 316–333. The latter paper is reprinted as a biographical appendix in a photographic reprint of De Moivre's *The Doctrine of Chances*, brought out by the Chelsea Publishing Company, New York, 1967. Articles on the Bernoulli Family, Gauss, Laplace, De Moivre, Pearson, and Quetelet will be found in the 1968 edition of the *International Encyclopedia of the Social Sciences*.

[4] Conjecto = to throw together, therefore, to gamble, therefore, to guess at or surmise.

The probability curve is often called the Gaussian curve, because until recently it was supposed that Gauss had been the first person to make use of its properties. However, in 1924, Karl Pearson discovered a hitherto unknown derivation by Abraham De Moivre.[5]

The idea that this curve could be used to describe data other than errors of observation in the physical sciences seems to have originated with the great Belgian statistician Adolphe Quetelet (1796–1874), who first popularized the idea that statistical method was a fundamental discipline adaptable to any field of human interest in which mass data were to be found. He was convinced that the measurement of mental and moral traits waited only for the collection of sufficient and trustworthy data and was sure that when such measurement was feasible, the distribution of these traits would be found to be in accordance with the "law of error."

[5] Pearson, K., "Historical note on the origin of the normal curve of error," *Biometrika*, **16** (1924), 402–404.

9

INTRODUCTION
TO STATISTICAL
INFERENCE

In an investigation which deals with matters of general interest, the findings are usually applied to a much larger number of cases than were actually observed in the sample of cases. This use of samples in statistics is probably the most important and certainly the most intriguing aspect of this science. The idea that information obtained from the examination of a relatively small number of cases can be used to throw light on the characteristics of a vast and unexamined universe is an exciting idea. That the sample not only furnishes an estimate of some characteristic of an unknown population but also furnishes a measure of how much faith can be placed in that estimate is still more remarkable.

Situations in which Samples Are Employed. Statistical sampling plays a part in almost all statistical studies on which decisions for future action are to be based. Sometimes sampling is necessary because the universe, or population, about which information is desired is infinite and could never be examined in its entirety. This situation is typical of most psychological experiments. Some populations, even though finite, are so vast or so inaccessible that it would not be feasible to examine all their members. Sometimes, as in Example 1 below, it is important to process the results more quickly than would be possible if a complete count (or census) of the entire population were made. In certain kinds of research, measurement means the destruction of the individual measured. This is the typical situation in industrial studies where the strength of material or the length of life of some product is measured. To measure the tensile strength of cloth, pieces of cloth must be destroyed; to measure the breaking load of steel, steel must be broken; to measure the effective life of a battery or a light bulb or an automobile tire, it must be used until it is worthless. Consequently, in much industrial research the idea of studying an entire population is fantastic, and even the use of large samples is prohibitive.

Population.   A sampling study is characterized by the fact that its objective is not simply to describe those individuals observed in the sample but rather to search for information about the population from which the sample was drawn.

The statistical concept of a *population*, also called a *universe*, is an abstract concept fundamental to statistical inference. "It refers to the totality of numbers that would result from indefinitely many repetitions of the same process of selecting objects, measuring or classifying them, and recording the results. A population is, thus, a fixed body of numbers about which we would like to know. What we actually know is the numbers of a *sample*, a group selected from the population."[1] The reader must not think of a population in this sense as necessarily consisting of people or even of measurements of people. The three examples which follow happen all to be situations in which the population consists of observations on persons; they were chosen not because of but in spite of that peculiarity. A population might consist of observations on schoolhouses, on books, on cities, on shoes, on carloads of grain, on manufactured items, and so on.

To appreciate the nature of such studies, let us examine a few examples more closely.

1.   *Estimating the number of unemployed in the United States.* In order to obtain current information on the state of the labor market, the United States Bureau of the Census conducts periodic sampling studies from which it calculates an estimate of the number of persons unemployed in various kinds of work and in different parts of the country. For the purpose of this investigation, each person in the labor market is placed in one of two classes: employed and unemployed. A two-class population is called a *dichotomous* population. The sample is necessarily obtained by selecting individual persons from the total number in the labor market. For the purpose of this investigation, when John Jones is drawn into the sample, we are not interested in all the personal attributes of John Jones as a human being, but only in whether he belongs in the class of the employed (E) or the unemployed (U). Thus the population is the totality of all observations U and E, and the sample consists of the number of

---

[1] W. Allen Wallis and Harry V. Roberts, *Statistics, a New Approach*, The Free Press, Glencoe, Ill., 1956, p. 126.

observations of U and of E among the individuals selected. This particular example was chosen because it illustrates a dichotomous population and because it illustrates a situation in which the processing of a sample can be done rapidly enough to provide information while it is fresh and useful. The processing of a complete census would require so long that, by the time the findings could be published, they might no longer be applicable.

2. *Standardizing a test.* Standardized tests of achievement, of ability, of attitudes, or of preferences are used to determine the status of an individual person relative to a population or of a subgroup relative to a population. (Such tests are also used to compare subgroups, but for this purpose it is not usually necessary that they be standardized.)

Consider for example the use of the Stanford Binet Test to provide information concerning the intelligence of an 8-year-old child. The child's test score would be useless in this connection without information about the distribution of the Binet scores of 8-year-old children in general, in order to provide a standard against which an individual score could be rated. But the administration of the Binet test to *all* 8-year-old children is manifestly impossible, partly because of expense and partly because the class of persons who are ultimately to be measured by this test extends on into the future and includes children who are not now members of the population of 8 year olds. The desired norms can be developed only on the basis of sample data.

In this example the population of Binet scores for 8 year olds may be presumed to have a normal distribution, whereas in the first example the population was composed of two classes. While in this example it is impossible to obtain a complete census, in the first example it was merely inexpedient.

3. *Testing the effectiveness of a new drug.* When a new drug or vaccine is developed in a laboratory, its effectiveness must be tested by using it on live subjects. The first tests may be made on animal subjects, but, if it is intended for human use, later tests of its effectiveness must be made on human subjects.

A naive research worker might assume that all he needs to find to prove the effectiveness of a new drug is the proportion of subjects who recover from the malady in question when they are treated with the given drug. Thus he would be thinking in terms of a two-class population of observations; each person who had had the malady and been treated by the drug would be labeled either "recovered" or "did not recover." The proportion who recover under the given treatment would indeed provide some measure of the effectiveness of the drug but would not provide a basis for deciding whether to advocate its use. That decision can be made only after the results are compared with some alternative treatment, which may be some other drug or no drug at all.

The research worker must, therefore, consider two populations: (1) observations on persons who have the malady and are treated by drug A, and (2) observations on persons who have the malady and receive treatment B. Each of these populations is composed of two classes, those cured and those not cured. Presumably the decision to use drug A would be made only if the first population has a larger proportion of cures than the second.

To sum up, we speak of *a population as the collection of all possible measures of a certain kind, and we associate with that population a frequency distribution.* For convenience, the individuals themselves are often spoken of as a population, but this is a loose use of the term. Each individual may be measured on many different traits so that the collection of individuals is related to many different populations of measures. It is not the individuals themselves but some measure of those individuals which has a frequency distribution. Consider, for example, all the colleges in the United States. When these are classified as men's colleges, women's colleges, or coeducational, there is a population of three classes; when they are rated according to the state (or District of Columbia) in which they are located, there is a population of 51 classes; when they are scored as to age or present enrollment or endowment or rate of tuition, there are populations of quite different natures. Clearly then it is not the colleges themselves which form a population but rather the set of measures on those colleges.

Randomness.   It should be fairly clear at this stage of the discussion that it is ordinarily not possible to study a population in its entirety. Usually information about the population can be obtained only by the selection of a limited group of individuals from the population. The manner in which these individuals are selected is a very important consideration.

All statistical theory involving the use of samples is based on the notion of randomness. *By randomness we mean that any group of individuals is as likely to be chosen as any other group of the same size.* Randomness is required because this procedure permits us to apply the laws of chance to the theory of samples. This matter will be discussed more fully below.

Achieving randomness in the selection of subjects is often one of the most difficult aspects of a statistical investigation. Such investigations have probably been criticized more often for failing to achieve randomness than for any other shortcoming. The well-known Kinsey studies of sexual behavior are an example of studies which have been criticized severely in this respect. The problem of achieving randomness is solved differently in each field of endeavor and becomes an element of study in that field. However, some indications of methods for achieving randomness will be pointed out here.

Suppose, for example, it is proposed to set up an experiment to study three methods of teaching statistics. For the purpose of the study, suppose the 98 students listed in Appendix Table VIII are to be divided into three groups, each of which is to be taught by one of these methods. It becomes, obviously, a matter of considerable importance to make sure that the groups to be taught by different procedures do not basically differ in other ways which will influence their learning. Assigning students to the different groups in a random manner is the best way to eliminate this difficulty.

To make such random assignment, slips of paper may be numbered 1 through 98, thoroughly shuffled, and then divided into three groups, just as playing cards might be dealt to three players.

A less cumbersome way to achieve the same result is the use of a table of random numbers like the one in Appendix Table XI. This table consists of digits which have been previously randomized, so that their use has the same effect as the shuffling just described. To facilitate keeping track of one's position while using the table, the subdivisions of the numbers in Table XI are presented in blocks of five rows and five columns. Random numbers of any size can be obtained from this table.

For making a selection among the 98 students, we find their code numbers by choice of consecutive pairs of digits beginning with any row or any column. If we use the first 20 digits in line 1, we find the following pairs of digits: 10, 48, 01, 50, 11, 53, 60, and 20. This selection gives us 8 students after eliminating duplicates. If we continue in this way until 33 students have been selected, the first list of randomly selected students is available. Should the number 99 occur, it would be passed over because no student has that code number. The table may be entered at any point and read in any direction—to right, to left, up the column, down the column, or diagonally. Use of a different starting point and reading plan will produce a different but equally good random selection. Had there been, for example, 4200 students in the population to be sampled, it would have been necessary to read four-digit instead of two-digit numbers and to discard all such larger than 4200.

EXERCISE 9-1

1. Suppose you want to choose 10 students at random from a group of 260. You decide to begin in the upper left-hand corner of Appendix Table XI,

reading downward. Verify that you would select the code numbers 104, 223, 241, 94, 103, 71, 23, 10, 70, and 24.

2. Suppose you had decided to begin in row 31 and block 6 and to read downward. Verify that you would have selected code numbers 183, 170, 257, 244, 45, 219, 214, 130, 221, and 25.

3. Suppose you have 18 cases which you want to divide randomly into two groups of 9. You decide to begin at the extreme left of row 40 and read to the right. That row would yield numbers 14, 12, 7, 18, 1, and 3, all other numbers being larger than 18. You should have decided in advance what to do in case this row did not provide enough numbers. One course would be to go to row 41, reading from the left and picking up numbers 4, 2, and 9. These 9 numbers then form one group and the remaining 9 the other,

> Group I: Students 14, 12, 7, 18, 1, 3, 4, 2, and 9
> Group II: Students 5, 6, 8, 10, 11, 13, 15, 16, 17

Sample. The group of subjects selected from a population for the purpose of obtaining information about that population is called a sample. For the purpose of the analysis which is described below, it is necessary that the sample be selected by a process involving randomness. When the population is considered as a single group and a sample is selected from it by some random method, as by use of a table of random numbers, the sample is called a *simple random sample*.

Sometimes it is more appropriate to subdivide a population into several subpopulations, say into occupational or geographic groups, and then to sample randomly from the subpopulations. The subpopulations are called *strata*, and the sample selected from them is called a *stratified random sample*. Other variations on random sampling are described in advanced statistical literature. The expression *parent population* is often used in describing the relationship of a population to a sample drawn from it.

What is wanted, of course, is a *sample of measures drawn from a population of measures*, but usually the only practicable way of obtaining such measures is to obtain a sample of subjects from the collection of subjects loosely called a population of subjects and then to measure the subjects in this sample. If the subjects have been selected at random, the measures of those subjects will constitute a random sample from the population of measures.

Parameter and Statistic. A statistical population always has a definite distribution but ordinarily the characteristics of that distribution are unknown. If the observations are on a scaled trait, the distribution has a mean, a median, percentiles, a standard deviation, and so forth. If the distribution is not scaled, it is characterized by the proportion of individuals in its various classes. The characteristics of a population are called *parameters;* the corresponding characteristics of a sample are called *statistics.*

The symbols used in this text for the parameters which we shall discuss and the related statistics are as follows:

Characteristic	Population Parameter	Sample Statistic
Mean	μ (mu)	$\bar{x}$
Standard deviation	σ (sigma)	s
Variance	σ^2	s^2
Correlation coefficient	ρ (rho)	r
Proportions in the two classes of a dichotomous distribution	P, Q	p, q

Statistical Inference.   Ordinarily, the only information available to an investigator is contained in the statistics which he obtains from a sample. From this information he makes inferences about the population parameters.

Statistical inferences are usually of two kinds. One is *estimation of parameters* of a single population. The other is *comparison of populations.* Simple examples of both kinds of inference will be described below.

Law of Large Numbers.   The basic justification for statistical inference is that the distribution which is obtained from a random sample tends to resemble the distribution of the population from which it was drawn. This tendency increases as the size of sample increases. Consequently, certain statistics computed from the sample tend toward corresponding values of parameters of the parent population. Thus the mean of a large random sample tends to be close to the mean of its parent population. This tendency of distributions of random samples to resemble the distributions of their parent population more closely as sample size increases is called the *law of large numbers.*

The importance of randomness in the operation of the law of large numbers can hardly be exaggerated. Mere size without randomness may not provide the necessary approximation of sample to population.

Sampling Variability.   Two extreme views, both incorrect, are often expressed by statistically untrained persons in respect to inferences from samples to population. One view is to regard the sample as an exact reproduction of the population and to speak of statistics computed from the sample as if they were actual values of parameters. The other view is to hold that information from samples is quite valueless as a basis for inference about population values because so much depends upon what sample one happens to have chosen. It is the purpose of the discussion which follows to evaluate the reliability, or credibility, of information obtained from samples.

As a first step in this discussion, we shall set up an artificial situation in which we have a population with a fully known distribution. From this population we shall draw random samples and see how the statistics computed from these samples agree with the corresponding known parameters.

For the artificial population with which we are going to work, we have chosen a set of scores on the Cooperative Service English Test of 447 college students. The scores have a distribution which is very nearly normal, with parameters $\mu = 121.6$ and $\sigma = 37.15$. The list of scores, together with code numbers of students to whom each applies, appears in Appendix Table X. However, understanding of what follows is in no way dependent on reference to the set of original scores.

We now continue the fiction by supposing that 20 people separately seek information about the parameter μ by random sampling. Each of these people draws randomly a sample of 1, a sample of 5, and a sample of 25, and computes the sample mean $\bar{x}$ for each sample.

In Table 9-1, the column headed $n = 1$ shows the distribution of the means of 20 random samples of 1 case each. Obviously, the mean of a sample of 1 case is merely the value of the observation itself. The mean and standard deviation of this column would therefore be expected to approximate μ and σ but to differ somewhat from them because of random sampling errors. You should note that this is what happened. However, an attempt to estimate μ from information about a single individual would expose the investigator to risk of a very large error. Although the mean of the column for $n = 1$ comes close to μ, one sample had a value of only 36, and one had a value of 169. The range of estimates of μ based on these 20 samples was therefore $169 - 36 = 133$.

Now examine the column headed $n = 5$. The observed means of the 20 samples recorded here have considerably less variability than those in the column $n = 1$. Ungrouped data not presented here show the lowest as 84.4 and the highest as 144.2, so that the range is $144.2 - 84.4 = 59.8$. Obviously, the error made in estimating μ from 1 sample of 5 cases is likely to be numerically smaller than the error made in estimating μ from a single case. However, a person might happen to draw a single case with score almost exactly equal to μ and then might happen to draw a sample of 5 cases with a fairly extreme value of $\bar{x}$.

Table 9-1 Distributions of Means of 20 Samples Drawn from a Normal Population with $\mu = 121.6$ and $\sigma = 37.15$, when $n = 1$, $n = 5$, and $n = 25$

	Number of Samples		
Score Interval	$n = 1$	$n = 5$	$n = 25$
165–169	2		
160–164	1		
155–159	1		
150–154	3		
145–149	1		
140–144	2	3	
135–139		1	
130–134	1	2	1
125–129		2	5
120–124	1	1	6
115–119		2	5
110–114	1	1	3
105–109	1	4	
100–104	2	2	
95–99			
90–94			
85–89		1	
80–84		1	
75–79			
70–74	1		
65–69	1		
60–64			
55–59	1		
50–54			
45–49			
40–44			
35–39	1		
Total	20	20	20
* Mean of 20 means	122.5	119.0	121.7
* Standard deviation of 20 means	39.5	18.0	6.1

* Computed from ungrouped values.

Look now at the column headed $n = 25$. From the ungrouped data in Table 9-3 on page 155 we see that the smallest mean in the 20 samples was 111.0 and the largest was 133.6, so that the range was only 22.6.

The three distributions in Table 9-1 suggest certain generalizations about the validity of an estimate of μ made from a sample value of $\bar{x}$. These generalizations can be obtained by mathematical reasoning and are supported by many empirical studies. They apply only on the assumption that sampling is random. These generalizations are:

1. The value of $\bar{x}$ obtained from a sample can be expected to differ somewhat from μ.
2. In computations from many samples, the errors $\bar{x} - \mu$ are both positive and negative and the average error tends toward zero. In other words, the average value of $\bar{x}$ in many samples tends toward μ, regardless of sample size.
3. As sample size increases, the variability of the distribution of $\bar{x}$ decreases, and, consequently, there is smaller probability of making a large error in using $\bar{x}$ as an estimate of μ.

The third generalization is suggested both by the ranges and by the standard deviations of $\bar{x}$ in the three distributions of Table 9-1, which are:

	$n = 1$	$n = 5$	$n = 25$
Range of $\bar{x}$	133	59.8	22.6
Standard deviation of $\bar{x}$	39.5	18.0	6.1

Sampling Distributions.   The 20 individual observations in Table 9-1 had a mean and a standard deviation similar to those of the parent population, μ and σ. If sampling were continued indefinitely, that mean and standard deviation would tend to approach μ and σ more and more closely. The 20 means of samples of 5 cases also had a mean (119.0) similar to the mean of the parent population, and if more and more samples were drawn, the mean of their means would tend to approach μ more and more closely. A similar statement can be made for the samples of 25 cases. However for samples larger than one, the standard deviation of the set of means was smaller than the standard deviation σ of the parent population and would remain so even if the number of samples were vastly increased.

The mean actually has a distribution of its own which can be ascertained if the distribution of the parent population is known. In the previous paragraph we have noted that the variability of the distribution of the mean appears to depend on sample size, and that relationship will presently be made explicit.

For the samples used to obtain Table 9-1, many other statistics might have been computed and recorded. Thus, a distribution of medians might have been obtained, a distribution of variances, a distribution of ranges, or a distribution of standard deviations. In referring to such distributions, it is customary to refer to the statistic in the singular. Thus, the "distribution of the means of indefinitely many random samples of size n" is called "the distribution of the mean," just as the distribution of a set of heights is called the distribution of height rather than heights. Every statistic has a distribution which depends on the distribution of the parent population and on sample size, but the distributions of different statistics are not the same. *The distribution of a statistic is called the sampling distribution of that statistic.*

A sampling distribution is itself a population. It differs from the parent population by the fact that its elements are values of a statistic rather than single observations. The individuals on which a sampling distribution is based are samples rather than individuals of the parent population. A sampling distribution has parameters which are related to the parameters of the population of observations from which the sampling distribution is derived and to the number of cases in the samples.

The Normal Distribution for Statistics.   When the parent distribution is normal, the distribution of the mean is also normal. When the parent distribution is not normal, the sampling distribution of means is not normal for small samples but becomes very nearly normal as sample size increases. For many (but not for all) statistics the sampling distribution is approximately normal for large samples. For this reason the normal distribution is very important in statistical inference. However, for many purposes the normal distribution is insufficient for the needs of statistical inference, and three very important non-normal distributions will be introduced in later sections of this text. Even the elementary problems considered in this chapter will require the use of certain non-normal distributions.

When the sampling distribution of a statistic is normal, or approximately normal, the mean of that sampling distribution is either a parameter of the parent population or an expression mathematically derivable from the parameters of that population. The same is true of the standard deviation of the distribution of the statistic.

The Standard Error.   *The standard deviation of the distribution of a statistic is called the standard error of the statistic.* The standard error is especially important in dealing with statistics which are normally distributed. For such statistics the standard error provides information concerning the probability that a statistic will deviate from its parameter by a specified amount. The standard error of the mean will be denoted $\sigma_{\bar{x}}$ and is given by the formula

(9-1) $$\sigma_{\bar{x}} = \frac{\sigma}{\sqrt{n}}$$

To illustrate the meaning of $\sigma_{\bar{x}}$ let us look again at the data of Table 9-1. Since the standard deviation of the parent population was known to be $\sigma = 37.15$, we can compute $\sigma_{\bar{x}}$ for each of the three sample sizes and compare that value with the observed standard deviation of the 20 samples, thus:

Size of Sample	$\sigma_{\bar{x}}$	Observed Standard Deviation
$n = 1$	$\dfrac{37.15}{\sqrt{1}} = 37.15$	39.5
$n = 5$	$\dfrac{37.15}{\sqrt{5}} = 16.6$	18.0
$n = 25$	$\dfrac{37.15}{\sqrt{25}} = 7.4$	6.1

Even for as few as 20 samples, the correspondence between the expected value $\sigma_{\bar{x}}$ and the observed value is notable. Now let us see what we might have predicted about the three distributions of Table 9-1 from knowledge of the fact that $\mu = 121.6$ and $\sigma = 37.15$ for the parent population. Since the parent population was approximately normal, the distribution of the sample mean may be assumed to be normal. We know that 95 percent of the area under a normal curve lies between -1.96σ and $+1.96\sigma$. Therefore, if a great many samples are drawn from this parent population, we should expect about 95 percent of them to have means lying between $\mu - 1.96\sigma_{\bar{x}}$ and $\mu + 1.96\sigma_{\bar{x}}$. Now let us compare the observed distributions with these expected values.

Size of Sample	Value of $1.96\sigma_{\bar{x}}$	Range from $\mu - 1.96\sigma_{\bar{x}}$ to $\mu + 1.96\sigma_{\bar{x}}$	Observed Means Outside Range Number	Percent
$n = 1$	72.8	48.8 to 194.4	1	5
$n = 5$	32.5	89.1 to 154.1	2	10
$n = 25$	14.5	107.1 to 136.1	0	0

EXERCISE 9-2

Assume that random samples are drawn from a normal population for which μ and σ have the values shown below. For samples of the size indicated, what would be the mean and the standard error of the sampling distribution of $\bar{x}$? Within what range would the middle 95 percent of the sample means fall? The middle 50 percent?

Population Parameters			Distribution of $\bar{x}$		Range of Middle 95 Percent of Sample Means	Range of Middle 50 Percent of Sample Means
μ	σ	n	Mean	Standard Error		
1. 60	12	9	60	4	52.2 to 67.8	57.3 to 62.7
2. 60	12	64				
3. 40	12	36				
4. 40	18	36				
5. 92	8	100				
6. 53	25	100				
7. 39	15	900				
8. 39	15	144				

Estimation by a Single Value.   Previously in this chapter we indicated two extreme views toward statistical inference, one that a sample provides an exact reproduction of the population and the other that a sample provides little or no information about a population.

We have now reached a stage where we can discuss this problem somewhat rationally. We have found that the mean of a random sample is likely to be not far from the population mean, provided the sample is not too small. On this basis we feel justified in calling the sample mean an estimate of the population mean. We have used here the expression "estimation by a single value" in order to distinguish it from estimation by an interval which we shall describe later on. An estimate by a single value is also called a *point estimate*, because it represents a point on a scale of possible estimates.

The reasoning by which we have justified using $\bar{x}$ as an estimate of μ can also be used to justify using s^2 as an estimate of σ^2, using a sample percentile as an estimate of a population percentile and using other statistics as estimates of other parameters.

Some questions about point (single-value) estimates remain to be answered. We note, for example, that for a normal population μ is not only the mean but is also the median. One may, therefore, raise the question as to the possibility of using the sample median as a basis for estimating μ for a normal population. Another question of some interest is the proper estimate for σ^2. Shall it be the estimate $\Sigma(x - \bar{x})^2/(n - 1)$ which is used in this text, or the estimate $\Sigma(x - \bar{x})^2/n$ used by most of the older books and by a few of the newer ones?

Two criteria for choosing estimates will now be discussed.

Unbiased Estimate.   *An estimate is unbiased if the mean of its distribution is equal to the value of the parameter which it estimates.* Thus, no matter what the size of a sample may be, its mean is an unbiased estimate of the population mean. For the normal population, the sample median is also an unbiased estimate of the population mean. This relation is to be expected, since in a normal population the mean and the median are the same.

The formula $s^2 = \Sigma(x - \bar{x})^2/(n - 1)$ gives an unbiased estimate of σ^2, but the alternative formula $s^2 = \Sigma(x - \bar{x})^2/n$ does not. This is the justification for the use of $n - 1$ rather than n in the denominator. For example, suppose a great many samples of 9 cases each were drawn from a normal population in which $\sigma^2 = 144$. If for each sample s^2 were computed with $n - 1 = 8$ as the denominator, the mean of all the sample variances would be approximately equal to $\sigma^2 = 144$. However, if $n = 9$ were used as the denominator, the mean of the sample variances would be approximately $\sigma^2(n - 1)/n = 8(144)/9 = 128$ instead of 144.

Table 9-2 presents distributions of sample means and sample medians for 20 samples of $n = 5$. Means and medians were computed for the same samples. We find that the means of the two distributions of statistics are very nearly the same and are close to μ. This illustrates the statement that the sample mean or the sample median is each an unbiased estimate of μ for a normal population.

Efficient Estimate.   An interesting feature of the two distributions in Table 9-2 is that the medians are much more variable than the means. This is a reflection of the fact that the standard error of the median is greater than the standard error of the mean.

Because of this difference in standard errors, the mean of a sample is likely to be nearer than the median to μ, or, in other words, the error in using the sample mean as an estimate of μ is likely to be less than the error in using the sample median for this purpose.

Since the standard error of the mean is less than the standard error of the median, the sample mean is said to be a more *efficient statistic* for estimating μ in a normal population. When two statistics are available for estimating the same parameter, the one with the lesser standard error is the more efficient.

It should be pointed out that in a population in which the mean and median are presumed to differ, the sample median should be used for estimating the population median, and the sample mean used for estimating the population mean. This conclusion is based on consideration of bias.

Table 9-2   Distributions of Means and Medians of Random Samples of Five from a Normal Population with $\mu = 121.6$ and $\sigma = 37.15$

Score Interval	Number of Sample Means	Number of Sample Medians
160–164		1
155–159		1
150–154		1
145–149		1
140–144	3	1
135–139	1	1
130–134	2	
125–129	2	2
120–124	1	2
115–119	2	1
110–114	1	2
105–109	4	2
100–104	2	1
95–99		
90–94		2
85–89	1	1
80–84	1	
75–79		
70–74		1
* Mean of distribution	119.0	119.5
* Standard deviation of distribution	18.0	25.3

* Computed from ungrouped data.

The Standard Score for Means.   In Chapter 6 the standard score for an observation was defined as

(9-2)
$$z = \frac{x - \bar{x}}{s}$$

This expression is entirely in sample terms.

In Chapter 8 the corresponding expression in terms of parameters was given as

(9-3)
$$z = \frac{x - \mu}{\sigma}$$

We learned that, if x comes from normal population, the latter expression has the standard (or unit) normal distribution.

Corresponding to formula (9-3) for observations, the standard score for means is

(9-4)
$$z = \frac{\bar{x} - \mu}{(\sigma/\sqrt{n})}$$

Regardless of the size of n, the standard score in formula (9-4) is a standard normal deviate if the sample is drawn from a normal population with mean μ and standard deviation σ. Unless n is very small, this standard score is distributed approximately normally even if the population distribution is not normal. It may be helpful to the reader to notice that for means based on 1 observation ($n = 1$) formula (9-4) would be identical with formula (9-3).

Because of the normality of distribution of the standard score for means, we can apply the table of the normal distribution to answer questions like the following:

1. Assume that we are drawing samples of size 25 from a population with $\mu = 60$ and $\sigma = 8$. For what percent of these samples will $\bar{x}$ exceed 61 if the number of samples is very large? Solution:

$$z = \frac{61 - 60}{(8/\sqrt{25})} = \frac{1}{1.6} = .625$$

In a normal distribution, the area above the point $z = .625$ is $.5 - .2340 = .2660$. Hence $\bar{x}$ may be expected to exceed 61 in about 27 percent of samples. A better statement is that there is probability .27 that $\bar{x}$ will exceed 61.

2. Assuming the same situation as in the preceding question, between what two limits will the value of $\bar{x}$ for the middle 80 percent of samples be contained?

Solution:

The middle 80 percent of the area under a normal curve is contained between $z_{.10} = -1.282$ and $z_{.90} = +1.282$. For the sampling distribution of $\bar{x}$, the corresponding points are obtained from the two equations $(\bar{x} - 60)/(8/\sqrt{25}) = -1.282$ and $(\bar{x} - 60)/(8/\sqrt{25}) = 1.282$. The solutions of these equations for $\bar{x}$ are 57.95 and 62.05, which are the required limits, and we may say there is probability .80 that $\bar{x}$ will not be less than 58 nor more than 62.

The expression for the standard score for means (9-4) which uses the population value σ is of great theoretical interest. Its direct application to data is limited to those situations where an estimate of σ is available from earlier experience. In the most common situations, σ is not known, and s must be used as a substitute. The standard score for means is then called t rather than z.

$$(9\text{-}5) \qquad t = \frac{\bar{x} - \mu}{(s/\sqrt{n})}$$

This t is not normally distributed, but for large samples its distribution is sufficiently like the normal so that the tables of the normal distribution may be used. Questions as to how large a sample should be in order to make the normal approximation applicable and what to do when it is not applicable will be taken up on page 152. The denominator in formula (9-5) is the point estimate for the standard error of the mean.

EXERCISE 9-3

1. If random samples of size 36 are drawn from a normal population for which $\mu = 57$ and $\sigma = 12$, what proportion of such samples will have a mean
   (a)   Greater than 57?
   (b)   Greater than 60?
   (c)   Smaller than 59?
   (d)   Between 55 and 59?

2. In the situation described in Question 1, what is the probability that the mean of a random sample will be
   (a)   Greater than 53?
   (b)   Between 54 and 60?

3. Suppose random samples of 64 cases are drawn from a normal population for which $\mu = 90$ and $\sigma = 18$. Consider the 5 percent of samples with means which deviate most widely from μ. Outside what values will these means fall?

   Solution: For half of them $(\bar{x} - \mu)/(\sigma/\sqrt{n})$ will be larger than $z_{.975}$, and for half it will be smaller than $z_{.025}$.

   $$\frac{\sigma}{\sqrt{n}} = \frac{18}{\sqrt{64}} = 2.25$$

   $$z_{.025} = -1.96 \quad \text{and} \quad z_{.975} = 1.96$$

   If $\dfrac{\bar{x} - 90}{2.25} = -1.96$, $\bar{x} = 90 - 1.96(2.25) = 85.6$

   If $\dfrac{\bar{x} - 90}{2.25} = 1.96$, $\bar{x} = 90 + 1.96(2.25) = 94.4$

Therefore the most extreme 5 percent of the samples will have $\bar{x}$ either smaller than 85.6 or larger than 94.4.

4. Suppose random samples are drawn from a normal population for which $\mu = 75$ and $\sigma = 12$. Between what two values will the middle 99 percent of sample values of $\bar{x}$ fall if n is
   (a)   9?   (c)   36?   (e)   400?
   (b)   16?   (d)   100?   (f)   900?

5. Suppose a random sample of 200 cases is drawn from a population for which $\mu = 75$. On the scale of $\bar{x}$ find two values symmetrically placed about μ such that the probability of obtaining a value of $\bar{x}$ between them is .95 if σ is
   (a)   4?   (c)   12?   (e)   20?
   (b)   10?   (d)   15?   (f)   24?

The Confidence Interval.   Estimation of a parameter by a single number is valuable because the estimate is likely to be close to the value of the parameter. Because that estimate will almost never be exactly the same as the value of the parameter, however, it is somewhat unsatisfactory. Such point estimation does not sufficiently tie down our information about the parameter.

Another type of estimate is obtained by computing from the same sample two statistics and using the interval between these statistics as an interval estimate for the parameter. These two statistics are computed in such a way that the interval between them usually contains the parameter. By "usually" we mean that for some preassigned percentage of samples the interval between the computed values contains the parameter. In other words, *for any given sample the interval may or may not contain the parameter, but the computation of the interval is of such nature that for some preassigned percent of samples the interval contains the parameter.* An interval obtained in this way is called an *interval estimate* or a *confidence interval;* the two statistics which determine the limits of the interval are called *confidence limits;* and the preassigned percentage of samples is called a *confidence coefficient.* These concepts will be clarified and made more precise when we develop them for the problem of obtaining a confidence interval for μ.

Confidence Interval for the Mean when the Sample Is Large.   The method of obtaining this interval for a sample of 50 or more cases will first be described, and then a justification will be given for the method.

Suppose we have drawn a random sample of 125 cases and we wish to compute a confidence interval for μ with confidence coefficient .99. In other words, we wish to compute a confidence interval of such nature that in 99 percent of similar computations the interval will contain the parameter. We compute $\bar{x}$ and s from the sample and find $z_{.995}$ and $z_{.005}$ from the table of the normal distribution. From these values we compute the two statistics, $\bar{x} + z_{.995}(s/\sqrt{125})$ and $\bar{x} + z_{.005}(s/\sqrt{125})$. These are the two *confidence limits*. The interval between them is the *confidence interval*.

Suppose, in the above example, $\bar{x} = 93.5$ and $s = 14.2$. In Appendix Table II of the normal distribution, we find $z_{.995} = 2.576$ and $z_{.005} = -2.576$. Then the limits are

$$93.5 - 2.576 \left(\frac{14.2}{\sqrt{125}} \right) = 93.5 - 3.3 = 90.2$$

and

$$93.5 + 2.576 \left(\frac{14.2}{\sqrt{125}} \right) = 93.5 + 3.3 = 96.8$$

Consequently the interval between 90.2 and 96.8 is the required confidence interval, and we state, with confidence coefficient .99, that $90.2 < \mu < 96.8$. This is read "μ is greater than 90.2 and less than 96.8" or "μ is between 90.2 and 96.8."

In more general terms, suppose the sample size is sufficiently large to permit use of the normal approximation and the confidence coefficient is c, then confidence limits for μ are

(9-6) $\bar{x} + z_{\frac{1-c}{2}} \dfrac{s}{\sqrt{n}}$ and $\bar{x} + z_{\frac{1+c}{2}} \dfrac{s}{\sqrt{n}}$

The following justification for the limits in (9-6) is somewhat mathematical and may be skipped by a reader who wishes to do so. We know from the discussion of the tables of the normal distribution that the percentage of samples between $z_{(1-c)/2}$ and $z_{(1+c)/2}$ is $[(1 + c)/2] - [(1 - c)/2] = c$. In terms of the standard score for means, the percentage of samples for which $(\bar{x} - \mu)/(s/\sqrt{n})$ is between $z_{(1-c)/2}$ and $z_{(1+c)/2}$ is c. The statement can be put mathematically by saying that the inequality

(9-7) $z_{\frac{1-c}{2}} < \dfrac{\bar{x} - \mu}{(s/\sqrt{n})} < z_{\frac{1+c}{2}}$

is true for a proportion c of samples. By simple algebra, inequality (9-7) can be transformed into inequality,

(9-8) $\bar{x} + z_{\frac{1-c}{2}} \dfrac{s}{\sqrt{n}} < \mu < \bar{x} + z_{\frac{1+c}{2}} \dfrac{s}{\sqrt{n}}$

and this latter inequality is true whenever the former is true. Therefore c is also the proportion of samples for which inequality (9-8) is true.

It is important for the reader to realize that the probability expressed by (9-8) is about the intervals containing μ. It is not a statement about possible value of μ in the given interval. In other words, the inequality (9-8) might mislead a reader to think that there is a collection of populations, each with some value of μ and that a proportion c of these populations has values of μ in the interval indicated by the inequality. This is an incorrect interpretation of the inequality. The correct interpretation is that there is one population with a unique μ. There are, however, as many possible intervals as there are possible samples. Some of these intervals satisfy the inequality because they actually contain μ between their limits. Some do not. By the theory stated, the proportion of intervals which satisfy (9-8) is c.

Student's Distribution. We have stated above that, if the population has a normal distribution, the sample values $z = (\bar{x} - \mu)/(\sigma/\sqrt{n})$ also have a normal distribution, and that if n is sufficiently large, the values $t = (\bar{x} - \mu)/(s/\sqrt{n})$ have a distribution which approximates the normal. For small samples, say $n = 40$ or less, the normal approximation for t is inadequate, and its exact distribution must be used. This distribution is called either "Student's"[3] distribution or the t distribution. It is a symmetrical distribution, with mean zero, but its form differs for different values of n.

A difficulty in using Student's distribution, one not met in using the normal distribution, is that we need a separate probability table for each sample size. The situation is somewhat further complicated by the fact that the tables are not stated directly in terms of numbers of cases, but in terms of a related value called *degrees of freedom* which we shall denote by ν, which is the small Greek letter *nu*. The relationship between number of cases and number of degrees of freedom differs from one type of problem to another and must be ascertained separately for each type. The concept of degrees of freedom is discussed further on pages 250 and 296.

A general understanding of that concept is not needed at this time. It is enough now to know that for $t = (\bar{x} - \mu)/(s/\sqrt{n})$, the number of degrees of freedom is one less than the number of cases.

Tables of areas for Student's distribution are found in Appendix Table

---

[3] "Student" was a well-known British statistician named William Sealy Gosset (1876–1947) who was adviser to the Guinness brewery in Dublin. A ruling of that firm forbidding their employees to publish the results of research was relaxed to allow him to publish mathematical and statistical research under a pseudonym. His paper on "The Probable Error of a Mean," published in *Biometrika* in 1908, which first called attention to the fact that the normal curve does not properly describe the distribution of the ratio of mean to standard error in small samples, is now a classic.

III. Each row is really a separate table corresponding to the number of degrees of freedom indicated in the left-hand column with the heading ν. Tabular entries are values of t. In the column headed $t_{.90}$, each entry is the 90th percentile of the distribution to which it belongs. A similar description applies to the other columns. Since the distributions are symmetrical with mean zero, it is not necessary to tabulate percentiles below the 50th. Instead we use the relationships $t_{.10} = -t_{.90}$, $t_{.005} = -t_{.995}$, and so forth. The entries in the bottom row are percentiles of the normal distribution.

EXERCISE 9-4

1. By examination of Appendix Table III, verify enough of the following statements to be sure you are reading the table correctly.
   (a) The 95th percentile of Student's distribution is
      (i) $t = 2.35$, when $\nu = 3$.
      (ii) $t = 1.94$, when $\nu = 6$.
      (iii) $t = 1.74$, when $\nu = 17$.
   (b) The 5th percentile of Student's distribution is
      (i) $t = -2.35$, when $\nu = 3$.
      (ii) $t = -1.94$, when $\nu = 6$.
      (iii) $t = -1.74$, when $\nu = 17$.
   (c) For Student's distribution with $\nu = 5$,
      (i) 10 percent of the area lies to the right of $t = 1.48$.
      (ii) 25 percent of the area lies to the right of $t = .73$.
      (iii) 1 percent of the area lies to the right of $t = 3.36$.
      (iv) 80 percent of the area lies to the right of $t = -.92$.
      (v) 99.5 percent of the area lies to the right of $t = -4.03$.
   (d) For Student's distribution with $\nu = 12$,
      (i) 50 percent of the area lies between $t = -.70$ and $t = .70$.
      (ii) 80 percent of the area lies between $t = -1.36$ and $t = 1.36$.
      (iii) 90 percent of the area lies between $t = -1.78$ and $t = 1.78$.
      (iv) 95 percent of the area lies between $t = -2.18$ and $t = 2.18$.
      (v) 98 percent of the area lies between $t = -2.68$ and $t = 2.68$.
      (vi) 99 percent of the area lies between $t = -3.05$ and $t = 3.05$.

2. In the following questions you may assume that we are considering $t = (\bar{x} - \mu)/(s/\sqrt{n})$ and that the number of degrees of freedom is $\nu = n - 1$.
   (a) If $n = 10$, what is the point on the t scale such that
      (i) 5 percent of the area lies above it (that is, to the right)?
      (ii) 10 percent of the area lies below it (that is, to the left)?
      (iii) 80 percent of the area lies above it?
      (iv) 99 percent of the area lies above it?
   (b) Answer the same four questions, if $n = 16$.

(c) If $n = 7$, what two points on the t scale include between them
   (i) The middle half of the area?
   (ii) The middle 80 percent of the area?
   (iii) The middle 90 percent of the area?
   (iv) The middle 95 percent of the area?
   (v) The middle 99 percent of the area?
   (vi) The middle 99.9 percent of the area?

3. Answer the same six questions, if $n = 20$.

4. What percent of samples of size $n = 15$ would have values of $(\bar{x} - \mu)/(s/\sqrt{n})$ between -2.14 and 2.14? Does your answer depend on the value of μ? Of s? Of n?

Confidence Interval for the Mean when the Sample Is Small. The reasoning which led to the confidence limits given in formula (9-6) for large samples leads also to the confidence limits for small samples, with the only modification being that z is replaced by t and that reference is made to the t distribution instead of to the normal. For a small sample, the confidence limits for μ can, therefore, be obtained as follows:

Suppose that a random sample of size n is drawn from a normally distributed population and that the confidence coefficient is to be c, then the confidence limits for μ are

$$(9\text{-}9) \qquad \bar{x} + t_{\frac{1-c}{2}} \frac{s}{\sqrt{n}} \quad \text{and} \quad \bar{x} + t_{\frac{1+c}{2}} \frac{s}{\sqrt{n}}$$

Values of $t_{(1-c)/2}$ and $t_{(1+c)/2}$ are read from Appendix Table III with $\nu = n - 1$.

In Table 9-3 is shown the confidence interval for each of the 20 samples of 25 cases which have been previously discussed and for which the distribution of means was given in Table 9-1. We recall that these samples were drawn randomly from a population for which μ and σ are known, which is, of course, a fictitious situation because parameters are ordinarily not known.

Entries in the column at the extreme right show that 18 of the 20 confidence intervals include $\mu = 121.6$ and 2 do not. Inclusion of μ in the interval is indicated by "Yes" and noninclusion by "No." Here the proportion of samples ($\frac{18}{20} = .90$), in which μ is included between the computed confidence intervals, is exactly equal to the confidence coefficient of .90. However, one must not expect such perfect agreement of experience and theory every time an interval estimate is made. The value $c = .90$ applies to the totality of computations from a very large number of samples rather than to any fixed number of them. It is interesting to

Table 9-3   Confidence Interval for μ with Confidence Coefficient .90 for Each of 20 Random Samples of 25 Cases from a Normal Population with $\mu = 121.6$ and $\sigma = 37.15$

Sample Number	$\bar{x}$	s	Lower Limit $\bar{x} + t_{.05}\dfrac{s}{\sqrt{n}}$	Upper Limit $\bar{x} + t_{.95}\dfrac{s}{\sqrt{n}}$	Includes μ?
1	123.7	36.9	111.1	136.3	Yes
2	116.0	31.9	105.1	126.9	Yes
3	117.9	25.6	109.1	126.7	Yes
4	126.7	35.9	114.4	139.0	Yes
5	128.0	41.4	113.8	142.2	Yes
6	124.5	38.1	111.5	137.5	Yes
7	112.9	38.7	99.7	126.1	Yes
8	123.8	34.3	112.1	135.5	Yes
9	118.8	39.0	105.5	132.1	Yes
10	127.6	39.7	114.0	141.2	Yes
11	128.3	37.0	115.6	141.0	Yes
12	122.1	41.7	107.8	136.4	Yes
13	125.6	36.0	113.3	137.9	Yes
14	111.0	30.1	100.7	121.3	No
15	111.4	39.6	97.9	124.9	Yes
16	123.8	31.8	112.9	134.7	Yes
17	117.8	37.2	105.1	130.5	Yes
18	117.3	37.8	104.4	130.2	Yes
19	133.6	34.7	121.7	145.5	No
20	123.0	41.4	108.8	137.2	Yes

note in Table 9-3 that, even when the confidence interval missed the true value of μ, it missed it only slightly.

It will be helpful to the reader to check some of the computations in Table 9-3. We shall describe the first of these:

As the confidence coefficient is stated to be $c = .90$

$$\frac{1 - c}{2} = \frac{.10}{2} = .05 \quad \text{and} \quad \frac{1 + c}{2} = \frac{1.90}{2} = .95$$

As we must enter Table III with the appropriate value of ν, we first note that $\nu = 25 - 1 = 24$.

Referring to Student's distribution with $\nu = 24$, we read $t_{.05} = -1.71$

and $t_{.95} = 1.71$. Hence the confidence limits for sample 1 are

$$\bar{x} + t_{.05} \frac{s}{\sqrt{n}} = 123.7 + (-1.71) \frac{36.9}{\sqrt{25}} = 111.1$$

and

$$\bar{x} + t_{.95} \frac{s}{\sqrt{n}} = 123.7 + (1.71) \frac{36.9}{\sqrt{25}} = 136.3$$

EXERCISE 9-5

1. A sample of 16 cases has yielded $\bar{x} = 23.5$ and $s = 12$.
   Make an interval estimate for μ with confidence coefficient
   (a) $c = .99$.   (b) $c = .95$.   (c) $c = .90$.

2. A sample of 400 cases has yielded $\bar{x} = 23.5$ and $s = 12$.
   Make an interval estimate for μ with confidence coefficient
   (a) $c = .99$.   (b) $c = .95$.   (c) $c = .90$.

3. A sample of 16 cases has yielded $\bar{x} = 23.5$ and $s = 4$.
   Make an interval estimate for μ with confidence coefficient
   (a) $c = .99$.   (b) $c = .95$.

10

TESTING HYPOTHESES

That aspect of statistical inference which deals with estimati[on of] unknown parameters of a population was discussed in Chapter [9. Of] the three examples involving the use of samples described at the beg[inning] of that chapter, the first two illustrate problems in which the ma[in task] is the estimation of parameters. The problem presented in th[e third] example differs from those in the first two in that it requires a [choice] between two alternatives: (1) continued use of the medical treatmen[t now] in use, or (2) substitution of the new kind of treatment which is [under] consideration. These alternatives can be reduced to statistical con[sider]ations if we say that the old method will be continued if the new m[ethod] does not cure a larger proportion of sick persons than the old. Howe[ver,] if the new method does produce a larger proportion of cures, then it will replace the old method.

The approach to this problem is to formulate a *hypothesis* that some particular relation between the proportions of cures is true and then to test the hypothesis experimentally. On the basis of the outcome of the experiment, a conclusion is drawn about the relative effectiveness of the two methods of treatment.

Methods of statistical inference involving proportions will be described in Chapter 11. In this chapter we shall develop methods required in dealing with hypotheses about means. Two applications will be discussed.

A Problem Involving a Hypothesis about Means. Suppose the question has been raised as to whether people who are deaf and those with normal hearing differ in motor abilities. By now the student should have developed far enough in statistical thinking to be unwilling to accept a generalization about this matter based only on a few personal experiences, "I know several deaf persons who . . . ." He should also be unwilling to accept a generalization based only on general theories of physiology. Although one should recognize that either abstract theorizing or personal experience may lead to a formulation of a useful though tentative generalization, an appeal to empirical data is essential to establish that generalization.

In a research conducted by Long,[1] deaf and girls having normal hearing were compared on a number of motor abilities. Inasmuch as motor ability increases with age, if one group had been older than the other, a difference might erroneously have been interpreted as related to deafness. For fear that the motor traits studied might be related to intelligence or that performance might be affected by the ability to concentrate and directions, it also seemed important to control intelligence. There veral ways to arrange an experiment for the purpose of controlling fect of background traits which might bias the outcome.[2] Long each deaf girl with a girl having normal hearing of the same age elligence quotient. Such pairing is usually laborious and not always st efficient device, but it is very frequently employed. The data ong reported on 37 pairs are presented in Table 11-1 on pages 185, the purpose of the present discussion, only the first ten pairs en used. Long presented findings on several traits of which only will be reported here. The method of obtaining a score for each n a test of balance is important for the research but extraneous analysis of the data, which is our present concern.

each pair, the score of the deaf child is subtracted from that of her h g mate, leaving a difference score which may be denoted x. It sh u d be recognized that x is a measure *of the pair*, that the matched pair is the individual in our sample, and that we shall generalize to a *population of matched pairs*.

The work of pairing has presumably eliminated gross differences related to age and intelligence. Differences related to training or experience will presumably be eliminated by selecting for measurement those tasks the children are not likely to have practiced specifically. It is never feasible to control *all* sources of variation, but sources which might seriously bias the outcome must be controlled. Differences related to individual performance will always remain. Deaf children are not uniform in motor ability nor are children with normal hearing. Even if deaf

[1] John A. Long, *Motor Abilities of Deaf Children*, Teacher's College, Columbia University, Bureau of Publications, New York, 1932.

[2] See, for example, H. M. Walker and J. Lev, *Statistical Inference*, Holt, Rinehart and Winston, Inc., 1953, p. 382.

and hearing children are generally equal in balance, very few pairs will have zero difference. Sometimes a deaf child whose balance is better than average is mated with a hearing child whose balance is poorer than average, producing a negative difference score. Sometimes the accidents of pairing will produce a positive difference. The population of such differences may be assumed to have a normal distribution with mean μ. If there is no difference between deaf and hearing in general, $\mu = 0$. If hearing children have better balance than deaf, $\mu > 0$, although for many pairs the difference x may be negative. If hearing children have poorer balance than deaf, $\mu < 0$, although for many pairs the difference x may be positive.

A Statistical Hypothesis.   A hypothesis about μ is a tentative statement that μ has some particular value or that μ has some one of a set of values. After it is formulated, the hypothesis is subjected to an experimental check known as a *test of the hypothesis.*

Suppose your knowledge of physiology inclines you to believe that μ is greater than zero; however, you are not inclined to guess how much greater. You might set up the statistical hypothesis $\mu = 0$ and test it against the alternative $\mu > 0$. If then the statistical hypothesis $\mu = 0$ is found to be unacceptable, you would have grounds for asserting your original belief; if the statistical hypothesis is found to be acceptable, you would be warned that you could not safely assert your belief that $\mu > 0$.

Suppose that a researcher is unaware of the connection between balance and the semicircular canals in the ears and has no prior opinion as to the value of μ. He merely wants to find out whether any general difference exists. He will then set up the hypothesis $\mu = 0$ and test it against the alternative "μ is not 0."

The statistical hypothesis tested is often called the *null hypothesis.* A hypothesis is always a hypothesis about a parameter (or sometimes about more than one parameter) of a population, never about a statistic of a sample. It is stated in the present tense, that some general state of affairs is now existent. It should not be stated in future tense ("the mean will be zero," "the correlation will be positive," and so forth), because that phraseology suggests a kind of fortune telling about the outcome of a particular sample.

Test of Hypothesis that $\mu = 0$.   To test this hypothesis a random sample must be drawn; in the present problem that will be a sample of matched pairs. The data for 10 such pairs will be found in Table 10-1. Each pair-difference has been denoted x, and the statistics $\bar{x} = 2.29$,

Table 10-1   Scores on Balance for 10 Pairs of Deaf and Hearing Girls*

| Pair Number | Balance Score for | | x |
	Hearing	Deaf	Hearing-Deaf
1	2.3	2.0	0.3
2	1.0	2.0	−1.0
3	3.7	2.7	1.0
4	3.3	2.7	0.6
5	10.0	3.0	7.0
6	2.7	2.7	0
7	8.3	3.7	4.6
8	6.0	1.3	4.7
9	4.3	2.0	2.3
10	7.7	4.3	3.4
Total	49.3	26.4	22.9
$\bar{x}$			2.29
s			2.56
$s/\sqrt{n}$			0.810

* Source of data: Long, *op. cit.*

$s = 2.56$, and $s/\sqrt{n} = 0.810$ have been computed. The standard score for means is

$$t = \frac{\bar{x} - \mu}{s/\sqrt{n}} = \frac{2.29 - 0}{0.810} = 2.83$$

with $n - 1 = 9$ degrees of freedom. Referring $t = 2.83$ to Student's distribution, we find it slightly larger than $t_{.99}$. This means that, if the hypothesis $\mu = 0$ is true, we have accidentally drawn a sample which is in the 1 percent of samples with the largest values of t. We are inclined, therefore, to say that this sample is exceptional, or unusual, for samples from a population in which $\mu = 0$.

Now μ is not known. Only $\bar{x}$, s, and n are known, and from these statistics evidence must be extracted about the reasonableness of the hypothesis, $\mu = 0$. Since that hypothesis leads to a value of t that would occur only rarely in a random sample, it does not seem a reasonable hypothesis to entertain. Hence we *reject* the hypothesis that $\mu = 0$. When we have made this decision, our test of the hypothesis is completed.

The reader may feel that the hypothesis $\mu = 0$ should be rejected whenever a value of $\bar{x}$ different from zero is obtained in a sample. Such a policy would be mistaken because it fails to take account of sampling variability. Even if μ is actually zero in the population from which samples are drawn, the sample values of $\bar{x}$ will vary so that almost none of them will be precisely zero; most of them will deviate from zero by a small amount in one direction or the other, and only a few will show large deviations from zero. Consequently, values of $\bar{x}$ near zero do not throw suspicion on the hypothesis, $\mu = 0$.

In a general sense (this point will be clarified in further discussion), one tends to accept a hypothesis about μ whenever the value of $\bar{x}$ computed from a sample is close to the value specified by the hypothesis and to reject that hypothesis whenever the value of $\bar{x}$ is far from the specified value. In either accepting or rejecting a hypothesis, an investigator may make the wisest decision possible in view of the data and still be in error. However, the nature of the procedures which will be now described is such that at least the probability of making large errors is kept at a low level.

Level of Significance.   To this point the word "exceptional" applied to the size of $\bar{x}$ has been treated rather vaguely. Actually, the decision as to what is to be regarded as exceptional is arbitrary. For a small sample like the one we have been considering, it is customary to regard the 5 percent of samples having the most extreme values of $\bar{x}$ as exceptional and to reject the null hypothesis when one of these samples has been drawn. This arbitrary value (here selected as 5 percent) is called a *level of significance*, because it sets apart certain differences between sample mean and hypothetical population mean as being sufficiently significant to lead to rejection of the hypothesis.

Having decided upon a level of significance, it is still necessary to decide how to use it. Shall we apply it to the right tail of the distribution and reject the hypothesis $\mu = 0$ only when $\bar{x}$ is exceptionally large or to the left and reject only when $\bar{x}$ is exceptionally small, or shall we regard both extremely large and extremely small values of $\bar{x}$ as exceptional?

The answer to these questions depends on what alternatives seem particularly important to guard against in view of the problem situation. If the investigator has started out with no opinion concerning the relative balance scores of the deaf or hearing children, he will wish to reject the hypothesis that $\mu = 0$ should he find a value of t that is numerically very large, whether it is positive or negative. In other words, the alternative he will wish to consider is simply: μ is different from zero. Suppose he has selected .05 as the level of significance at which he will work. He will then decide to reject the hypothesis, $\mu = 0$, if he obtains a sample for which t exceeds $t_{.975}$ or a sample for which t is less than $t_{.025}$. The decision then is to reject the hypothesis $\mu = 0$ if a sample of 10 cases shows a value of $t = \sqrt{n}\,(\bar{x} - \mu)/s$ which is smaller than -2.26 or larger than 2.26. This set of values of t is called the *region of rejection* or the *region of significance* or the *critical region*. The probability .05 corresponding to that region is the *level of significance*. The test we have just used is called a *two-sided test* or a *two-tail test* because the region of rejection is on both sides of the probability curve (or in both tails of the curve).

One-sided Test. Now if the investigator has enough prior information to entertain a strong belief that the mean balance score for hearing children exceeds that for deaf children, he may wish to reject the hypothesis $\mu = 0$ should his sample show a large positive value of t. He will not consider it important, however, to reject the hypothesis if a large negative value appears in the sample. (You will recall that in Table 10-1 x is positive for a pair if the score of the hearing child exceeds that of the deaf child.) Therefore, his region of rejection will consist of those values of t larger than $t_{.95} = 1.83$. This region is a one-sided or one-tail region, and the test is a one-sided test. Note that the level of significance is .05 in both situations but that the location of the critical region differs.

The steps in formulating and testing a statistical hypothesis are stated in Table 10-2, first in general terms and then in terms specific to the problem of comparing deaf and hearing persons in regard to balance.

Comparing the Means of Two Populations. Pairing cases as described for deaf and hearing children is very laborious and time consuming and often requires elimination of cases for which no mates can be found. It is employed only when the investigator fears that a random sample from each of the two populations he wishes to compare might differ in respect to some extraneous variable which would introduce bias into the comparison. Fortunately, in many problems this danger is not crucial, and it is satisfactory merely to draw a separate random sample from each population.

Suppose, for example, the question is whether sophomore men in a certain university are more familiar with contemporary science than sophomore women. A random sample of men is drawn from the sophomore class and given the science section of the *Contemporary Affairs Test* of the Cooperative Test Service. By this device what may be presumed to be a random sample of science scores of sophomore men is obtained. Let us use n_1, $\bar{x}_1$, and s_1 to denote the number of cases, the mean and the standard deviation of this sample.

A random sample of sophomore women is also drawn and given the

Table 10-2   Steps in Formulating and Testing a Statistical Hypothesis

Step	General Procedure	Application to Comparison of Deaf with Hearing
1	Select the measures on which the investigation will be based.	Choose as a measure the balance score of a hearing girl minus the balance score of the deaf girl with whom she is paired.
2	Specify the general nature of the population and the parameter or parameters needed for the investigation.	The population of differences can be assumed to be approximately normal, and the parameter in question is its mean μ.
3	Formulate a hypothesis about the population and decide on the alternatives.	The hypothesis is $\mu = 0$, and the alternatives are μ is not equal to 0.
4	Determine a statistic by which the hypothesis is to be tested.	$$t = \frac{\bar{x} - 0}{(s/\sqrt{n})}$$
5	Ascertain the distribution of the statistic.	Student's distribution with $\nu = n - 1$ degrees of freedom.
6	Choose a level of significance.	5 percent.
7	Determine the region of rejection on the basis of the level of significance and the alternatives to the hypothesis.	Reject $\mu = 0$ when t is larger than $t_{.975}$ or is smaller than $t_{.025}$.
8	Draw a random sample of size n from the population.	Determine differences in balance for a random sample of matched pairs.
9	Compute for this sample the value of the previously specified statistic.	Compute t from the sample in Table 10.1 obtaining $t = 2.83$.
10	Determine whether the computed value of the statistic is in the region of rejection.	Ascertain that t is in the region of rejection because $t_{.975} = 2.26$ and the sample value $t = 2.83$ is larger than $t_{.975}$.
11	Reject the hypothesis if the value of the statistic is in the region of rejection; otherwise accept it.	Reject the hypothesis $\mu = 0$.

same test. Let us use n_2, $\bar{x}_2$, and s_2 to denote the corresponding statistics of this sample.

If μ_1 represents the mean of the population of science scores achieved by sophomore men and μ_2 that of science scores of sophomore women, the hypothesis to be tested is $\mu_1 - \mu_2 = 0$.

The statistic by means of which this hypothesis is to be tested is the standard score for the mean difference $\bar{x}_1 - \bar{x}_2$. To obtain this standard score, the standard error of the mean difference is required. There are two basic formulas for this standard error, one to be used when no assumption is made about the equality of σ_1^2 and σ_2^2, the other when it seems reasonable to assume that $\sigma_1^2 = \sigma_2^2$. (A test for the hypothesis that two variances are equal will be presented on page 260.)

1. Method when there is no assumption that $\sigma_1^2 = \sigma_2^2$ and samples are large. In this case the formula for the standard error of $\bar{x}_1 - \bar{x}_2$ is

$$(10\text{-}1) \qquad s_{\bar{x}_1 - \bar{x}_2} = \sqrt{(s_1^2/n_1) + (s_2^2/n_2)}$$

and the standard score for the mean difference is

$$(10\text{-}2) \qquad t = \frac{(\bar{x}_1 - \bar{x}_2) - (\mu_1 - \mu_2)}{\sqrt{(s_1^2/n_1) + (s_2^2/n_2)}}$$

This method should be used only when the samples are fairly large; neither n should be less than 25.

Aside from the fact that the statistic of formula (10-2) is used instead of that of (9-5) on page 149 and that we have two independently chosen samples instead of one sample of matched pairs, the test of the hypothesis $\mu_1 - \mu_2 = 0$ proceeds by the same eleven steps which were listed in Table 10-2.

2. Method when it is not reasonable to assume $\sigma_1^2 = \sigma_2^2$ and samples are small. Fortunately this situation does not occur very often, because when samples are small a very large disparity between the two sample variances may not be inconsistent with equality of the population variances. If you do find yourself in this uncomfortable—but unusual—situation, you would be wise to seek advice from an expert.

3. Method when it can be assumed that $\sigma_1^2 = \sigma_2^2$. One great advantage of making this assumption is that the resultant formula is appropriate for samples of any size. When $n_1 + n_2$ is small, the standard score is referred to Student's table with $n_1 + n_2 - 2$ degrees of freedom. Whenever $n_1 + n_2 - 2$ is larger than the values of ν shown in the column at the left of Appendix Table III, the standard score can be referred to a normal probability table.

Another advantage of making the assumption that $\sigma_1^2 = \sigma_2^2$ is that the resultant procedure is a special case of analysis of variance as discussed in Chapter 17, which provides a test for the equality of several means, $\mu_1 = \mu_2 = \cdot\,\cdot\,\cdot = \mu_k$.

If it can be assumed that $\sigma_1^2 = \sigma_2^2 = \sigma^2$ the common variance σ^2 is estimated by pooling the variances of the two samples

$$(10\text{-}3) \qquad s^2 = \frac{(n_1 - 1)s_1^2 + (n_2 - 1)s_2^2}{n_1 + n_2 - 2}$$

then the standard error of the mean difference is

$$(10\text{-}4) \qquad s_{\bar{x}_1 - \bar{x}_2}^{\,2} = \frac{s^2(n_1 + n_2)}{n_1 n_2}$$

or its equivalent

$$(10\text{-}5) \qquad s_{\bar{x}_1 - \bar{x}_2}^{\,2} = \frac{(n_1 - 1)s_1^2 + (n_2 - 1)s_2^2}{n_1 + n_2 - 2} \times \frac{n_1 + n_2}{n_1 n_2}$$

and the standard score is

$$(10\text{-}6) \qquad t = \frac{(\bar{x}_1 - \bar{x}_2) - (\mu_1 - \mu_2)}{s_{\bar{x}_1 - \bar{x}_2}}$$

Formulas (10-6) and (10-2) are algebraically identical whenever $n_1 = n_2$ or $s_1^2 = s_2^2$. If n_1 is very different from n_2 and s_1^2 is very different from s_2^2, the two formulas may yield quite different results.

If you are working from raw data you will find it convenient to use formula (10-7) instead of (10-3).

$$(10\text{-}7) \qquad s^2 = \frac{\Sigma x_1^2 + \Sigma x_2^2 - [(\Sigma x_1)^2 / n_1] - [(\Sigma x_2)^2 / n_2]}{n_1 + n_2 - 2}$$

Now let us return to the problem suggested at the beginning of this section, and let us suppose that the pertinent sample data are as follows:

	Men	*Women*
n	120	130
$\bar{x}$	7.57	6.40
s^2	23.05	21.49
s^2/n	.1921	.1653

By formula (10-1)

$$s_{\bar{x}_1 - \bar{x}_2}^{\,2} = .1921 + .1653 = .3574$$

By formula (10-5)

$$s_{\bar{x}_1-\bar{x}_2}^2 = \frac{119(23.05) + 129(21.49)}{120 + 130 - 2} \cdot \frac{250}{(120)(130)} = .356$$

In this case, because neither the variances nor the sample sizes are very different and the latter are not small, the two formulas give similar results.

Suppose now that a two-sided test and significance level .01 have been chosen. Then

$$t = \frac{7.57 - 6.40 - 0}{\sqrt{.356}} = \frac{1.17}{.597} = 1.96$$

At the .05 level this would have been barely significant. At the .01 level it is not significant, and the hypothesis $\mu_1 = \mu_2$ would be considered tenable.

However let us imagine a situation in which the samples are small and unequal in size and the variances unequal, as follows:

	Men	Women
n	36	6
$\bar{x}$	12.51	9.46
s^2	18.6	10.2

Using formula (10-1) we obtain

$$s_{\bar{x}_1-\bar{x}_2}^2 = \frac{18.6}{36} + \frac{10.2}{6} = 2.217$$

Using formula (10-5) we obtain

$$s_{\bar{x}_1-\bar{x}_2}^2 = \frac{35(18.6) + 5(10.2)}{40} \cdot \frac{42}{216} = 3.4125$$

By the first method $t = 3.05/\sqrt{2.217} = 2.05$, which is greater than $t_{.975}$ for 40 degrees of freedom. By the second method $t = 3.05/\sqrt{3.4125} = 1.65$, which is less than $t_{.95}$. By one method at the .05 level we would reject and by the other accept the hypothesis $\mu_1 = \mu_2$. Which computation is correct? The number of cases is too small to justify using (10-1). The test for the hypothesis $\sigma_1^2 = \sigma_2^2$ described in Chapter 15 shows that this hypothesis may be accepted, and therefore the latter method is preferable, and we accept the hypothesis $\mu_1 = \mu_2$.

A word of caution is needed at this point. To retain a hypothesis does not prove it true but merely indicates that it is not inconsistent with the observed data of a sample. Similarly, to reject a hypothesis does not prove it false but merely indicates that the sample data obtained are improbable if the hypothesis is true. Furthermore, if the hypothesis is not rejected, it is inappropriate for the investigator to discuss why one sample had a higher mean than the other. The failure to reject the hypothesis $\mu_1 - \mu_2 = 0$ means simply that the difference in sample means may well have arisen through accidents of sampling from populations with the same mean. It is wrong to say, "There was a tendency for men to have the higher mean, but the difference was not significant." Failure to reject the null hypothesis means that there is no convincing evidence of a tendency for the scores of men and women to be different.

EXERCISE 10-1

1. The same 10 pairs of deaf and hearing girls whose balance scores were presented in Table 10-1 had scores on strength of grip as follows:

Pair	1	2	3	4	5	6	7	8	9	10
Hearing	16	12	17	18	15	21	16	22	18	16
Deaf	10	14	12	21	19	12	19	17	19	25

Test the hypothesis that deaf and hearing do not differ in respect to strength of grip, the alternative being that they do differ, and .05 the level of significance.

2. For all 37 pairs in Long's study, $\Sigma x = 87.1$ and $\Sigma x^2 = 363.17$, where x is the balance score of a hearing child minus the balance score of her deaf mate. Test the hypothesis $\mu = 0$, the alternative being $\mu > 0$, and .01 the level of significance.

3. For all 37 pairs, the corresponding data for strength of grip are $\Sigma x = 86$ and $\Sigma x^2 = 2072$. Test the null hypothesis, $\mu = 0$, when the alternative is $\mu > 0$ and .01 the level of significance.

4. For all 37 pairs, the corresponding data for rate of tapping with best hand are $\Sigma x = 5.9$ and $\Sigma x^2 = 982.55$. Test the null hypothesis, $\mu = 0$, the alternative being $\mu \neq 0$ and .02 the level of significance.

5. For all 37 pairs, the corresponding data for rate of tapping with right hand are $\Sigma x = 28.8$ and $\Sigma x^2 = 1595.52$. Test the null hypothesis $\mu = 0$ against the alternative $\mu \neq 0$, using the .05 level of significance.

Effectiveness of a Test of Hypothesis. In this section the various problems involved in testing a hypothesis such as choice of level of significance and choice of region of rejection will be reconsidered from the point of view of the relation of the hypothesis to its alternatives.

The reader will recall that the decision to reject or to accept a hypothesis involves a degree of arbitrariness. There is always a risk that a hypothesis may be rejected when it is true or accepted when it is false. The error in rejecting a hypothesis when it is true is known as an error of the first kind or a Type I error. The error of accepting a hypothesis when some other hypothesis is true is known as an error of the second kind or a Type II error.

What is called the level of significance is the probability of rejecting a hypothesis when it is, in fact, true. In the long run, during the course of a great many experiments in which he is testing a true hypothesis, an investigator who is employing the .05 level of significance will falsely reject 5 percent of those true hypotheses. If he is employing the .01 level of significance, he will falsely reject 1 percent of the true hypotheses he tests. The level of significance, denoted $\propto$, is the probability of a Type I error. Naturally one would like to make the probability of a Type I error as low as possible; in other words, to make the level of significance as low as possible. At first thought this seems easy because the choice of this level is arbitrary. However, the possibility of reducing the level of significance is limited by the need for rejecting the hypothesis when it is false. To see this need, suppose the level of significance is set at zero, so that there is no possibility of an error of Type I. If this is done, however, there is no possibility of rejecting the hypothesis even when it is false. There is then no point to the study at all, because the essence of hypothesis testing is the utilization of experimental evidence as a basis for a choice between the acceptance of a hypothesis and its rejection.

In addition to the choice of level of significance for a test of a hypothesis, the statistical analyst is concerned with the location of the region of rejection. Shall it be in one tail of the sampling distribution or the other or in both tails?

The choice of *level of significance* and the choice of location of the *region of significance* (which is also the region of rejection) can be evaluated in terms of the need for rejection of the hypothesis when it is false.

The problems just posed can be approached directly in terms of the need for reducing the probability of a Type II error; that is, of reducing the probability of accepting a hypothesis when it is false. In the following discussion we approach the matter from the mathematically equivalent view of increasing the probability of rejecting the hypothesis when it is false.

The Power of a Test.   This discussion will involve comparison of statistical tests. It will be necessary to define the sense in which one test may be said to be better than another. To make this discussion clear, it will be necessary to state explicitly what is meant by a test of hypothesis. For the purpose of the following discussion, a test is defined when the following have been specified:

1. A statistic
2. Sample size
3. Level of significance
4. Region of rejection.

The reader should note that a test thus defined is only one of a variety of test types. It is known as a fixed-sample-size, two-decision test. One different type of test is the sequential type in which sample size is not specified in advance of the experiment. In another type of test, known as a multiple-decision test, there are more decisions than the two of acceptance and rejection considered here.

For the fixed-sample-size, two-decision test with which we are dealing, a change in any one of the four elements constitutes a change in the test. Consider now a hypothesis and an alternative; for example:

Hypothesis H: $\mu = 0$
Alternative A: $\mu = 1$.

Suppose a test has been selected to test hypothesis H. It is possible to determine the probability that this test will lead to the rejection of H in case H is false and A is true, that is, in case $\mu = 1$. Now consider a second test obtained by changing the level of significance or changing sample size or both. It is also possible to compute the probability that this second test will lead to the rejection of $H: \mu = 0$ if $\mu = 1$. If the second probability is greater than the first, we say that the second test is more *powerful* than the first for rejecting H when A is true. For each test separately, the probability of rejecting H when A is true is called the *power* of the test in that relationship between H and A.

Table 10-3 shows the power of eight different tests for rejecting $\mu = 0$ when $\mu = 1$ and of the same eight tests for rejecting $\mu = 0$ when $\mu = -1$. Consider first the alternative $\mu = 1$ and the four tests with sample size 100. Of the two-tail tests, the one with level of significance .05 is more powerful than the one with level of significance .01; the situation is similar for the two tests with critical region in the upper tail. These comparisons illustrate the principle that an increase in probability of a Type I error leads to an increase in power. In other words, a decrease in the probability of a Type I error can be bought at the price of an increase in the probability of a Type II error.

Another principle can be discovered by comparing the power of the two upper-tail tests with the power of the corresponding two-tail tests for testing the hypothesis $\mu = 0$ against the alternative $\mu = 1$. In this situation the upper-tail tests are clearly the more powerful. The two-tail tests are equally powerful for the alternatives $\mu = 1$ and $\mu = -1$, but the upper-tail tests have almost no power against the alternative $\mu = -1$. In fact, if the hypothesis $\mu = 0$ is tested by use of this upper-tail test when μ is actually equal to -1, the false hypothesis $\mu = 0$ is almost sure to be accepted. This important point will be developed more fully below.

Table 10-3 Power of Several Tests of the Hypothesis $\mu = 0$ against the Alternatives $\mu = 1$ and $\mu = -1$ ($\sigma = 10$)

Alternative	Level of Significance	Power of the test			
		Two-tail Test		Upper-tail Test	
		$n = 100$	$n = 400$	$n = 100$	$n = 400$
$\mu = 1$	.05	.17	.52	.26	.64
	.01	.06	.28	.09	.37
$\mu = -1$	.05	.17	.52	.00	.00
	.01	.06	.28	.00	.00

If sample size is increased from 100 to 400, the power is increased regardless of level of significance and regardless of the alternative value of μ. However, for the alternative $\mu = -1$, the power is still so small that several more decimal places would be required to distinguish the value from zero.

Power Function.   The relationships in Table 10-3 are developed more fully in Figures 10-1 and 10-2. From these figures it is possible to read the probability of rejecting the hypothesis $\mu = 0$ if some particular alternative is true. For each curve, possible alternative values of μ are given on the horizontal scale. For each such alternative (as, for example, the alternatives $\mu = 1$ or $\mu = -1$ shown in Table 10-3) the probability of rejecting the hypothesis $\mu = 0$ has been computed, and a point has been plotted with the alternative value of μ as abscissa and the corresponding probability of rejection (or power) as ordinate. Examining these graphs, the reader should note that on each of them, when $\mu = 0$, the probability of rejection is exactly equal to the level of significance.

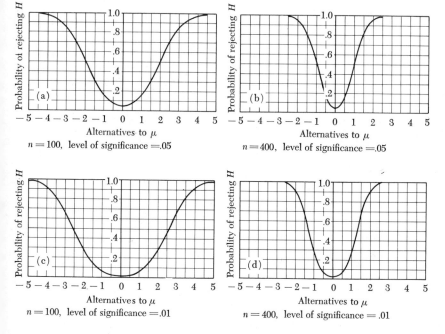

Figure 10-1 Power Functions of Two-tail Tests of Hypothesis H: $\mu = 0$ for Two Levels of Significance and Two Sizes of Samples, with $\sigma = 10$

He should also identify on the appropriate graph the point corresponding to each entry in Table 10-3.

Certain general principles which are of importance in deciding whether to use a larger or a smaller significance level, a larger or a smaller sample, or a one-tail or a two-tail test, can now be made clear through further examination of these graphs.

Look first at Figure 10-1, and verify the following probabilities of rejection when $n = 100$:

	For Test at .05 Level	For Test at .01 Level
If $\mu = 0$	.05	.01
If $\mu = .5$ or $\mu = -.5$	.07	.02
If $\mu = 1$ or $\mu = -1$	.17	.06
If $\mu = 1.5$ or $\mu = -1.5$	.30	.10
If $\mu = 2$ or $\mu = -2$	.52	.28
If $\mu = 3$ or $\mu = -3$	.85	.66
If $\mu = 4$ or $\mu = -4$	.98	.92

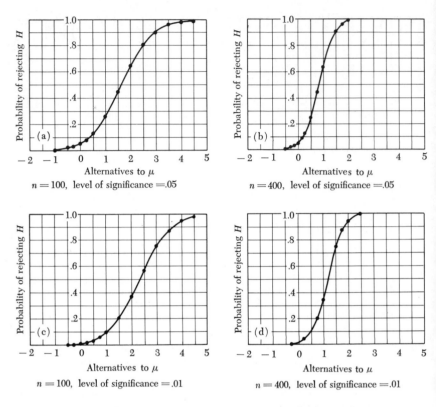

Figure 10-2   Power Functions of Upper-tail Tests of Hypothesis $H: \mu = 0$ for Two Levels of Significance and Two Sizes of Sample, with $\sigma = 10$

From the probabilities thus verified and from the corresponding probabilities for samples of 400, the following conclusion should now be clear. In general, when a two-tail test is used, *the probability of rejection increases as the true alternative deviates more widely from the value under the hypothesis.* If μ is near zero, though not precisely zero, there is small probability of rejecting $H: \mu = 0$, but, when the actual value of μ departs considerably from zero, the test makes rejection of $H: \mu = 0$ almost a certainty. This is a quite satisfactory state of affairs, for obviously in a practical situation it may be essential to reject a hypothesis about μ when the hypothesized value is very wrong and not essential to reject it when it is only slightly wrong. Now compare the probabilities of rejection shown above for the two levels of significance. For each alternative stated, the power is less when .01 is used as the level of significance than when .05 is used. In general, reducing the level of significance causes a reduction in the power of the test. In other words, an investigator who demands a very low probability of rejecting a true hypothesis must pay for that caution by incurring a larger probability of *not* rejecting it when some other hypothesis is true.

Now compare the probabilities of rejection listed above for $n = 100$ with corresponding probabilities read from the graphs for $n = 400$. *For a fixed level of significance, power is increased as sample size is increased*, and this relation holds for all alternatives to the hypothesis tested. In Figure 10-2 it is obvious that the probability of rejection is near one for alternatives much greater than zero, and for such alternatives the power of the one-tail test shown in Figure 10-2 is greater than the power of the two-tail tests shown in Figure 10-1. However, for alternatives less than zero, the upper-tail test has scarcely any power.

For the upper-tail test, as for the two-tail test, lowering the level of significance results in a reduction of power, and increasing sample size results in an increase of power for alternatives greater than $\mu = 0$.

The practical consequences of the preceding discussion for the selection of tests of hypotheses will now be outlined.

1. Sample size should be made as large as feasible without sacrificing the needs of good experimentation, to assure good power.

2. If sample size is small because of the nature of the study, then the level of significance should not be very small. The justification for this statement is that small sample size provides low power for a test. Since a low level of significance further contributes to reduction of power, the combination of small sample size and a low level of significance is not desirable.

3. The use which is to be made of the information gained from an investigation plays a part in the selection of a test of hypothesis. Ordinarily, the hypothesis which is being tested is rather noncommittal. It may consist, for example, of a statement that two groups do not differ. If the purpose of the study is to arrive at a scientific truth, one would be reluctant to reject a noncommittal hypothesis and thereby accept a more positive alternative, unless the evidence is strongly in favor of the alternative. In these circumstances, one is likely to adopt a low level of significance, say, .01 rather than .05.

However, if the purpose of the study is to choose some remedial action, the investigator would not wish to reject a method which may be helpful, even if the evidence in favor of the method is slight. Under such circumstances, the investigator would use a higher level of significance—for example, .05, in preference to .01.

4. In the choice between a one-tail test and a two-tail test, it in necessary to weigh the gains and losses involved from this choice. On the one hand, the one-tail test gives greater power against certain alternatives. On the other hand, the two-tail test provides assurance that the hypothesis $\mu = 0$ will be rejected for all alternatives differing considerably from zero, whenever the one-tail test does not provide such assurance.

The one-tail test is chosen when certain alternatives are either of no importance or no interest. For example, suppose a new medical treatment is being compared with one already in general use. The hypothesis "old and new treatments are equally effective" should certainly be rejected if the alternative "new treatment is more effective" is true, because a change of practice would then be indicated. However, the hypothesis and the alternative "old treatment is more effective" call for the same action—namely, retaining the old treatment. Consequently, this alternative is of no importance, and a one-tail test may be used. However, if two new treatments are under comparison, a difference in favor of either one is of interest, and the two-tail test should be used.

Computing the Power of a Test. This section is intended for those readers who want to know how the power of a test is computed, and it may be disregarded by others. In Figure 10-3, the solid line gives the sampling distribution of t under the hypothesis $\mu = 0$, when $n = 100$ and $\sigma = 10$. The region of rejection has been placed in the upper tail.

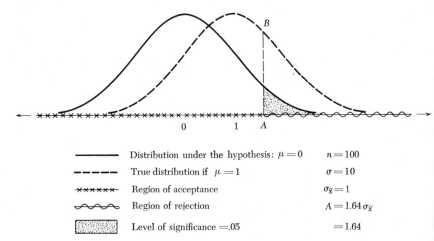

	Distribution under the hypothesis: $\mu = 0$	$n = 100$
— — — —	True distribution if $\mu = 1$	$\sigma = 10$
✕✕✕✕✕	Region of acceptance	$\sigma_{\bar{x}} = 1$
∿∿∿	Region of rejection	$A = 1.64\,\sigma_{\bar{x}}$
▓	Level of significance $= .05$	$= 1.64$

Figure 10-3 Distribution of $t = (\bar{x} - \mu)/(s/\sqrt{100})$ Under the Hypothesis that $\mu = 0$ and True Distribution if $\mu = 1$. Region of Rejection Based on a One-tail Test, $n = 100$ and $\sigma = 10$

In the figure, the shaded area represents the level of significance .05; the segment of line extending to the right from $A = 1.645$ and marked $\sim\!\sim$ includes values of $(\bar{x} - 0)/(10/\sqrt{100}) = \bar{x}$ which are larger than 1.645 and is the region of significance; the segment extending to the left from $A = 1.645$ and marked $\times\times\times$ includes all values smaller than 1.645 and is the region of acceptance.

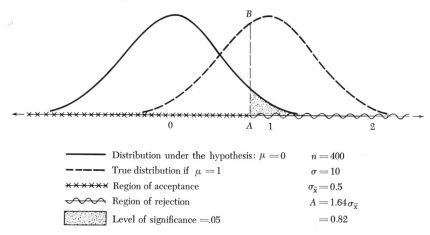

—————— Distribution under the hypothesis: $\mu = 0$	$n = 400$
— — — — True distribution if $\mu = 1$	$\sigma = 10$
×××××× Region of acceptance	$\sigma_{\bar{x}} = 0.5$
$\sim\!\sim\!\sim\!\sim$ Region of rejection	$A = 1.64\,\sigma_{\bar{x}}$
▨ Level of significance $= .05$	$= 0.82$

Figure 10-4   Distribution of $t = (\bar{x} - \mu)/(s/\sqrt{n})$ Under the Hypothesis that $\mu = 0$ and True Distribution if $\mu = 1$. Region of Rejection Based on a One-tail Test, $n = 400$ and $\sigma = 10$

Suppose, however, that the true value of μ is actually $\mu = 1$ and not, as hypothesized, $\mu = 0$. Then the actual sampling distribution is the distribution of $(\bar{x} - 1)/(10/\sqrt{100}) = \bar{x} - 1$ and is given by the broken line in Figure 10-3. The regions of rejection and acceptance, being based on the hypothetical distribution, are the lines marked $\sim\!\sim$ and $\times\times\times$ as already described. The probability that a sample will show a value of $\bar{x}$ in the region of rejection is represented by the area under the actual sampling distribution (the broken line) to the right of A. But the point A is $1.64 - 1.0 = .64$ units above $\mu = 1$, and so the probability sought is the area under a normal curve to the right of $z = .64$. (The sample size is large enough to make the use of the normal probability distribution appropriate.) This area is .26. Hence, the probability of rejecting the hypothesis $\mu = 0$, when μ is actually 1, is .26. This is the power of the one-sided test against the alternative $\mu = 1$. In Figure 10-2, the test for $\mu = 0$ at significance level .05, when $n = 100$ and $\sigma = 10$, is shown in Section A. The ordinate at $\mu = 1$ can be read as .26.

In similar fashion, the power against the alternative $\mu = 1.5$ can be computed as .44, against the alternative $\mu = 2.0$ as .64 and against $\mu = 3.5$ as .97.

The computation shown above was made particularly simple because $\sigma/\sqrt{n}$ came out so conveniently as $10/10 = 1$, but in Section B of Figure 10-2, $s = 10$ and $n = 400$. Then the region of significance consists of values of $(\bar{x} - 0)/(10/\sqrt{400}) = \bar{x}/.5$ greater than 1.64, which means values of $\bar{x}$ greater than .82. The hypothetical and actual distributions now appear like Figure 10-4, and the probability sought is the area under the actual curve to the right of A. But the point A is $(.82 - 1.00)/(10/\sqrt{400}) = -.18/.5 = -.36$ standard deviation units from the alternative $\mu = 1$. The area under a normal curve to the right of $z = -.36$ is .64. Compare this result with the ordinate at $\mu = 1$ of the power curve in Section B.

For that same one-tail test of $\mu = 0$ ($n = 400$, $\sigma = 10$, and significance level of .05), you may wish to verify by computation that the power against the alternative $\mu = .25$ is .13; against the alternative $\mu = .75$ is .44; against the alternative $\mu = 1.5$ is .91.

11

INFERENCES ABOUT
PROPORTIONS

Methods of statistical inference for means have been developed in Chapters 9 and 10. Now we shall consider similar procedures applicable to proportions.

Meaning of Proportion. The word *"proportion"*[1] is used here to mean the fractional part of a group of *discrete individuals*. Thus, if a survey of 800 dwelling units should find 128 units substandard, the proportion of substandard units could be expressed as $\frac{128}{800}$ or as any other equivalent fraction such as $\frac{4}{25}$ or $\frac{16}{100}$ or .16. The percent of substandard units, however, can be expressed in only one way, namely as 16 percent, since percent means literally per hundred (per centum).

In ordinary speech the term "proportion" is also applied to the fractional part of a continuous quantity, as, for example, the proportion of milk sold in New York City which is supplied by a particular distributor or the proportion of a teacher's time which is spent on routine tasks. The methods described in this chapter *are not applicable* to proportions of this sort.

One other correct use of the term "proportion" to which the methods of this chapter cannot be correctly applied should be mentioned. Consider a psychological test in which subjects make free responses which

---

[1] The word "proportion" is used here with a meaning somewhat different from that given to it in textbooks in mathematics where it is defined as the equality of two ratios. The meaning attributed to the word here has become standard usage in statistical writing.

are then classified by judges into categories. Suppose these results have been obtained:

Number of subjects examined	150
Number of responses made in all categories	3000
Number of responses in category A	420
Number of subjects who gave no responses in category A	48

It is possible to make inferences about the proportion of subjects who give no responses in category A and to use as datum the statistic $\frac{48}{150} = .32$. It is not possible to use the methods of this chapter to make inferences about the proportion of responses falling into category A based on the statistic $\frac{420}{3000} = .14$, because such responses are not independent individuals subject to the processes of random sampling. Samples are not drawn from a population of responses but from a population of subjects.

Conversion of Dichotomous Data to Measurement Data.   Methods of statistical inference for proportions are very similar to those for means, which are already familiar. The reason for this similarity will be clarified if we note how a dichotomy can be transformed into a special kind of measurement by assigning numerical values to the two classes in the dichotomy.

Consider, for example, the data reported in Appendix Table VIII. The six test scores reported are scaled data, and sex is a dichotomy. We shall assign the numbers 0 and 1 to the two sex categories. It is immaterial which number is assigned to which category, but, as some choice must be made, we shall score women as 0 and men as 1. The value 1 indicates presence in a class (males) and 0 means absence from that class. For the first 10 students from Table VIII we have the following tabulation:

Code number of student:	1	2	3	4	5	6	7	8	9	10
Sex of student:	M	M	M	M	F	F	M	M	M	F
Value of x:	1	1	1	1	0	0	1	1	1	0

By counting entries in the second line we see that there are 7 males among the 10 students so the proportion of males is .7. By adding entries in the third line we obtain $\Sigma x = 7$ and $\bar{x} = \Sigma x/n = \frac{7}{10} = .7$. Clearly the arbitrary assignment of the numbers 0 and 1 to the two categories of this dichotomy shows that a proportion can be regarded as analogous to a mean.

The reader will notice that the proportion of women among the 10 students is .3, and that if female sex is scored $x = 1$ and male $x = 0$, then $\Sigma x = 3$ and $\bar{x} = .3$.

Symbolism for Proportions. The symbol p will be used here to represent the proportion of individuals in a sample which have a specified characteristic (that is, the proportion which are in a given class) and $1 - p$ to represent the proportion which do not have that characteristic (are not in the given class). The letter q is often used in place of $1 - p$. If there are n individuals in a sample, then np is the *number* which have the given characteristic and $n(1 - p)$ or nq is the number which do not have it.

For the population from which the sample is presumed to be drawn, we shall use capital P to represent the proportion in the given class and $1 - P$ or Q to represent the proportion not in the class. The Greek letter π which is analogous to p will not be used for two reasons: (1) it is in familiar use to mean the number 3.1416 and, (2) the Greek alphabet has no letter analogous to our Q.

Statistical Inference about Proportions. The parameter P of a dichotomous population is analogous to the parameter μ of a population defined on a continuous variable. Therefore the four main problems of inference about μ described in Chapters 9 and 10 suggest the following analogous problems about P:

1. To find a point estimate for P
2. To find an interval estimate, or confidence interval, for P
3. To test the hypothesis that P has some specified value
4. To test the hypothesis that two populations have the same value of P, regardless of what that value may be.

Population Distribution. In making inferences about means, it was assumed that there was a population distribution of some variable X with population parameters μ and σ. The sample statistic $\bar{x}$, which will vary from sample to sample, must provide the basis for inferences about μ, as μ itself is almost always unknown. In the problems of Chapters 9 and 10, the variable was always scaled and always treated as if continuous.

In the problems of our present discussion, the population distribution is a dichotomy in which the proportion of individuals in one class is P and in the other, $1 - P$. This population has a mean and a standard deviation also, and it can be shown by a little mathematics that these are

(11-1)   Mean $= \mu = P$

(11-2)   Variance $= \sigma^2 = P(1 - P)$

(11-3)   Standard deviation $= \sigma = \sqrt{P(1 - P)}$

One notable difference appears here between the populations considered previously and the two-class populations which we are now considering. In the former, μ and σ are quite unrelated, knowledge of one gives no information as to the value of the other. However, in these two-class populations, σ depends completely on P.

Figure 11-1 illustrates a dichotomous population in which $P = .5$ and consequently $\sigma = \sqrt{.5(.5)} = .5$. Figure 11-2 illustrates a dichotomous population in which $P = .8$ and consequently $\sigma = \sqrt{.8(.2)} = .4$.

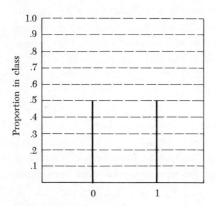

Figure 11-1 Population Distribution when $P = .5$ **Figure 11-2** Population Distribution when $P = .8$

Sampling Distribution of a Proportion.   Suppose we draw a random sample of 2 persons out of a population in which the proportion of men is $P = .5$. The sample may have no men, 1 man, or 2 men, and the corresponding proportions will be $p = \frac{0}{2} = 0$, $p = \frac{1}{2} = .5$, and $p = \frac{2}{2} = 1.0$. Thus if $n = 2$, p can take one of three different values, and the sampling distribution of p is a distribution with three classes. The probabilities attaching to those classes can be obtained mathematically from the value of P, either from a known value of P or a hypothesis about P.

In general, if there are n cases in a sample, p can take any one of $n + 1$ values ranging from $p = 0/n = 0$ to $p = n/n = 1.00$, and so the sampling distribution of p is a distribution of $n + 1$ discrete classes. The probabilities attaching to these classes can be obtained exactly by mathematical formulas which are available in all advanced texts but will not be reproduced here.[2]

Life is made much simpler for statisticians because the normal distribution provides an approximation to the sampling distribution of p.

[2] See the authors' more advanced text *Statistical Inference*, Holt, Rinehart and Winston, Inc., 1953, pages 19–30.

This approximation is good enough for most practical purposes unless n is small or P is near 0 or near 1. Figure 11-3 shows the exact sampling distribution for p when $P = .5$ and $n = 10$ and also shows a normal curve placed over the exact distribution. Figure 11-4 is similar except that $P = .8$. The relationship between the normal distribution and the

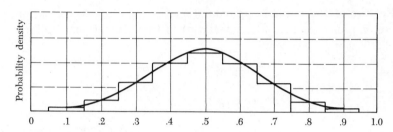

Figure 11-3   Exact Sampling Distribution for p when $P = .5$ and $n = 10$ and Normal Distribution with $\mu = .5$ and $\sigma = \sqrt{(.5)(.5)/10} = .158$

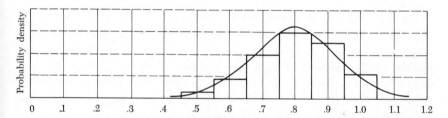

Figure 11-4   Exact Sampling Distribution for p when $P = .8$ and $n = 10$ and Normal Distribution with $\mu = .8$ and $\sigma = \sqrt{(.8)(.2)/10} = .126$

sampling distribution for proportions suggests that areas of the former may in some cases be used as approximations for areas of the latter, to simplify computation. Ways in which the approximation can be used effectively will be described below. Methods for exact calculation will also be developed.

Some features of the normal approximation need to be pointed out.

1. The normal approximation improves as sample size increases.
2. For given n the goodness of approximation depends on P. The normal approximation is useful for the distribution of p from samples of 10 when $P = .5$, but is very poor for samples of the same size when P is near 0 or near 1. A rough rule of thumb is that the normal approximation is satisfactory for given P if sample size n is such that $nP = 5$ or more when P is equal to or less than 0.5. Thus if P is 0.1,

then n should be at least 50. For P more than 0.5, apply the same rule to $Q = 1 - P$.

3. Because the normal approximation involves fitting a curve to a histogram, certain corrections are necessary to improve the fit. These will be introduced as computations are described below.

For use of the normal approximation the following formulas for the mean, variance, and standard deviation of the distribution of p for samples of size n are needed.

(11-4) $$\text{Mean of } p = \mu_p = P$$

(11-5) $$\text{Variance of } p = \sigma_p{}^2 = \frac{P(1 - P)}{n}$$

(11-6) $$\text{Standard deviation of } p = \sigma_p = \sqrt{\frac{P(1 - P)}{n}}$$

The standard deviation of p is usually spoken of as the *standard error of p*.

Point Estimation of a Proportion. When the unknown proportion, P, of a population is to be estimated, the sample value p provides the best single value available, just as a sample mean $\bar{x}$ provides the best single estimate of the parameter μ. Both p and $\bar{x}$ are *unbiased estimates* of their parameters. In other words, repeated samples will yield a variety of values of p, but the average value of p for many samples of the same size will be almost exactly equal to the population value P.

Interval Estimation of a Proportion. A confidence interval for P will be obtained by use of the normal approximation to the distribution of p. The procedure is similar to the one already described in obtaining a confidence interval for μ. The essential difference in the procedure to be described is that it requires a correction for fitting a normal curve to a histogram.

As before, we begin by adopting a confidence coefficient c and reading the values $z_{(1+c)/2}$ and $z_{(1-c)/2}$ from a table of normal probability. The confidence limits are then:

(11-7) $$\text{Upper limit} = p + \frac{1}{2n} + z_{\frac{1+c}{2}} \sqrt{\frac{p(1 - p)}{n}}$$

(11-8) $$\text{Lower limit} = p - \frac{1}{2n} + z_{\frac{1-c}{2}} \sqrt{\frac{p(1 - p)}{n}}$$

Here the quantity $1/2n$ is a correction for fitting the normal curve to a histogram, usually called a *correction for continuity*. The reader will also note that the quantity, $\sqrt{p(1-p)/n}$, is an estimate of the standard error of p obtained by substituting p for P in formula (11-6).

As an illustration, suppose a superintendent of schools wants an estimate of how many persons of voting age in his community would be in favor of a bond issue for a new school building, and suppose he selects a random sample of 200 such adults. (This selection is usually the hardest part of the whole undertaking and requires a great deal of careful planning.) Suppose an inquiry to these 200 voters reveals that 140 are in favor of the bond issue. Then the best single estimate of the proportion of voters who will vote affirmatively—assuming no major differences between what a voter says now and what he will do at a later date—is $\frac{140}{200} = .70$.

If the superintendent wishes to make an interval estimate, he must choose a confidence coefficient. Suppose he chooses to let $c = .95$. Then,

$$\text{Upper limit} = .70 + \frac{1}{2(200)} + 1.96 \sqrt{\frac{(.7)(.3)}{200}} = .766$$

$$\text{Lower limit} = .70 - \frac{1}{2(200)} - 1.96 \sqrt{\frac{(.7)(.3)}{200}} = .634$$

With confidence coefficient .95 he can make the statement that $.634 < P < .766$. If he wants to make a statement in which he has still greater confidence, he can use a larger value of c. If he uses $c = .99$, his statement would be $.614 < P < .786$. Certainly he can feel quite confident that, unless there is a definite shift of opinion before election day, the vote will carry with a good majority.

Test of a Hypothesis about a Proportion. To test a hypothesis about a proportion, we proceed in a manner similar to the one which was adopted for testing hypotheses about a mean. The procedure will be modified somewhat because of the necessary corrections for continuity. We choose a level of significance—for example, .05 or .01—and set up a region of rejection which may be in the upper tail, in the lower tail, or in both tails.

Suppose now that the hypothetical proportion is P, sample size is n and the level of significance is .05. Then, the region of rejection for an upper-tail test consists of values of p such that

$$(11\text{-}9) \qquad p > P + \frac{1}{2n} + z_{.95} \sqrt{\frac{P(1-P)}{n}}$$

where $z_{.95}$ is the usual value referred to the tables of normal probability. Similarly, for a lower-tail test, the region of significance would be

$$(11\text{-}10) \qquad p < P - \frac{1}{2n} + z_{.05} \sqrt{\frac{P(1 - P)}{n}}$$

For the two-tail region, we have both

$$(11\text{-}11) \qquad p > P + \frac{1}{2n} + z_{.975} \sqrt{\frac{P(1 - P)}{n}}$$

and

$$p < P - \frac{1}{2n} + z_{.025} \sqrt{\frac{P(1 - P)}{n}}$$

Exact procedures for testing hypotheses will be indicated below.

The Sign Test. An especially interesting and important way to apply tests of hypotheses about proportions is to test the hypotheses on the basis of scores achieved by matched pairs of individuals. The procedure is applicable to problems such as the comparison of deaf and hearing persons discussed in Chapter 10. The procedure, known as the *sign test,* provides a test of the hypothesis that the median of a population of differences of scores is zero.

The procedure will be described in relation to the scores on balance of deaf and hearing girls appearing in Table 11-1. For all 37 pairs of subjects let us examine the signs of the differences of hearing score minus deaf score. For pair number 1, this difference is plus. For pair number 2, it is minus. As we go through the 37 pairs, we find plus scores 33 times, minus twice, and zero twice. Ignoring the two zeros, the proportion of plus signs is $\frac{33}{35} = .94$.

On the assumption that, in general, the deaf and hearing do not differ in balance, half of the differences would be assumed to be plus and half minus. If P is the proportion of positive differences in the population, then $P = 0.5$. This is the hypothesis to be tested.

Using the .05 level of significance and an upper-tail region, the region of rejection is

$$p > .5 + \frac{1}{2(35)} + 1.645 \sqrt{\frac{(.5)(1 - .5)}{35}}$$

or

$$p > .65$$

Hence, the hypothesis $P = .5$ is to be rejected if the observed proportion

Table 11-1 Motor Abilities of 37 Pairs of
Deaf and Hearing Girls*

Pair Number	Tapping with Right Hand		Grip		Balance		Tapping with Best Hand	
	D	H	D	H	D	H	D	H
1	1.3	25.7	10	16	2.0	2.3	18.3	25.7
2	17.0	27.7	14	12	2.0	1.0	17.0	27.7
3	26.3	26.0	12	17	2.7	3.7	26.3	26.0
4	29.7	26.3	21	18	2.7	3.3	29.7	28.7
5	23.0	26.3	19	15	3.0	10.0	23.0	26.3
6	22.7	24.0	12	21	2.7	2.7	22.7	24.0
7	22.0	35.0	19	16	3.7	8.3	22.3	35.0
8	20.7	28.7	17	22	1.3	6.0	30.7	28.7
9	29.0	30.3	19	18	2.0	4.3	29.0	30.3
10	30.0	26.3	25	16	4.3	7.7	30.0	26.3
11	26.7	37.3	23	28	1.0	1.3	26.7	37.3
12	36.0	30.7	25	22	1.7	4.0	36.0	30.7
13	30.7	31.3	24	21	5.0	5.3	30.7	31.3
14	22.3	25.0	15	21	1.0	2.0	22.3	25.0
15	38.3	38.7	25	30	1.7	2.7	38.3	38.7
16	28.7	24.0	13	25	3.3	4.7	28.7	24.0
17	34.3	32.3	25	31	3.0	4.7	34.3	32.3
18	35.7	26.3	23	23	3.3	2.3	35.7	26.3
19	32.3	30.7	23	35	8.0	10.0	32.3	33.0
20	33.0	31.0	15	28	2.3	8.0	33.0	31.0
21	27.3	36.7	18	33	1.7	4.3	29.3	36.7
22	27.7	31.7	19	23	2.0	4.7	27.7	31.7
23	29.0	28.7	34	20	2.0	7.0	34.0	34.0
24	35.3	30.7	32	45	2.7	3.3	35.3	30.7
25	30.3	27.7	29	39	1.0	1.7	32.3	27.7
26	39.0	32.0	26	29	2.0	5.0	39.7	32.0
27	35.7	28.0	35	32	1.3	4.0	35.7	28.0
28	31.7	31.3	31	25	2.3	6.7	31.7	31.3
29	37.7	36.7	29	26	4.3	6.7	37.7	36.7
30	39.3	35.3	31	33	5.3	10.0	39.3	35.3
31	30.7	31.3	31	38	2.3	3.7	30.7	32.0
32	26.7	27.3	22	28	2.0	8.7	26.7	30.7
33	35.7	33.0	34	19	1.7	3.7	35.7	33.0
34	34.0	32.0	22	32	3.0	3.0	34.0	32.0
35	25.3	30.3	27	29	4.0	7.0	25.3	30.3
36	29.7	33.3	31	31	4.7	6.3	29.7	33.3
37	35.3	29.3	29	28	1.3	7.3	35.3	29.3

* Source of data: John A. Long, *Motor Abilities of Deaf Children*, Teacher's College,
Columbia University, Bureau of Publications, New York, 1932.

exceeds .65. Since the proportion in the sample is actually .94, the hypothesis is rejected.

Applying a two-tail test with a .05 level of significance, the region of rejection consists of all points p so that

$$p > .5 + \frac{1}{2(35)} + 1.96 \sqrt{\frac{(.5)(1 - .5)}{35}} = .68$$

and

$$p < .5 - \frac{1}{2(35)} - 1.96 \sqrt{\frac{(.5)(1 - .5)}{35}} = .32$$

Consequently, the region of rejection consists of all values of p greater than .68 and less than .32. Since the observed value of $p = .94$ is in the region of rejection, the hypothesis $P = .5$ is rejected on the basis of the two-tail test as well.

The sign test is an example of a wide variety of tests known as *nonparametric* or *distribution-free*. These tests have the great advantage that they make no assumption such as normality about the distribution of the parent population and are applicable to populations of any kind. The chi-square test discussed in Chapter 16 is a nonparametric test. Tests based on rank order are nonparametric. Other nonparametric methods may be found in advanced texts.

Another advantage, though of lesser importance, is that nonparametric tests give, with relatively little work, results closely approximating those obtained by tests based on the assumption of normality.

A Table for Testing Hypotheses about a Proportion. As was pointed out, the methods for testing hypotheses about proportions described to this point have been approximate. Exact methods are available but require extensive tables and mathematical insight. Some help in obtaining exact results is available in Appendix Table XII. This table consists of three parts: A, for the two-tail test; B, for the upper-tail test; and C, for the lower-tail test. All parts are suitable for testing at the 5-percent level of significance only. Each entry in the table indicates a region of rejection corresponding to the number of cases at the left of the row and the value of P at the head of the column in which the entry is located. The entries are stated in terms of number of cases rather than in terms of proportion. Thus, for the sign test one would use the number of differences having a plus sign rather than the corresponding proportion of differences.

As an example of the use of this table, let us apply it to make the sign test of the previous section. The data are $n = 35$, $k = 33$, and $P = .5$.

To make a two-tail test we refer to part A, the column headed .50, and the row for which $n = 35$. The corresponding pair of entries is 11 and 24. This means that the hypothesis $P = .50$ is rejected if the number of positive differences is either as small as, or smaller than, 11 or is as large as, or larger than, 24. Since the observed value of k is 33, the hypothesis is rejected. If the numbers 11 and 24 are reduced to proportions, we have $\frac{11}{35} = .314$, which approximates the value .320 previously obtained, and $\frac{24}{35} = .686$, which approximates the value .680.

To develop the corresponding test using a region of rejection in the upper tail, we refer to part B of Table XII. The entry in the column headed .50 and the row for which $n = 35$ is 23. Hence, the hypothesis is again rejected.

Hypothesis that P Is the Same for Two Populations. Suppose that 105 of the 200 adults in the sample described earlier in this chapter have children in school and 95 do not; and suppose that 90 of the 105 parents are in favor of the bond issue, and only 50 of the 95 nonparents are in favor. The situation can be summarized thus:

	Parents	Nonparents	Combined Group
Approve	90	50	140
Disapprove	15	45	60
Total	105	95	200

Two populations are under consideration here, a population of parents and one of nonparents. To distinguish these two populations and the samples from them, we shall use the following symbols:

P_1 = proportion of the population of parents who approve the issue
P_2 = proportion of the population of nonparents who approve the issue
p_1 = proportion of the sample of parents who approve the issue; $p_1 = 90/105 = .857$
p_2 = proportion of the sample of nonparents who approve the issue; $p_2 = 50/95 = .526$.

The question to be answered is whether the difference $p_1 - p_2 = .857 - .526 = .331$ is so great that it cannot reasonably be presumed to be an outcome of random sampling from two populations which have the same value of P. In other words, do the sample data contradict the hypothesis $P_1 = P_2$?

To answer this question, we need the standard error of the difference $p_1 - p_2$, which is

$$(11\text{-}12) \qquad \sigma_{p_1-p_2} = \sqrt{\frac{P_1(1 - P_1)}{n_1} + \frac{P_2(1 - P_2)}{n_2}}$$

In the present situation there is no hypothetical value for either P_1 or P_2. Each of them must be estimated from the data. On the hypothesis that they are equal, so $P_1 = P_2 = P$, the best sample estimate of P is the proportion in the combined group. If we call this estimate p, then for the given data

$$p = \frac{90 + 50}{105 + 95} = \frac{140}{200} = .70$$

Under the hypothesis that $P_1 = P_2$, the standard error is estimated by the value

$$(11\text{-}13) \qquad s_{p_1-p_2} = \sqrt{\frac{p(1 - p)}{n_1} + \frac{p(1 - p)}{n_2}}$$

If neither n_1 nor n_2 is very small, tests of the hypothesis $P_1 = P_2$ can be based on the standard normal distribution.

When the alternative is $P_1 > P_2$, use the statistic

$$(11\text{-}14) \qquad z = \frac{p_1 - p_2 - \dfrac{n}{2n_1n_2}}{\sqrt{\dfrac{p(1 - p)}{n_1} + \dfrac{p(1 - p)}{n_2}}}$$

and set the region of rejection in the upper tail. When the alternative is $P_2 > P_1$, use the statistic

$$(11\text{-}15) \qquad z = \frac{p_2 - p_1 - \dfrac{n}{2n_1n_2}}{\sqrt{\dfrac{p(1 - p)}{n_1} + \dfrac{p(1 - p)}{n_2}}}$$

and set the region of rejection in the upper tail.

When the alternative is $P_1 \neq P_2$, use a two-tail region. For the upper tail, use the statistic given by formula (11-14). For the lower tail, use the statistic

$$(11\text{-}16) \qquad z = \frac{p_1 - p_2 + \dfrac{n}{2n_1n_2}}{\sqrt{\dfrac{p(1 - p)}{n_1} + \dfrac{p(1 - p)}{n_2}}}$$

The term $n/2n_1n_2$ is known as *Yates' correction* or the *correction for continuity* and is usually presented somewhat differently, as in formula (16-11) on page 282. Here, $n = n_1 + n_2$.

In the numerical problem introduced earlier in this section, we found $p_1 = .857$, $p_2 = .526$, $p = .700$, $n_1 = 105$, $n_2 = 95$, $n = 200$. Then, if the alternative is $P_1 > P_2$, we compute

$$z = \frac{.857 - .526 - \dfrac{200}{2(105)(95)}}{\sqrt{\dfrac{(.7)(.3)}{105} + \dfrac{(.7)(.3)}{95}}} = \frac{.321}{.0649} = 4.9$$

This computed z is larger than $z_{.99} = 2.33$, and the hypothesis is rejected. It must be concluded that the parents are more favorable to the bond issue than the nonparents.

EXERCISE 11-1

1. Apply the sign test to the data of Table 10-1 on page 160, using an upper-tail test and the .05 level of significance. Use both the normal approximation and Appendix Table XII.

2. Apply the sign test to the difference in the scores of grip recorded in Table 11-1. Are the results consistent with the answer to Question 3 in Exercise 10-1?

3. A group of 50 laboratory animals was divided into two random groups of 25 each, and each group was trained in a complex task by a different method. At the end of a training period, 18 animals in group A and 12 in group B had mastered the task. Test the hypothesis that the two methods are equally effective. (Consider that the two methods of training generate two populations.)

12

REGRESSION

The law of Regression tells heavily against the full hereditary transmission of any gift. Only a few out of many children would be likely to differ as widely as the more exceptional of the two Parents. The more bountifully the Parent is gifted by nature, the more rare will be his good fortune if he begets a son who is as richly endowed as himself, and still more so if he has a son who is endowed yet more largely. But the law is even-handed; it levies an equal succession tax on the transmission of badness as of goodness. If it discourages the extravagant hopes of a gifted parent that his children will inherit all his powers; it no less discountenances extravagant fears that they will inherit all his weakness and disease.

It must be clearly understood that there is nothing in these statements to invalidate the general doctrine that the children of a gifted pair are much more likely to be gifted than the children of a mediocre pair. They merely express the fact that the ablest of all the children of a few gifted pairs is not likely to be as gifted as the ablest of all the children of a very great many mediocre pairs.

—Sir Francis Galton

It is a commonplace of even the most elementary scientific as well as practical effort to use a score of an individual on one variable as a means of *determining*, or *estimating*, a score of the same individual on another variable. Sometimes the determination is exact, as in finding the cost of a piece of cloth from its cost per yard. At other times, the determination is inexact, as in estimating the performance of a student in a course in college mathematics from the knowledge of his score in a college entrance test in the same subject.

Estimation (or Prediction) of One Variable from Another. When two variables are of such nature that knowledge of a value of one provides

190

information concerning the other, the variables are said to be *related*, or *associated*, and a *relationship* is said to exist between them. In such a matter as estimating a student's class performance from his score on a prognostic test, the test score comes first in point of time and so is said to be used to *predict* performance in the course. From such literal usage, the term *prediction* has spread to other situations in which, without any implication of futurity, a score on one variable is used to furnish information about another variable. Thus one may try to "predict" a child's weight from his height, or his mental age from his score on a test of paragraph meaning.

The purpose of this chapter is to consider inexact "prediction." However, the treatment of inexact prediction will be clearer if exact determination is discussed first.

The Exact Straight-line Relationship.   Exact determination can be readily illustrated by the relationship between salary rate and annual earnings. Table 12-1 contains a listing of monthly salary rates and annual incomes for five employees. Once the salary rate for an employee is known, the annual income is fully determined, assuming, of course, that the employee works a full year at the same rate. These data are presented graphically in Figure 12-1, where each point represents the paired monthly salary rate and annual income for one employee. The graph provides a ready means for ascertaining the annual income of an employee whose monthly salary is some amount other than those given in Table 12-1.

Table 12-1   Monthly Salary Rate and Annual
Income of Five Employees

Employee	Monthly Salary Rate	Annual Income
A	$400	$4800
B	425	5100
C	450	5400
D	500	6000
E	600	7200

To make a graph like that in Figure 12-1, horizontal and vertical scales for the two variables are drawn on coordinate paper. If no coordinate paper is at hand, a grid may be constructed by drawing horizontal and vertical lines at selected, usually equidistant, points on the two scales. These scales do not need to begin at zero. A point is plotted for each individual.

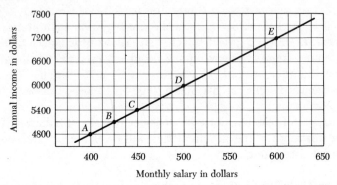

Figure 12-1 Relation of Annual Income to Monthly Salary Rate for Five Employees, A, B, C, D, and E

The Formula for an Exact Straight-line Relationship. The relationship of annual salary to monthly rate already shown by table and by graph can be succinctly stated in the word formula:

Annual salary = Twelve times monthly rate

To reduce this word formula to an expression more compact, more vivid to the eye, and more easy to manipulate algebraically, we agree to let some letter (say y) stand for "annual salary of an employee" and some other letter (say x) stand for "monthly rate for that employee" and write a symbolic formula as

$$y = 12x$$

The reader may wish to substitute pairs of values from Table 12-1 in this formula. Thus for employee C, $x = 450$, $y = 5400$, and $5400 = 12(450)$. Now consider a sixth employee not included in Table 12-1. If $x = 375$, what is y? $y = 12(375) = 4500$.

The graph of $y = 12x$ shown in Figure 12-1 is a straight line. When both variables are of the first degree, the graph of their relationship is a straight line and the relationship is called *linear* or sometimes *rectilinear*. (The terms x and y are of first degree; x^2, y^2, and xy are of second degree, x^3, y^2x, xy^2, and y^3 are of third degree.) Formulas showing the relationship between variables of degree other than the first will not be considered in this book. The graphs of such relationships are called nonlinear or curvilinear.

To introduce another form of the linear relationship, suppose that each of the five employees listed in Table 12-1 receives a year-end bonus of $300 in addition to his monthly salary. The annual income y then becomes

$$y = 300 + 12x$$

This formula contains two variables x and y (quantities which may change from individual to individual) and two constants, 12 and 300, which are the same for all individuals.

The general expression for a formula whose graph is a straight line has the form $y = a + bx$ where x and y change from individual to individual and where a and b represent numbers which are constant for any one set of data. This is a general algebraic expression of the *linear* relationship between two variables.

Independent and Dependent Variables.   The usefulness of such a relationship becomes clear when for one or more individuals the value of y is unknown and the formula is used to determine the unknown y from the known x. In obtaining the formula it was necessary to know values of both x and y for a group of subjects, but, once the relationship has been obtained, it can be used to determine y for other subjects for which only the values of x are known. The variable whose values are known is called the *independent variable;* the one whose values are determined from the formula is called the *dependent variable.*

In most practical situations, the logic of the situation makes clear which variable is considered independent and which dependent. Thus the familiar formula $C = 2\pi r$ suggests that the radius of a circle is known or can be obtained by direct measurement, whereas the circumference is to be obtained not by direct measurement but by multiplication of the measured radius. Because $C = 2\pi r$ is an exact relationship, it may be algebraically transformed into the equivalent form $r = C/2\pi$. Here the independent and dependent variables have interchanged their roles. The new formula suggests that the circumference of a circle is known or can be readily obtained by direct measurement and the radius is to be ascertained from the circumference, as in obtaining the radius of a circular pipe or cylinder where only external measurement is possible. *When the relationship is inexact, the two variables cannot be thus interchanged algebraically* but a second and distinct relationship is required, as will be discussed in Chapter 13.

Inexact Relationship.   In statistical work the relationships between two variables are almost never exact. The kind of relationships which are obtained in statistical investigations are well illustrated in the relationship between scores on the arithmetic test and on the midterm test for the data in Appendix Table VIII. As in the discussion of exact relationships, the inexact relationship will be described by a table, a graph, and a formula.

The tabular statement of the relationship between scores on arithmetic and scores on the midterm test is provided by the two columns in Table VIII. A comparison of pairs of scores for the same students shows at once that no simple relationship exists between them. For example, students coded 1, 28, and 52 all have the same score on the arithmetic test (33) but have the scores 58, 47, and 62 on the midterm test. These differences may be accounted for by such considerations as that the score on the prognostic test may not reflect accurately the ability of a student or that students with the same ability may be motivated differently in doing their course work.

The reader may be inclined to ask whether in the midst of such confusion any relationship at all can be found between these two variables. This question will be discussed in some detail.

Bivariate Frequency Distribution.   To construct the bivariate frequency distribution shown in Figure 12-2 from the data of Appendix Table VIII, an appropriate step interval is selected for each of the two variables. The step intervals for the independent variable are laid off on a horizontal axis, those for the dependent variable on a vertical axis. A grid is obtained either by placing the axes on coordinate paper or by drawing in the network of horizontal and vertical lines.

Each of the 98 individuals in Table VIII is represented by one tally in Figure 12-2. Thus the student with code number 1 who has arithmetic score $x = 33$ and midterm test score $y = 58$ is represented by a tally in the cell for which the interval on the horizontal scale is 30–34 and the interval on the vertical scale is 55–59. There are 7 tallies in this cell.

Transferring the data from such a table as Table VIII to the bivariate distribution is probably the most tedious part of all the procedures described in this and Chapter 13, and mistakes in tabulation are easily made. The task is less boring and results are usually more accurate if one person reads while another tabulates. Cardboard guides can be used by both reader and recorder to reduce errors and prevent eyestrain. The reader needs some device to keep his eye from jumping to the wrong row or wrong column on the data sheet and to keep him from reading the same values twice or omitting others. The recorder will find it an advantage to prepare a strip of cardboard with an exact duplicate of the scale for the midterm test. As the arithmetic score is read, the recorder moves this strip so that the scale appears directly at the left of the appropriate column, with zero point properly aligned. It is a simple matter then to place a tally in the proper cell.

When a tally has been recorded for every individual, the marginal frequencies are obtained by adding across each row and down each column.

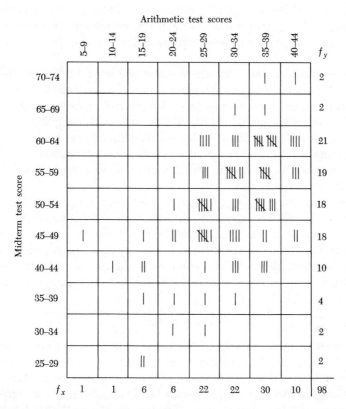

Figure 12-2   Midterm Test Scores Shown in Relation to Arithmetic Test Scores (98 cases) (Data from Appendix Table VIII)

The tallies jointly make up a *bivariate frequency distribution,* also called a *scatter diagram* and a *correlation chart.* The set of marginal frequencies at the right of this figure, obtained by adding across the rows, is the frequency distribution for the midterm test scores; and as we are using the letter Y to denote this variable, these frequencies are labeled f_y. The set of frequencies at the bottom of the figure, obtained by adding down the columns, are the frequencies related to intervals of the X scale and so are labeled f_x. Frequencies in the interior of the figure represent the joint occurrence of a particular value of X and a particular value of Y and so could be called f_{xy}.

A cursory examination of the bivariate distribution shows that high scores in the arithmetic test tend to be associated with high scores in the midterm test, and low scores in the arithmetic test with low scores in the midterm. Because of this immediate evidence of relationship, one may speak of the two variables as being co-related or correlated. A measure of correlation will be described in Chapter 13.

Mathematical Model. An examination of Figure 12-2 shows for each column, or each interval of arithmetic scores, the existence of a frequency distribution of midterm test scores. One readily observes in the figure, that the means of the distribution rise with increasing scores on the arithmetic test. The analysis of this relationship can be simplified by use of a mathematical model based on a two variable population. The model to be described is based on the straight-line relationship which was previously developed.

As was done in preceding paragraphs, the independent variable will be denoted by X and the dependent variable by Y. It will be supposed that for each value x in the population there is a distribution of y values, called the conditional distribution of y for a given x. Furthermore, it will be supposed that the mean of a conditional distribution, denoted by the symbol $\mu_{y.x}$ is related to the x value with which it is associated by the formula

$$(12\text{-}1) \qquad\qquad \mu_{y.x} = a + bx$$

Formula (12-1) states that in the population the means of the conditional distribution all lie on the same straight line. Therefore, the relationship described by this formula is called linear. For reasons to be stated later, the straight line in formula (12-1) is called a line of regression. Since regression lines need not necessarily be straight, this formula is further specified to represent linear regression.

To determine the conditional distribution fully, it is necessary to provide a measure of variability. Such a measure is the conditional standard deviation denoted by the symbol $\sigma_{y.x}$. Although the symbolism permits $\sigma_{y.x}$ to vary from one distribution to another, it is ordinarily assumed that the standard deviation is the same or constant for all distributions, that is, for all values of x.

For the purposes of statistical inference, it is commonly assumed that each conditional distribution is normal with mean $= a + bx$ and standard deviation $= \sigma_{y.x}$.

In succeeding paragraphs formulas will be given for obtaining estimates of the parameters a, b, and $\sigma_{y.x}$ from data. These estimates will be denoted $\hat{a}$, $\hat{b}$, and $s_{y.x}$.

The Problem in Miniature. Before using the entire 98 cases in a study of the relationship between scores on the two tests, working out that relationship for a very small number of students may serve to clarify principles. Nine students have been chosen. Their code numbers and scores on the two tests have been recorded in the first three columns of Table 12-2. The remainder of that table shows computations based on these scores for the purpose of obtaining a formula to express the manner of estimating midterm test score from the arithmetic score. You should read the ensuing discussion before attempting to follow these computations.

In Figure 12-3 the nine pairs of scores have been plotted as a bivariate distribution. There appears to be a tendency for higher scores on one

Table 12-2 Computation of Regression Equation to Predict Midterm Test Score from Arithmetic Test Score for Nine Cases from the Data in Appendix Table VIII

Code Number of Student	x Arith- metic Score	y Mid- term Score	x^2	y^2	xy	y_x	$y - y_x$	$(y - y_x)^2$
10	37	44	1369	1936	1628	55.2	−11.2	125.44
20	17	37	289	1369	629	44.5	−7.5	56.25
30	25	52	625	2704	1300	48.8	3.2	10.24
40	32	58	1024	3364	1856	52.5	5.5	30.25
50	18	47	324	2209	846	45.0	2.0	4.00
60	25	52	625	2704	1300	48.8	3.2	10.24
70	27	50	729	2500	1350	49.9	0.1	.01
80	36	53	1296	2809	1908	54.7	−1.7	2.89
90	36	61	1296	3721	2196	54.7	6.3	39.69
Sum	253	454	7,577	23,316	13,013	454.1	−0.1	279.01

$$\bar{x} = \frac{253}{9} = 28.1 \qquad \hat{b} = \frac{9(13{,}013) - (253)(454)}{9(7{,}577) - (253)^2} = .539$$

$$\bar{y} = \frac{454}{9} = 50.4 \qquad \hat{a} = 50.4 - (.539)(28.1) = 35.3$$

$$y_x = 35.3 + .539x$$

variable to accompany higher scores on the other, but the relationship is far from close, and most people would have no clear conviction as to where to draw the line of relationship. It will be necessary in this case to obtain the formula first and to draw the line of relationship afterward as a graph of that formula.

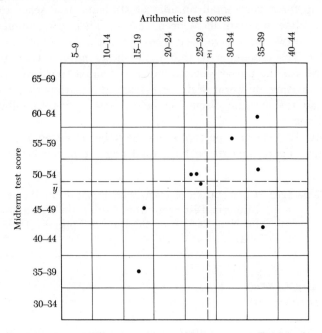

Figure 12-3   Midterm Test Scores Shown in Relation to Arithmetic Test Scores (9 selected cases) (Dotted lines pass through means of the two variables.)

This line is to be placed in a position which will best fit the nine points. But what is meant by "best fit"? Obviously, no matter where the line is drawn most, if not all, the points will deviate from it to some extent, and there will be a discrepancy between the actual midterm score of an individual (which has been denoted y) and the midterm score estimated from his x score. We shall denote the estimated score as y_x and the discrepancy as $y - y_x$. If the relation were perfect, every point would lie on the line and $y - y_x$ would be zero for every subject. In any reasonably good position of the line, some of these discrepancies will be positive and some negative, and so it is not enough to make the sum of all discrepancies zero. In fact, the sum of discrepancies from the horizontal line through the mean of the Y scale would be zero; yet that line does not represent the trend of the data at all well. The criterion commonly adopted for good fit is to make *the sum of the squares of the discrepancies* $(y - y_x)$ *as small as possible.* This is the principle or the method called *least squares.* By means of it the numbers $\hat{a}$ and $\hat{b}$ in the formula

$$(12\text{-}2) \qquad\qquad y_x = \hat{a} + \hat{b}x$$

are computed. We are now writing y_x instead of y because the relationship is between the known arithmetic score (x) and the best estimate which can be made of the midterm score (y_x), not the actual midterm score $(= y)$. With calculus it can be proved that the values of $\hat{b}$ and $\hat{a}$ which give the best position of the line—that is, "best" by the criterion stated—can be computed by formulas (12-3) and (12-4).

$$(12\text{-}3) \qquad\qquad \hat{b} = \frac{n\Sigma xy - (\Sigma x)(\Sigma y)}{n\Sigma x^2 - (\Sigma x)^2}$$

$$(12\text{-}4) \qquad\qquad \hat{a} = \bar{y} - \hat{b}\bar{x}$$

In Table 12-2 the sums of the columns headed x, y, x^2, and xy furnish the values which must be substituted in formulas (12-3) and (12-4). The result is the equation to predict y from x:

$$y_x = 35.3 + .539x$$

Let us now apply this equation to predict the y score of the first subject listed in Table 12-2 from his x score. As that x score is 37, we have

$$y_x = 35.3 + .539(37) = 55.2$$

The value of y_x has been computed for each of the nine subjects in this fashion and the results listed in the column head y_x. You should verify enough of these computations to be sure you understand the use of the regression equation.

If you should plot, for each of the nine students, the point representing his x and y_x scores, you would have nine points lying on one straight line, and that line would pass through the point whose paired values are $\bar{x}$ and $\bar{y}$. Therefore, you could locate this line by plotting any two of these points.

Figure 12-4 shows the regression line superimposed on the scatter diagram of Figure 12-3 and also shows for each of the nine subjects the amount $(y - y_x)$ by which the regression value was in error. The discrepancy $y - y_x$ is a deviation from the regression line sometimes called an *error of estimate*, an *error of prediction*, and sometimes *a residual error.* The reader should check some of these discrepancies and compare them with the corresponding vertical lines in Figure 12-4.

For historical reasons which will be explained on page 210, the line drawn on Figure 12-4 is called a *regression line,* and the value of $\hat{b}$ in the equation is called the *regression coefficient.*

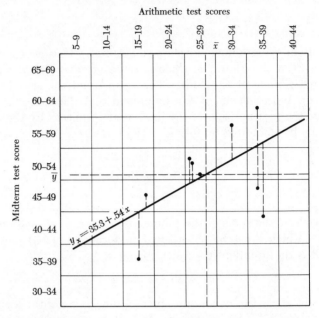

Figure 12-4   Regression Line $y_x = 35.3 + .539x$ and Errors of Estimate for the Nine Cases of Table 12-2

Before leaving Table 12-2, we should take notice of certain relations among its figures:

1. The sum of predicted values of y should equal the sum of actual values except for rounding errors. Here we have $\Sigma y = 454$ and $\Sigma y_x = 454.1$.
2. Because of the relationship just stated, the sum of the errors of estimate should approximate zero. Here we have $\Sigma(y - y_x) = -0.1$.

A General Routine of Computation for the Regression Formula.   The procedures illustrated for the 9 cases will now be applied to all 98 cases of Appendix Table VIII. It is unnecessary to copy down the scores as in Table 12-2, but sums can be obtained directly on a computing machine. With a wide-carriage desk calculator, all the sums needed for the formula can be obtained in a single operation. A method of getting these same sums from a scatter diagram with or without a machine will be described in Chapter 13.

The sums obtained from the raw data of Table VIII have been entered as Step I in Table 12-3. In subsequent steps in that table the computations are shown in convenient form. The reader will find it helpful to check these computations.

For the entire set of 98 cases, the regression equation is now found to be

$$y_x = 31.6 + .66x$$

whereas for the 9 selected cases it was

$$y_x = 35.3 + .539x$$

The regression coefficient for the 98 scores is .66 and for the 9 selected scores is .54. If the reader should now take another small sample and carry out the same computations, he will probably get an equation with constants still different from these. If, for example, he should use the first nine cases in Table VIII, he would find $\Sigma x = 316$, $\Sigma y = 504$, $\Sigma x^2 = 11390$, $\Sigma xy = 17762$, and $y_x = 48.1 + .22x$. These differences in the values of the coefficients of the regression formula illustrate the variation which results from using different samples. Some aspects of the problem of sampling variability have been discussed in earlier chapters.

Checks on Computation.   Even when a desk calculator is used, a check on the accuracy of the sums entering into the regression formula is essential. (In Chapter 13 you will see that these same sums are used in computing a correlation coefficient.) Recomputing is tedious and not very satisfactory because a person may make the same mistake twice. In Table 12-4 there is illustrated one way of obtaining a foolproof verification of the correctness of these sums. After $x + y$ and $x - y$ are written down for each individual, the sums and sums of squares can be readily obtained on a machine. The method would be laborious for hand work.

Computation by Hand or by Electronic Computer.   Most persons engaged in research of any magnitude have access to an electronic computer which can be programed to supply not only the relevant intermediate data but also regression equations, correlation coefficients, and individual deviations from regression estimates. Even though sophisticated computers are now widely available, there are several reasons why a textbook—particularly an elementary textbook emphasizing concepts—cannot omit all discussion of methods of hand computation.

The person who has little or no experience in this area needs especially to be able to look at a scatter diagram in order to grasp the basic concepts, and if he never plots a bivariate distribution his thinking is likely to remain unclear. Even the experienced person often finds that visual examination of a two-way distribution tells him something useful about his data, such as its apparent nonlinearity or the position of an aberrant individual.

Table 12-3   Steps in the Computation of the Regression Formula with Application to the Data of Appendix Table VIII (x = Arithmetic Score, y = Midterm Score, n = 98)

Step	x	y	xy
1.	$\Sigma x = 3{,}075$ $\Sigma x^2 = 101{,}581$	$\Sigma y = 5{,}141$ $\Sigma y^2 = 278{,}051$	$\Sigma xy = 164{,}694$
2.	$\bar{x} = \dfrac{3{,}075}{98} = 31.38$	$\bar{y} = \dfrac{5{,}141}{98} = 52.46$	
3.	$n\Sigma x^2 - (\Sigma x)^2$ $= 98(101{,}581) - (3{,}075)^2$ $= 499{,}313$	$n\Sigma y^2 - (\Sigma y)^2$ $= 98(278{,}051) - (5{,}141)^2$ $= 819{,}117$	$n\Sigma xy - (\Sigma x)(\Sigma y)$ $= 98(164{,}694)$ $\quad - (3{,}075)(5{,}141)$ $= 331{,}437$
4.	$s_x{}^2 = \dfrac{499{,}313}{98(97)} = 52.52$	$s_y{}^2 = \dfrac{819{,}117}{98(97)} = 86.17$	
5.	$s_x = \sqrt{52.52} = 7.2$	$s_y = \sqrt{86.17} = 9.3$	
6.	$\hat{b} = \dfrac{n\Sigma xy - (\Sigma x)(\Sigma y)}{n\Sigma x^2 - (\Sigma x)^2} = \dfrac{331{,}437}{499{,}313} = 0.664$		
7.	$\hat{a} = \bar{y} - \hat{b}\bar{x} = 52.46 - 0.664(31.38) = 31.62$		
8.	$y_x = 31.62 + 0.664x$		
9.	$\Sigma(y - y_x)^2 = \dfrac{1}{n}\left\{ n\Sigma y^2 - (\Sigma y)^2 - \dfrac{[n\Sigma xy - (\Sigma x)(\Sigma y)]^2}{n\Sigma x^2 - (\Sigma x)^2}\right\}$ $= \dfrac{1}{98}\left\{819{,}117 - \dfrac{(331{,}437)^2}{499{,}313}\right\} = 6113.4$		
10.	$s_{y.x}{}^2 = \dfrac{\Sigma(y - y_x)^2}{n - 2} = \dfrac{6113.4}{96} = 63.68$		
11.	$s_{y.x} = \sqrt{63.68} = 7.980$ or 8		

A teacher often wishes to study the relation between two or more measures of his own students or to explore the usefulness of a prognostic test he has administered. For this purpose he probably has only a small number of cases, and programing the data for a computer would not be worthwhile. He needs to know how to make his computations by hand or with only the help of a desk calculating machine.

The teacher of a statistics course often improvises small problems to illustrate a point. For these it would make little sense to halt the class, write a program, punch cards, and apply for time on the schedule of the nearest computer.

For reasons such as the preceding we have adopted the following philosophy and practice concerning computational problems:

1. Recognition of the fact that the availability of electronic computers has vastly reduced the need for hand computation on any large scale.
2. Omission of any discussion of methods appropriate to electronic computers because the primary emphasis of this text is on concepts and statistical thinking rather than on expert performance and because many good manuals telling how to program material for a computer are now available.
3. Presentation of computational routines to which the teacher of a statistics class may give light or heavy emphasis as the needs of his particular students seem to indicate or to which a student may refer when confronted with the need to make a computation by hand.
4. Provision of small problems with artificial data from which computations can be made with quickness and ease.

Estimation of the Conditional Standard Deviation. The conditional mean $\mu_{x.y}$ has already been estimated by the coefficients $\hat{a}$ and $\hat{b}$. It is now necessary to develop an estimate for $\sigma_{y.x}$. Under the assumption that $\sigma_{y.x}$ is the same for all values of x, the squares of deviations of the y scores from their respective regression values can be combined to provide an estimate of the standard deviation. Using these deviations, the formula for the estimate denoted as $s_{y.x}$ is given as

$$(12\text{-}5) \qquad s_{y.x} = \sqrt{\frac{\Sigma(y - y_x)^2}{n - 2}}$$

The symbol $s_{y.x}$ is read "s sub y point x" or "s y point x." $\Sigma(y - y_x)^2$ could be obtained by computing y_x for each case in the distribution, subtracting it from the corresponding y, squaring the remainder, and summing the squares, but it is much more easily obtained by the algebraically equivalent formula

$$(12\text{-}6) \quad \Sigma(y - y_x)^2 = \frac{1}{n}\left\{ n\Sigma y^2 - (\Sigma y)^2 - \frac{[n\Sigma xy - (\Sigma x)(\Sigma y)]^2}{n\Sigma x^2 - (\Sigma x)^2} \right\}$$

Because $s_{y.x}$ is the standard deviation of the extent to which actual scores differ from their regression estimates and therefore measures the errors one would make in estimating the y score of an individual by use of y_x, it is usually called the *standard error of estimate* or the *standard error of prediction*.

The standard error of estimate was calculated in Table 12-3, and the value $s_{y.x} = 8.0$ was obtained as the standard deviation of scores about the regression line, whereas $s_y = 9.3$ was the standard deviation of scores

Table 12-4 Checks on Computation of Sums Needed for Regression and Correlation, Applied to the Data of Table 12-2

Code Number	x	y	$x + y$	$x - y$
10	37	44	81	-7
20	17	37	54	-20
30	25	52	77	-27
40	32	58	90	-26
50	18	47	65	-29
60	25	52	77	-27
70	27	50	77	-23
80	36	53	89	-17
90	36	61	97	-25
Sum	253	454	707	-201
Sum of squares	7,577	23,316	56,919	4,867

$\Sigma x + \Sigma y = \Sigma(x + y)$	$253 + 454 = 707$
$\Sigma x - \Sigma y = \Sigma(x - y)$	$253 - 454 = -201$
$\Sigma(x + y) + \Sigma(x - y) = 2\Sigma x$	$707 - 201 = 506 = 2(253)$
$\Sigma(x + y) - \Sigma(x - y) = 2\Sigma y$	$707 + 201 = 908 = 2(454)$
$\Sigma(x + y)^2 + \Sigma(x - y)^2 = 2(\Sigma x^2 + \Sigma y^2)$	$56,919 + 4,867 = 61,786$ and
	$2(7,577 + 23,316) = 61,786$
$\Sigma(x + y)^2 - \Sigma(x - y)^2 = 4\Sigma xy$	$56,919 - 4,867 = 52,052$
	$4(13,013) = 52,052$
$\Sigma(x + y)^2 = \Sigma x^2 + 2\Sigma xy + \Sigma y^2$	$56,919 = 7,577 + 2(13,013)$
	$+ 23,316$

about the mean of the distribution. Obviously the scores cluster more closely about the regression line than about the mean.

Variation about the Regression Line. Computations near the end of Table 12-3 show the standard error of estimate to be $s_{y.x} = 8.0$, whereas the standard deviation of y had already been found as $s_y = 9.3$. Common sense suggests that because the regression line describes the trend of the

y scores more closely than the mean does, the variation of scores around the regression line should be less than the variation of those same scores about the general mean of the table. In fact, the sum of the squares of the deviations of scores from the mean $\Sigma(y - \bar{y})^2$ is made up of two parts, one of which is the sum of the squares of the deviations of scores from the regression line, $\Sigma(y - y_x)^2$ and the other is the sum of the squares of the deviations of regression values from the mean, $\Sigma(y_x - \bar{y})^2$. In symbols,

(12-7) $$\Sigma(y - \bar{y})^2 = \Sigma(y - y_x)^2 + \Sigma(y_x - \bar{y})^2$$

To verify this relation, return now to Table 12-2. Compute $\Sigma(y - \bar{y})^2 = \Sigma y^2 - (\Sigma y)^2/n = 414.2$. Compute $\Sigma(y_x - \bar{y})^2$ by subtracting $\bar{y} = 50.4$ from each of the 9 values of y_x, squaring the remainders and summing. The result is 133.8. We have already found that $\Sigma(y - \bar{y})^2 = 414.2$, and $\Sigma(y - y_x)^2 = 279.0$. Hence we have $133.8 + 279.0 = 412.8$, which is very close to 414.2. Rounding errors will usually prevent the verification from being perfect. By formula (12-7) $\Sigma(y - \bar{y})^2$ can never be less than $\Sigma(y - y_x)^2$.

Standard Error of Regression Coefficient.   One problem is to ascertain whether regression really plays a role in prediction in the sense that the parameter b in the regression line is something other than zero. The estimate $\hat{b}$ may, of course, be nonzero even when b is actually zero. What is needed, therefore, is a method of testing the hypothesis $b = 0$.

For tests of hypothesis and confidence intervals for b there is available the standard error of $\hat{b}$

(12-8) $$s_{\hat{b}} = \frac{s_{y.x}}{s_x \sqrt{n - 1}}$$

Then the ratio

(12-9) $$t = \frac{\hat{b} - b}{s_{\hat{b}}}$$

has Student's distribution with $n - 2$ degrees of freedom and can be used to test the hypothesis $b = 0$.

For the data in Table 12-3

$$s_{\hat{b}} = \frac{8.0}{7.2 \sqrt{97}} = .113$$

and

$$t = \frac{.664 - 0}{.113} = 5.88$$

It may be judged that t is significant at the .001 level, and therefore it cannot be said that arithmetic scores have no value for predicting midterm scores. Whether that prediction is precise enough to furnish dependable judgments concerning individuals in a practical situation is another issue.

Applications of the Regression Formula.   The concept of conditional distribution of y for a given x can now be used in the problem of estimation or prediction which was raised at the beginning of this chapter. The problem raised there was, "Given a value x for an individual, what is the best estimate of y for the same individual?"

The concept of conditional distribution means that y_x for a given x is the best estimate of this y mean of individuals with that same value of x. For this reason it is a best estimate of the y score of an individual with given x.

The concept of conditional distribution is useful beyond mere estimation since it provides a method of placing an individual with respect to an average, or expectation. Thus an individual with a given x who scores higher than the corresponding y_x exceeds his expectation, and he falls below his expectation if his score is below the appropriate y_x.

To illustrate these concepts consider the regression of midterm scores on scores in arithmetic which was computed in Table 12-3 as $y_x = 31.6 +$ .66x. Suppose an individual has an x score of 33, as is the case of Student 1 in Appendix Table VIII. By the regression formula the best estimate of the corresponding y score is 53.4. Since his actual score was 58, he did better than he might have been expected to do in view of his mathematical ability.

A similar analysis shows that Student 2 performed more poorly than might have been expected.

Let us now apply the regression formula to the study of the groups of students in sections I, II, and III. For each of these groups we can compute the mean of the arithmetic test scores. Call this mean $\bar{x}_i$. The resulting estimate $y_{\bar{x}_i}$ computed by the formula

(12-10) $$y_{\bar{x}_i} = \hat{a} + \hat{b}\bar{x}_i$$

is an estimate of the means of y values. In Table 12-5 are shown the means on the arithmetic test, the estimated means of midterm scores, and, for comparison, the means of actual midterm scores. The closeness with which the estimates of mean scores agree with actual means of scores is impressive. This approximation is far greater than for individuals. One might conclude, therefore, that the groups in the three sections performed on the midterm test very nearly in accordance with relative ability as shown on the arithmetic test.

Table 12-5   Means of Arithmetic Test Scores, Midterm Test Scores, and Estimated Midterm Test Scores for Students in Sections I, II, and III

$$y_{\bar{x}_i} = 31.6 + .66\bar{x}_i$$

Student Group	Mean of Arithmetic Test Scores $\bar{x}_i$	Estimated Mean of Midterm Test Scores $y_{\bar{x}_i}$	Actual Mean of Midterm Test Scores $\bar{y}_i$
Section I	35.85	55.26	55.05
Section II	30.07	51.44	52.78
Section III	26.32	49.0	48.18
All Students	31.38	52.3	52.46

Graphic Representation of the Regression Coefficient.   The regression coefficient $\hat{b}$, together with the two means $\bar{x}$ and $\bar{y}$, fully determine the line of regression. Geometrically, $\hat{b}$ is the *slope* of the line of regression. If we take any two points on the line of regression, draw a horizontal line through the lower and a vertical line through the upper point and continue these lines until they meet, a right triangle will be formed. The ratio of the vertical segment to the horizontal is the *slope of the line* and is $\hat{b}$.

$$(12\text{-}11) \qquad \hat{b} = \frac{\text{Vertical segment}}{\text{Horizontal segment}} = \frac{v}{h}$$

Figure 12-5 shows two regression lines and two such triangles drawn on each. For the left-hand figure, both h and v are considered positive, so $\hat{b}$ is positive, and the line has *positive slope*. When $\hat{b}$ is positive, the regres-

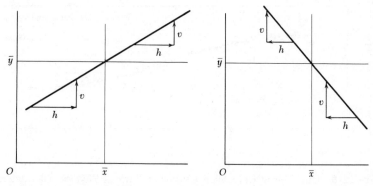

Figure 12-5   Geometric Representation of the Slope of a Regression Line, $\hat{b} = v/h$.

sion line is directed in a general southwest to northeast direction. For the right-hand figure, v is considered positive and h negative, so the quotient v/h is negative, and the line has *negative slope*. When $\hat{b}$ is negative, the regression line has a general northwest to southeast direction. Readers unfamiliar with directed lines and slope may find help in Chapters 8 and 10 of Walker, *Mathematics Essential for Elementary Statistics*.

The Assumption of Linearity. The discussion in this chapter has restricted itself to fitting a straight line to a set of data, and the reader is no doubt thinking of situations in which the trend is not linear. The graphs in Figure 12-6 are interesting in this respect. These curves were

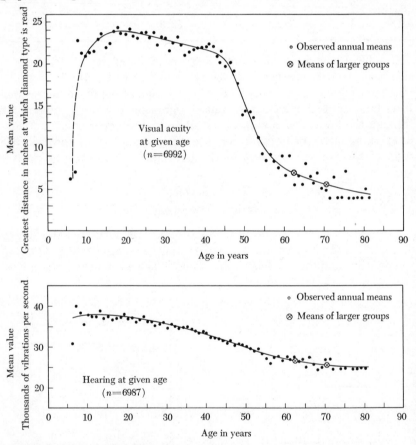

Figure 12-6 Curvilinear Regression Curves of Various Physical Characters on Age for Males (From H. A. Ruger, and B. Stoessiger, "On the growth curves of certain characters in man (males)," *Annals of Eugenics*, II, April, 1927, 76–110.)

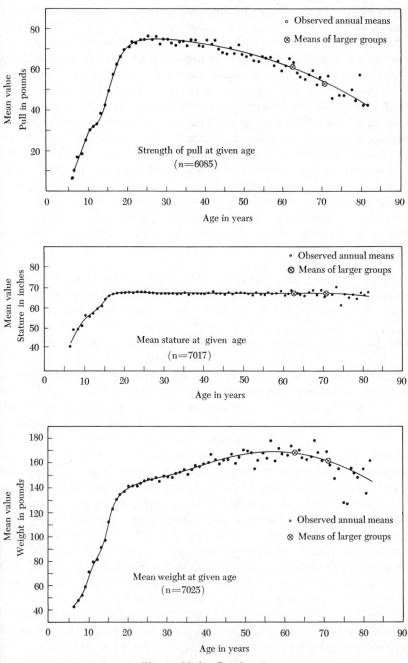

Figure 12-6—*Continued*

fitted by Henry A. Ruger to data originally collected by Sir Francis Galton and show for about 7000 adult males the way in which certain physical measures change with age. Each of the little black dots shows the mean value of the measure for a group of men of the specified age. Consider the first of these, showing the trend of visual acuity. Apparently if we had a group with age range from about 20 to about 40 years, the regression would be nearly linear and with a slightly negative slope indicating a slow decrease in visual acuity with advancing age. But if the age range were from about 20 to 80 years, no straight line would fit the data very well.

A discussion of nonlinear curves is beyond the scope of this book. A more or less intuitive decision as to the appropriateness of fitting a straight line can be made from a scatter diagram in which there has been some grouping. If the means of the columns cluster about a straight line, then a linear relationship may be presumed and a straight line should be fitted.

Origin of the Term "Regression." Most people think "regression equation" and "regression coefficient" rather strange terms unless they know the peculiar situation in which they were first used. Francis Galton, being profoundly impressed by *Origin of the Species* which his cousin Charles Darwin published in 1859, set himself to explore various aspects of heredity. In the process he gathered data on the stature of parents and their adult offspring and plotted a bivariate frequency distribution in which each entry represented one family, one axis being scaled for the average height of the two parents and the other for the average height of all the adult offspring. As might be anticipated, this chart showed clearly that taller parents tended to have taller children than did shorter parents. However, Galton also noted the surprising result that the children of very tall parents were, on the average, not quite so tall as their parents, and the children of very short parents were, on the average, not quite so short as their parents. In other words, extreme parents have children more mediocre than themselves. What is it, pondered Galton, which is dragging the human race back toward mediocrity? He commented that, "However paradoxical it may appear at first, it is theoretically a necessary fact, and one that is clearly confirmed by observation, that the stature of the adult offspring must on the whole, be more *mediocre* than the stature of the parents; that is to say, more near the *M* [the median] of the general population." In 1885 Galton published his famous paper on "Regression toward Mediocrity in Hereditary Stature" and introduced the term *lines of regression*.[1] The term still holds.

[1] For further discussion, see the following references: Francis Galton, *Memories of My Life*, London: 1908. Karl Pearson, "Notes on the history of correlation,"

Not only is historical interest and clarification of meaning to be found in Galton's use of the term regression, but there is also an important corrective for a certain kind of loose thinking. People are very prone to expect those who excel in one trait to excel in all others, whereas a proper understanding of the concept of regression would help them to be more realistic. There is, let us recognize, a tendency for high ability in reading to be associated with a high intelligence quotient. Therefore, if we select a group of children with high IQ, they will on the average read better than children with lower IQ, but their reading will not on the average exceed the mean reading score of their group by as many standard deviations as their intelligence quotients exceed the mean intelligence quotient of their group. This is an important lesson which many teachers fail to comprehend. And failing, they set unrealistic goals for their pupils. A clearer understanding of the concept of regression would make wiser teachers.

Routine for Computing Regression from a Scatter Diagram. This topic will be considered in Chapter 13.

EXERCISE 12-1

1. Examine each of the curves in Figure 12-6, and describe relationships which seem to you to be nearly linear, stating for each the age range over which linearity appears to hold. Describe situations in which a relationship appears not to be linear. Interpret what you see in terms of the way physical change appears to be related to increasing age. Do you find any variable which cannot be estimated from age?

2. (a) From the data of Appendix Table VIII, the student with code number 30 has the scores $x = 25$ and $y = 52$. Is his y value above or below what you would estimate by the regression equation $y_x = 31.6 + .66x$? By how much?

(b) Answer the same questions for student number 68 for whom $x = 39$ and $y = 44$.

3. Is there any fallacy in the comment that the general level of intelligence appears to be rising because in several families where the parents are of very low intelligence the children are considerably brighter than their parents?

Biometrika, **13** (1920), 25–45; *The Life, Letters and Labours of Francis Galton*, 3 vols., Cambridge University Press, 1914, 1924. Helen M. Walker, *Studies in the History of Statistical Method*, The Williams and Wilkins Co., 1929. The original works by Francis Galton are: "Regression towards mediocrity in hereditary stature," *Journal of the Anthropological Institute*, **15** (1885–86), 246–263; "Family likeness in stature," *Proceedings of the Royal Society*, **40** (1886), 42–63; "Co-relations and their measurement, chiefly from anthropological data," *Proceedings of the Royal Society*, **45** (1880), 135–145; and *Natural Inheritance*, 1889.

13

CORRELATION

In the previous chapter, the relationship between two variables was studied from the point of view of the line of regression and the variation about this line. In the present chapter, this relationship will be considered by use of the concept of the correlation coefficient. This coefficient can be viewed on one hand as providing a measure of the precision with which prediction is achieved by use of the line of regression and, on the other hand, as providing a measure of the mutuality of relationship between two variables.

The Correlation Coefficient. In Chapter 12 prediction was achieved by means of the regression equation. When all actual values are on the regression line, prediction is exact, and the relation between the two variables is perfect. We shall presently see that in such situations the correlation coefficient is either $+1$ or -1. When actual values are not identical with regression estimates, the prediction is not precise, and the relation between the variables is not perfect. In such situations, the correlation coefficient is between -1 and $+1$.

We have seen in Chapter 12 that the degrees of precision in estimating values of Y from values of X depends upon the sum of the squares of the residual errors, namely

$$\Sigma(y - y_x)^2$$

However, this sum of squares cannot be used as a measure of the precision of estimate because its value depends not only on the closeness of the relationship of X and Y but also on the variability of Y and on the number of cases involved. To obtain a measure which would be free of this dependence on n and on s_y, we might divide $\Sigma(y - y_x)^2$ by $\Sigma(y - \bar{y})^2$. The ratio

$$\frac{\Sigma(y - y_x)^2}{\Sigma(y - \bar{y})^2}$$

is zero when prediction is exact, for then every $y - y_x$ is zero, and so the numerator is zero. When knowledge of x contributes no information about y, the numerator of the ratio is equal to the denominator, and the value of the ratio is one. It should be noted that the numerator represents the sum of squares of the deviations of y values from the regression line, and the denominator represents the sum of squares of deviations from the mean; and the numerator will *never* be larger than the denominator.

This ratio could serve as a measure of precision of regression estimates, but, for reasons which are partly historical, the closely related coefficient of correlation is more commonly used for that purpose. The correlation coefficient, commonly denoted r, will now be defined by four algebraically equivalent formulas, each of which serves to reveal certain important aspects of this measure.

First, a formula which relates r to the sum of squares of residuals is

(13-1a)
$$r^2 = 1 - \frac{\Sigma(y - y_x)^2}{\Sigma(y - \bar{y})^2}$$

From this formula it can be seen at once that, when prediction is exact, $r^2 = 1$ because the ratio

$$\frac{\Sigma(y - y_x)^2}{\Sigma(y - \bar{y})^2}$$

is 0; and when prediction fails entirely, $r^2 = 0$ because the ratio is 1. In other words, $r^2 = 0$ means no correlation and $r^2 = 1$ means perfect correlation. It can also be seen that r^2 can never be larger than 1 because the ratio cannot be negative. Thus r cannot be numerically larger than 1 but can have values between -1 and $+1$. This formula gives no clue as to whether correlation is positive or negative. It does not provide a convenient pattern of computation.

As correlation is a mutual relationship between two variables, you are probably thinking that if x had been predicted from y, r should also be related to the precision of that prediction. You are correct, and the formula

(13-1b)
$$r^2 = 1 - \frac{\Sigma(x - x_y)^2}{\Sigma(x - \bar{x})^2}$$

is algebraically identical with (13-1a).

The best-known formula for the correlation coefficient is

(13-2)
$$r = \frac{\Sigma(x - \bar{x})(y - \bar{y})}{\sqrt{\Sigma(x - \bar{x})^2 \Sigma(y - \bar{y})^2}}$$

which can also be written in standard score form as

(13-3)
$$r = \frac{\Sigma z_x z_y}{n - 1}$$

Obviously, this formula has the same numerator as the formula for the regression coefficient. The denominator for r is positive, and the denominator for $\hat{b}$ is positive, and therefore r and $\hat{b}$ must have the same sign. When the regression line has positive slope, r and $\hat{b}$ are both positive, and there is a tendency for large values of the two variables to occur together and for small values to occur together. When the regression line has negative slope, r and $\hat{b}$ are both negative, and there is a tendency for large values of one variable to occur in connection with small values of the other.

Formula (13-2) does not provide a convenient routine for computation. The formula which provides the *easiest routine for computation* is considerably longer and looks more complicated. This formula can be derived from formula (13-2) by easy algebraic manipulation:

(13-4a)
$$r = \frac{n\Sigma xy - (\Sigma x)(\Sigma y)}{\sqrt{\{n\Sigma x^2 - (\Sigma x)^2\}\{n\Sigma y^2 - (\Sigma y)^2\}}}$$

The relation of this formula to formula (12-3) for $\hat{b}$ is obvious. It will be used to provide the computational pattern in later sections of this chapter.

Formula (13-4a) may be written in terms of coded deviations (x' and y') from an arbitrary origin instead of gross scores (x and y), the general pattern of the formula remaining unchanged. The formula is then written as

(13-4b)
$$r = \frac{n\Sigma f_{xy}x'y' - (\Sigma f_x x')(\Sigma f_y y')}{\sqrt{\{n\Sigma f_x(x')^2 - (\Sigma f_x x')^2\}\{n\Sigma f_y(y')^2 - (\Sigma f_y y')^2\}}}$$

This is the formula employed in the computational routine of Table 13-1 and Table 13-2. The symbols used in this formula will be defined in the discussion of the computational procedure.

A fourth formula equivalent to those already presented shows the relationship between the correlation and regression coefficients:

(13-5)
$$r = \hat{b}\frac{s_x}{s_y}.$$

Since s_y and s_x are both positive, the formula shows once more that r and

b must have the same sign. A formula equivalent to formula (13-5) is

$$(13\text{-}6) \qquad\qquad \hat{b} = r \frac{s_y}{s_x}$$

When the two variables have equal variability so $s_y = s_x$, the correlation coefficient r is identical with the regression coefficient $\hat{b}$. This was the situation in Francis Galton's study of hereditary stature. He used the letter r for the regression coefficient, but later it became almost universally used for the correlation coefficient instead of for the regression coefficient. We have already seen that r cannot be numerically larger than 1. There is, however, no such limitation on the size of $\hat{b}$. If, for example, $r = .8$, $s_x = 3$, and $s_y = 12$, $\hat{b}$ would be 3.2.

Computation of the Correlation Coefficient without Plotting. The computation of the correlation coefficient is very similar to that of the regression coefficient, and the two are usually obtained together. For a first illustration, let us turn to the data of Table 12-2 on page 197, where we already have $\Sigma(y - y_x)^2 = 279.01$ and $\Sigma(y - \bar{y})^2 = 23{,}316 - [(454)^2/9] = 414.22$. Therefore, by formula (13-1a)

$$r^2 = 1 - \frac{279.01}{414.22} = .326$$

and r is either $\sqrt{.326} = +.571$ or $-\sqrt{.326} = -.571$. Whether r is positive or negative we cannot tell from formula (13-1a).

From Table 12-2 we also know that $n = 9$, $\Sigma xy = 13{,}013$, $\Sigma x = 253$, $\Sigma y = 454$, $\Sigma x^2 = 7{,}577$, and $\Sigma y^2 = 23{,}316$, and therefore by formula (13-4a) we have

$$r = \frac{9(13{,}013) - (253)(454)}{\sqrt{[9(7{,}577) - (253)^2][9(23{,}316) - (454)^2]}} = .571$$

Although the two procedures are algebraically identical, formula (13-4a) is better to use because it involves easier computation and less rounding error and because it indicates whether r is positive or negative.

To compute the correlation coefficient between midterm test scores and arithmetic test scores for all 98 cases, work in Table 12-3 on page 202 is extended to obtain the following values:

$$n\Sigma xy - (\Sigma x)(\Sigma y) = 331{,}437$$
$$[n\Sigma x^2 - (\Sigma x)^2][n\Sigma y^2 - (\Sigma y)^2] = (499{,}313)(819{,}117) = 408{,}995{,}766{,}621$$
$$\sqrt{[n\Sigma x^2 - (\Sigma x)^2][n\Sigma y^2 - (\Sigma y)^2]} = \sqrt{408{,}995{,}766{,}521} = 639{,}528$$
$$r = \frac{n\Sigma xy - (\Sigma x)(\Sigma y)}{\sqrt{[n\Sigma x^2 - (\Sigma x)^2][n\Sigma y^2 - (\Sigma y)^2]}} = \frac{331{,}437}{639{,}528} = .518$$

This value computed from the ungrouped scores may be compared with the value $r = .53$ obtained in Table 13-2 from grouped scores.

The Second Line of Regression. In Chapter 12 and in the earlier parts of this chapter, regression has been treated as though there were only one regression line, and y values were always estimated from x values. In many situations only one prediction makes much sense. For example, it is often important to estimate performance on a job from score on an aptitude test, but no one wants to estimate aptitude score from job performance. However, in other situations the mutuality of relation is such that it is just as important to be able to predict x from y as to predict y from x. Suppose, for example, that after the data for Appendix Table IX have been analyzed, two new pupils appear. One of them has taken only the arithmetic computation test, and the teacher wants to obtain an estimate of his score in arithmetic reasoning by a regression equation. The second has taken only the arithmetic reasoning test, and a regression equation is needed to estimate what his arithmetic computation score would have been if he had taken that test also. There is nothing whatever in this situation to suggest that one regression equation is more important or more meaningful than the other.

The equation $y_x = \hat{a} + \hat{b}x$, which we have been using, will now be changed slightly by attaching a subscript to each of the constants, making it

(13-7a) $$y_x = \hat{a}_{yx} + \hat{b}_{yx}x$$

The analogous equation to estimate x from y is

(13-7b) $$x_y = \hat{a}_{xy} + \hat{b}_{xy}y$$

The order of letters in the subscript is the same as the order of variables in the equation. The first letter in the subscript names the variable to be estimated; the second letter in the subscript names the variable by means of which the estimation is made.

The regression coefficient $\hat{b}_{yx}$ is already familiar, but its formula will be repeated here in order that the similarity of form of $\hat{b}_{yx}$ and $\hat{b}_{xy}$ may be recognized:

(13-8a) $$\hat{b}_{yx} = \frac{n\Sigma xy - (\Sigma x)(\Sigma y)}{n\Sigma x^2 - (\Sigma x)^2}$$

(13-8b) $$\hat{b}_{xy} = \frac{n\Sigma xy - (\Sigma x)(\Sigma y)}{n\Sigma y^2 - (\Sigma y)^2}$$

Note that the two numerators are identical with each other and with the numerator of r. Compare the two denominators, and note that the square root of their product is the denominator of r.

The value of $\hat{a}_{yx}$ is already familiar. Note its relation to $\hat{a}_{xy}$ in the formulas

$$(13\text{-}9) \qquad \hat{a}_{yx} = \bar{y} - \hat{b}_{yx}\bar{x} \quad \text{and} \quad \hat{a}_{xy} = \bar{x} - \hat{b}_{xy}\bar{y}$$

The two regression lines can be written directly in terms of the correlation coefficients by use of standard scores as

$$(13\text{-}10) \qquad \frac{y_x - \bar{y}}{s_y} = r\,\frac{x - \bar{x}}{s_x}$$

and

$$\frac{x_y - \bar{x}}{s_x} = r\,\frac{y - \bar{y}}{s_y}$$

These forms of the regression line hold because of the relationships

$$(13\text{-}11) \qquad \hat{b}_{yx} = r\,\frac{s_y}{s_x} \quad \text{and} \quad \hat{b}_{xy} = r\,\frac{s_x}{s_y}$$

Computation of the Correlation and Regression Coefficients from a Scatter Diagram. In Chapter 12 we considered one method for obtaining the sums of squares and sums of products needed in the regression formula, and these same sums are the values needed in the correlation formula. The method described there is convenient to use when data are given in a list such as the lists of Appendix Tables VIII and IX and when a computing machine is available. If no machine is at hand and n is large, that method would be very time consuming. It would be difficult to use when data are presented in a scatter diagram such as that of Table 13-2. We shall now consider a method of computation more appropriate for data displayed in a scatter diagram. This method provides a complete check on computation of Σx, Σy, and Σxy but does not check the computation of Σx^2 and Σy^2.

Table 13-2 shows a computation for the correlation coefficient between midterm test scores of 98 students in a first course in statistical method and scores on an arithmetic test taken at the beginning of the term. Table 13-1 describes in words the steps in that process. You should examine the two tables together, identifying in Table 13-2 the outcome of each step described in Table 13-1.

In Table 13-2 it happens that $i_x = i_y$. Usually these two step intervals are not equal, and care must be taken to multiply by the proper numbers in computing regression coefficients. A summary of computation formulas

Table 13-1   Steps in the Computation of Correlation and
Regression Coefficients from a Scatter Diagram

Step	Procedure
1	Plot the data on a bivariate frequency chart. This will ordinarily be done by tallies as in Figure 12-2 on page 195, and the result will be confusing to work from. If the number of cases is at all large, the tallies should be replaced by numerals before the next steps are taken. The symbol f_{xy} is used to represent the number of cases in a cell at the intersection of a row and a column.
2	Add across each horizontal row to obtain the marginal frequency for that row, and record the result at the right in a column labeled f_y. Add down each vertical column to obtain the marginal frequency for that column, and record the result below the chart in the row labeled f_x. Find Σf_y and Σf_x, and verify that they are equal. Record that value as n. In Table 13-2, $n = 98$.
3	In the column labeled y', record an arbitrary set of coded scores for the y intervals. In the row headed x', record an arbitrary set of coded scores for the x intervals. If you are using a machine, it is convenient to place the arbitrary origin (coded 0) in the lowest interval in which a frequency occurs as in Table 13-2, because then all deviation values will be positive. When working by hand, some computers prefer to place the arbitrary origin nearer the center of the distribution in order to have smaller numbers to work with. Then special attention must be paid to signs.
4	Compute $\Sigma f_y y'$ and $\Sigma f_y (y')^2$ exactly as you obtained similar values when computing a standard deviation. In Table 13-2, these are 497 and 2849, respectively, and for ease of identification have been designated A and B. Turn the table around and compute $\Sigma f_x x'$ and $\Sigma f_x (x')^2$ in the same way. In Table 13-2, these are 479 and 2551, respectively, and can be identified as C and E.
5	For each row separately, compute $\Sigma f_{xy} x'$, and enter the result in the column headed $\Sigma x'$. For example, the computation for the third row from the top is $4(4) + 3(5) + 10(6) + 4(7) = 119$. The sum of that column must agree with the $\Sigma f_x x'$ obtained in step 4 and will thus furnish a check on the correctness of the preceding step. In the example, the sum of this column is 479 and has been marked C to show its relation to the sum of the row C. In practice it is helpful to make a small movable scale exactly like the fixed horizontal scale of the correlation chart, with code numbers inserted, and to place this scale immediately below each row so you can easily multiply each f_{xy} by the corresponding x' and sum the products. Turn the table around and, in similar manner, find $\Sigma f_{xy} y'$ for each column and enter the results below the table in the row headed $\Sigma y'$. The sum of entries in this row must be identical with that previously obtained for $\Sigma f_y y'$. In the example, both sums are 497.
6	In the column $\Sigma x'$, multiply each entry by the corresponding y', and record in the final column of the table headed $y'\Sigma x'$. The first entry in the column $\Sigma x'$ is 13, actually obtained as $1(6) + 1(7)$. When this is multiplied by $y' = 9$, we have $9(13) = 117$ and $1(6)(9) + 1(7)(9) = 117$, which is the value of $\Sigma f_{xy} x' y'$ for the two cases in the top row of the table. The sum of all the entries in this final column will be $\Sigma f_{xy} x' y'$ for the entire table. Turn the table around. Multiply each entry in the row $\Sigma y'$ by the corresponding x', and enter in the final row of the table. The sum of the entries in this row will also be $\Sigma f_{xy} x' y'$ and thus provides a complete check on that part of the computation. In the example, these two numbers have been labeled D.
7	Complete the computation of $\bar{x}$, $\bar{y}$, s_x, s_y, r, b_{yx} or b_{xy} by inserting the obtained sums into the appropriate formulas.

Table 13-2　Computation of r from a Scatter Diagram

Arithmetic score

Midterm score	5-9	10-14	15-19	20-24	25-29	30-34	35-39	40-44	f_y	y'	$f_y y'$	$f_y(y')^2$	$\Sigma x'$	$y'\Sigma x$
70-74							1	1	2	9	18	162	13	117
65-69						1	1		2	8	16	128	11	88
60-64				4	3	10	4		21	7	147	1029	119	833
55-59			1	3	7	5	3		19	6	114	684	101	606
50-54			1	6	3	8			18	5	90	450	90	450
45-49	1		1	2	6	4	2	2	18	4	72	288	78	312
40-44		1	2		1	3	3		10	3	30	90	42	126
35-39			1	1	1	1			4	2	8	16	14	28
30-34					1	1			2	1	2	2	7	7
25-29		2							2	0	0	0	4	0
f_x	1	1	6	6	22	22	30	10	98		497	2849	479	2567
											A	B	C	D
x'	0	1	2	3	4	5	6	7						
$f_x x'$	0	1	12	18	88	110	180	70	479	C				
$f_x(x')^2$	0	1	24	54	352	550	1080	490	2551	E				
$\Sigma y'$	4	3	12	22	106	113	174	63	497	A				
$x'\Sigma y'$	0	3	24	66	424	565	1044	441	2567	D				

$$\bar{y} = 27 + 5\left(\tfrac{497}{98}\right) = 52.36$$

$$\bar{x} = 7 + 5\left(\tfrac{479}{98}\right) = 31.44$$

$$s_y^2 = 84.66 \qquad s_y = 9.2$$

$$s_x^2 = 54.06 \qquad s_x = 7.4$$

$$\hat{b}_{yx} = \frac{5}{5} \cdot \frac{13,503}{20,557} = .66$$

$$\hat{b}_{xy} = \frac{5}{5} \cdot \frac{13,503}{32,193} = .42$$

$$n\Sigma xy - (\Sigma x)(\Sigma y) = 98(2567) - (497)(479) = 13,503$$

$$n\Sigma y^2 - (\Sigma y)^2 = 98(2849) - (497)^2 = 32,193$$

$$n\Sigma x^2 - (\Sigma x)^2 = 98(2551) - (479)^2 = 20,557$$

$$(32,193)(20,557) = 661,791,501$$

$$r = \frac{13,503}{\sqrt{661,791,501}} = \frac{13,503}{25,725} = .53$$

using the letters A, B, C, D, and E to represent the results of specific operations performed in Table 13-2 may be helpful.

(13-12)
$$s_x = i_x \sqrt{\frac{nE - C^2}{n(n-1)}}$$

(13-13)
$$s_y = i_y \sqrt{\frac{nB - A^2}{n(n-1)}}$$

(13-14)
$$r = \frac{nD - AC}{\sqrt{(nE - C^2)(nB - A^2)}}$$

(13-15)
$$\hat{b}_{yx} = \frac{i_y}{i_x} \cdot \frac{nD - AC}{nE - C^2}$$

(13-16)
$$\hat{b}_{xy} = \frac{i_x}{i_y} \cdot \frac{nD - AC}{nB - A^2}$$

EXERCISE 13-1

1. Using the bivariate distribution of Table 13-2, place the arbitrary origins at $a'_x = 27$ and $a'_y = 57$ and verify the following:

$$\Sigma f_x x' = 87 \qquad \Sigma f_y y' = -91 \qquad \Sigma f_{xy} x' y' = 57$$
$$\Sigma f_x (x')^2 = 287 \qquad \Sigma f_y (y')^2 = 413$$

Complete the computations until you obtain the same values of r and b_{yx} as are found in Table 13-2.

2. With the same frequency distribution and with arbitrary origins at $a'_x = 32$ and $a'_y = 62$, carry out the same computations.

3. Verify the values of r indicated below Figures 13-1 to 13-6.

4. Find r and both regression equations if preliminary computations have yielded the following results:

$$
\begin{array}{ll}
n = 250 & \Sigma x' = 160 \\
a'_x = 15 & \Sigma y' = -125 \\
a'_y = 50 & \Sigma (x')^2 = 812 \\
i_x = 3 & \Sigma (y')^2 = 914 \\
i_y = 5 & \Sigma x' y' = -34
\end{array}
$$

Scatter Diagrams Illustrating Various Values of the Correlation Coefficient. No one can judge the size of a correlation coefficient precisely from the visual appearance of a scatter diagram, but persons who work extensively in correlational material acquire a background of experience which enables them to make a rough estimate. To help you acquire a

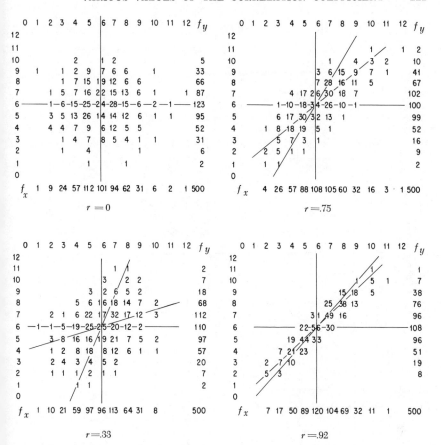

Figure 13-1   Scatter Diagrams and Regression Lines for Four Distributions with Different Sizes of r

little experience with correlations of various sizes, Figures 13-1 to 13-6 have been drawn up.

Look first at Figure 13-1 in which there are four distributions with correlation coefficients of 0, .33, .75, and .92. These are drawn from artificial data, primarily to show the positions of the two regression lines which are identical with the lines of the means when $r = 0$ and draw closer and closer together as r increases.

Figure 13-2 shows the joint distribution of language score and intelligence quotient for the 109 cases in Appendix Table IX. Here $r = .58$. To see more clearly the relation of the regression lines to the horizontal and vertical lines through the means of the marginal distributions (to which we shall refer briefly as the "lines of the means"), draw a red line

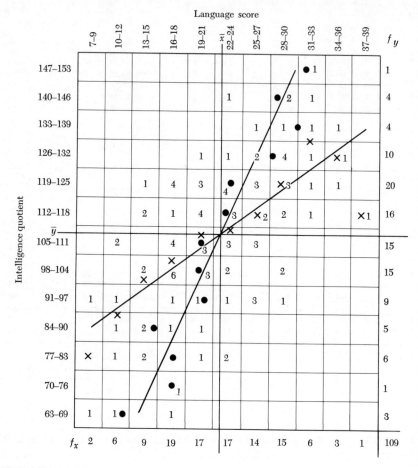

Figure 13-2   Language Score (x) and Intelligence Quotient (y). Bivariate Frequency Distribution with Two Lines of Regression Drawn

$$r = .583 \qquad x = 22.0 \qquad s_x = 6.51$$
$$y_x = 74.9 + 1.60x \qquad y = 110.1 \qquad s_y = 17.83$$
$$x_y = -1.45 + .213y$$

Circles mark means of horizontal rows. Crosses mark means of vertical columns.

horizontally across the chart at $\bar{y} = 110.1$. Then mark in red the small crosses which represent the means of the various columns, and also mark in red the straight regression line $y_x = 74.9 + 1.6x$. You will now see vividly how this regression line straightens out the irregularities of the set of observed column means.

Age to nearest month

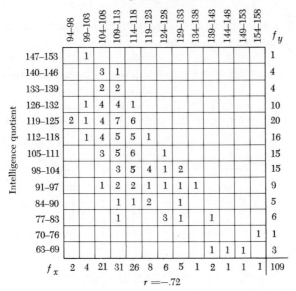

Intelligence quotient	94–98	99–103	104–108	109–113	114–118	119–123	124–128	129–133	134–138	139–143	144–148	149–153	154–158	f_y
147–153	1													1
140–146		3	1											4
133–139		2	2											4
126–132		1	4	4	1									10
119–125	2	1	4	7	6									20
112–118		1	4	5	5	1								16
105–111			3	5	6		1							15
98–104				3	5	4	1	2						15
91–97			1	2	2	1	1	1	1					9
84–90				1	1	2		1						5
77–83				1			3	1		1				6
70–76												1		1
63–69										1	1	1		3
f_x	2	4	21	31	26	8	6	5	1	2	1	1	1	109

$r = -.72$

Figure 13-3 Age and Intelligence Quotient. Bivariate Frequency Distribution for 109 Fourth-Grade Pupils

Age to nearest month

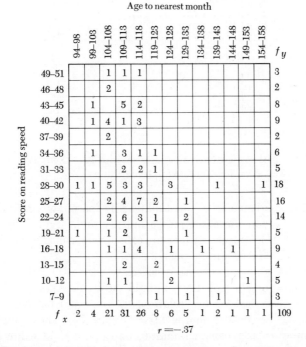

Score on reading speed	94–98	99–103	104–108	109–113	114–118	119–123	124–128	129–133	134–138	139–143	144–148	149–153	154–158	f_y
49–51			1	1	1									3
46–48			2											2
43–45		1		5	2									8
40–42		1	4	1	3									9
37–39			2											2
34–36		1		3	1	1								6
31–33				2	2	1								5
28–30	1	1	5	3	3		3			1			1	18
25–27			2	4	7	2		1						16
22–24			2	6	3	1		2						14
19–21	1		1	2				1						5
16–18			1	1	4		1		1		1			9
13–15				2		2								4
10–12			1	1			2					1		5
7–9					1		1		1					3
f_x	2	4	21	31	26	8	6	5	1	2	1	1	1	109

$r = -.37$

Figure 13-4 Age and Reading Speed. Bivariate Frequency Distribution for 109 Fourth-Grade Pupils

Now, with a blue pencil, mark the vertical line at $\bar{x} = 22.0$, the small circles which represent the means of rows, and the regression line $\bar{x}_y = -1.45 + .21y$. You will notice that of the two regression lines, the one to predict y is the one nearer to $\bar{y}$, and the one to predict x is the one nearer to $\bar{x}$.

Figure 13-3 shows that for the 109 fourth-grade pupils in Appendix Table IX, the correlation of age with IQ was negative and fairly high,

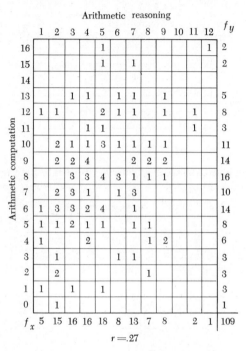

Arithmetic reasoning

Arithmetic computation	1	2	3	4	5	6	7	8	9	10	11	12	f_y
16				1							1		2
15				1	1								2
14													
13			1	1		1	1		1				5
12	1	1			2	1	1		1		1		8
11				1	1						1		3
10		2	1	1	3	1	1	1	1				11
9		2	2	4			2	2	2				14
8			3	3	4	3	1	1	1				16
7		2	3	1		1	3						10
6	1	3	3	2	4		1						14
5	1	1	2	1	1		1	1					8
4	1			2				1	2				6
3		1				1	1						3
2		2						1					3
1	1		1		1								3
0		1											1
f_x	5	15	16	16	18	8	13	7	8		2	1	109

$r = .27$

Figure 13-5 Arithmetic Reasoning and Arithmetic Computation. Bivariate Frequency Distribution for 109 fourth-Grade Pupils

$r = -.72$. Experience tells you that this is reasonable, because in any single school grade the older pupils are, by and large, duller than the younger ones. If all pupils in an entire school were considered, the correlation of age with IQ would be nearly zero.

Figure 13-4 shows a negative correlation ($-.37$) between age and reading speed for the 109 fourth-grade pupils in Appendix Table IX, and Figure 13-5 shows a low relationship ($r = .27$) between arithmetic reasoning and arithmetic computation. Figure 13-6 shows high positive correlation ($r = .66$) between midterm test and final examination for the 98 students of a first course in statistics.

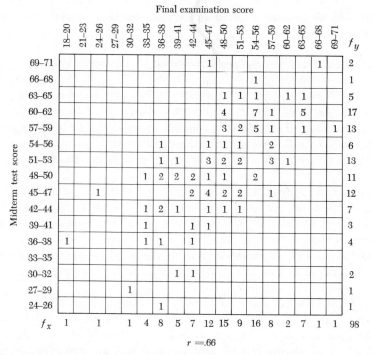

$r = .66$

Figure 13-6   Midterm and Final Examination Scores. Bivariate Frequency Distribution for 98 Students in a First-Term Course in Statistics

EXERCISE 13-2

Examine Figure 13-6 to obtain answers to the following questions:

1. Of the two persons scoring lowest on the midterm test, did either one make the lowest score on the final?

2. Compare the 8 persons with highest scores on the midterm and the 9 with highest scores on the final. How many persons were in both groups?

3. Do your answers to the preceding questions lead you to believe that, when $r = .66$, the persons who are at the extremes on one variable are almost certain to be equally extreme on the other?

4. Consider the 8 persons scoring lowest and the 8 scoring highest on the midterm. Do their scores on the final overlap at all?

5. Of the 30 persons who scored below 48 on the midterm, how many scored 48 or above on the final?

6. Of the 44 who scored above 50 on the final, how many scored 50 or below on the midterm?

Inferences about Correlation Coefficients. The correlation coefficient r is a statistic calculated from a sample just as $\bar{x}$, p, and $\hat{b}$ are statistics. Consequently, it is subject to sampling variability. In succeeding paragraphs, methods of making inferences about correlation coefficients will be described. The discussion will deal with the distribution of r, with tests of hypotheses, and confidence intervals. Comments will also be made on the interpretation of the size of correlation coefficients.

Nature of the Population. The population which is the source of the samples from which values of r are calculated is, of course, bivariate with variables, say X and Y. For each of these variables there is a mean and standard deviation denoted as μ_x and σ_x for X and μ_y and σ_y for Y. In addition, there is a fifth parameter namely the correlation in the population which is denoted as ρ or as ρ_{xy}. The Greek letter ρ (rho pronounced row) corresponds to the English r.

Sample estimates of these parameters are denoted as $\bar{x}$, s_x, $\bar{y}$, s_y, and r.

In the correlation model it is assumed that both x and y values differ from sample to sample. This assumption is in contrast with the view taken in the regression model, that the x values are fixed from sample to sample and only the y values vary.

Corresponding to the normality assumption in the univariate case, it is assumed that the joint distribution of the two variables is normal bivariate. Tables of the normal bivariate probability distribution are available, but these will not be used in this book. Because of the nature of the bivariate normal distribution, the values of either of the variables are distributed normally for a fixed value of the other variable.

Distribution of r. The distribution of r, like that of the proportion, is not normal but approaches normality as sample size increases. However, the approach to normality depends not only on sample size, but also on the value of ρ. If samples are drawn from a population for which ρ is at or near zero, the distribution of r is approximately normal for samples even as small as 30. However, when ρ is near $+1$ or -1, the distribution of r is quite skewed even for very large samples. We shall describe below a way of meeting this problem.

Standard Error of r. The standard error of r is given by the formula

$$(13\text{-}17) \qquad \sigma_r = \frac{1 - \rho^2}{\sqrt{n - 1}}$$

We may note that the standard error of r, like that of a proportion, depends on its population parameter. This fact and the non-normality of the distribution of r limit the usefulness of the standard error of r.

Tests of the Hypothesis that $\rho = 0$.   If $\rho = 0$, the standard error of r given in formula (13-17) reduces to $1/\sqrt{n-1}$ and the standard score for r, which is $(r - \rho)/\sigma_r$, becomes $r\sqrt{n-1}$. Then, if n is not small, for example, not less than 30, the distribution of $r\sqrt{n-1}$ will approximate the standard normal.

Suppose we wish to test the hypothesis $\rho = 0$ with a two-sided test and significance level .05. A table of the normal distribution indicates that the region of significance is $z > 1.96$ and $z < -1.96$. Now suppose a sample of 35 cases is drawn, and in this sample r is found to be .30. Therefore, $z = r\sqrt{n-1} = .30\sqrt{34} = 1.75$. As this value is not in the region of significance, the hypothesis $\rho = 0$ is tenable, and the observed value of r is *nonsignificant*.

The test just described is not exact because, even when $\rho = 0$, the distribution of r is not exactly normal. An exact test is available because

$$(13\text{-}18) \qquad\qquad t = \frac{r\sqrt{n-2}}{\sqrt{1-r^2}}$$

has the t distribution with $n - 2$ degrees of freedom. Thus for $n = 35$ and $r = .30$, $t = .30\sqrt{33}/\sqrt{1 - (.30)^2} = .30\sqrt{33/.91} = 1.81$. When we refer this value to Appendix Table III with 30 degrees of freedom (since Table III does not have entries for 33 degrees of freedom), we find it well below $t_{.975} = 2.05$. Again, we cannot reject the hypothesis $\rho = 0$.

Test of Hypothesis $\rho = 0$ by Use of Appendix Table VI.   Table VI enables one to test the null hypothesis about ρ directly from the known value of r without further computation. It is entered with degrees of freedom $\nu = n - 2$. (For a partial correlation, discussed in Chapter 18, ν would be something else.) By reference to Table VI verify the following:

1. $n = 22$. Then $\nu = 20$, $r_{.025} = -.423$, and $r_{.975} = .423$.
    In a sample of 22 cases, any value of r not between $r = -.423$ and $r = .423$ would be significant at the .05 level and would justify rejecting the hypothesis $\rho = 0$.
2. $n = 52$. Then $\nu = 50$, $r_{.025} = -.273$, and $r_{.975} = .273$.
    In a sample of 52 cases, the value of r which is numerically greater than .273 would be significant at the .05 level.

3. $n = 92$. Any value of r which is numerically greater than .205 would be significant at the .05 level; any value numerically greater than .242, significant at the .02 level; any value numerically greater than .338, significant at the .001 level.

The z_r Transformation. When ρ is different from zero, the sampling distribution of r is skewed, and neither of the tests described in the previous section nor Appendix Table VI can be used. A statistic which is usually called z but which we shall call z_r in order to distinguish it from the other measures which have already been denoted by that letter, is related to r by the formula

$$(13\text{-}19) \qquad z_r = \frac{1}{2} \log_e \frac{1 + r}{1 - r}$$

$$(13\text{-}20) \qquad = 1.1503 \log_{10} \frac{1 + r}{1 - r}$$

$$(13\text{-}21) \qquad = 1.1503[\log_{10} (1 + r) - \log_{10} (1 - r)]$$

This transformation, introduced by R. A. Fisher, has two very great advantages. Its distribution is approximately normal even for small samples in which ρ is near 1, and its standard error does not depend on the unknown population value but only on the size of the sample,

$$(13\text{-}22) \qquad \sigma_{z_r} = \frac{1}{\sqrt{n - 3}}$$

If ζ (zeta) is the population value of z_r corresponding to ρ, then $(z_r - \zeta) \sqrt{n - 3}$ may be treated as a normally distributed variable.

The transformation from r to z_r and vice versa can be made easily by reference to Appendix Table VII. For example, suppose $r = .85$. Look in the field of the table for the number nearest to .85. It is .8511, standing in the row headed 1.2 and in the column headed .06. The value of z_r corresponding to $r = .85$ is therefore read as 1.26. Or if $z_r = 2.09$, the corresponding r will stand in a cell in the row 2.0 and column .09. The value found in that cell is $r = .9699$. You will note that small values of r and z_r are very similar, but for large values, z_r is considerably larger than r.

Test of Hypothesis about ρ Using z_r Transformation. Suppose we have $r = .76$ for $n = 228$, and we wish to test the hypothesis that $\rho = .80$ at significance level .05, against the alternative $\rho < .80$. A correct procedure would be as follows:

1. From Appendix Table VII, we find that if $r = .76$, $z_r = 1.00$.

2. From the same table we find that if $\rho = .80$, $\zeta = 1.10$.

3. $\sqrt{n - 3} = \sqrt{225} = 15$.

4. Then the standard score is

$$z = \frac{1.00 - 1.10}{\dfrac{1}{\sqrt{225}}} = -1.50$$

5. As this is a one-sided test, the critical region is $z < z_{.05}$, which is $z < -1.64$.

6. Since the computed value $z = -1.50$ is not in the critical region, the hypothesis $\rho = .80$ is not rejected.

We may contrast with this the decision which might have been incorrectly obtained by computing $\sigma_r = .024$ and

$$\frac{r - \rho}{\sigma_r} = -1.67$$

The computations are not incorrect, but as we have no probability distribution to which the statistic -1.67 can be referred, they lead nowhere. If now we incorrectly refer it to the normal probability distribution and reject the hypothesis $\rho = .80$ because $-1.67 < z_{.05}$, we shall make an incorrect decision.

Confidence Interval for ρ. The z_r transformation is useful in obtaining a confidence interval for ρ. The procedure is as follows:

1. In Table VII, find the value of z_r corresponding to the sample value of r.

2. Compute a confidence interval for the population value ζ by the formula

(13-23)   Upper limit: $z_r + z_{\frac{1+c}{2}}\left(\dfrac{1}{\sqrt{n - 3}}\right)$

Lower limit: $z_r + z_{\frac{1-c}{2}}\left(\dfrac{1}{\sqrt{n - 3}}\right)$

Here one must remember that z_r and $z_{(1+c)/2}$ or $z_{(1-c)/2}$ are two quite different values represented by the same letter only because of tradition and the brevity of our alphabet.

3. In Table VII, read the value of ρ corresponding to each of the bounding values of ζ. From these form the interval for ρ.

Thus suppose we have obtained $r = .65$ in a sample of 40 cases, and we want an interval estimate for ρ with confidence coefficient .95. From Table VII we read $z_r = .78$. From the normal probability table we find

$$z_{.025} = -1.96 \quad \text{and} \quad z_{.975} = 1.96. \quad \sqrt{n-3} = \sqrt{37}.$$

Then

$$.78 - 1.96 \left(\frac{1}{\sqrt{37}} \right) < \zeta < .78 + 1.96 \left(\frac{1}{\sqrt{37}} \right) \quad \text{or} \quad .46 < \zeta < 1.10$$

and

$$.43 < \rho < .80$$

Notice that because of the skewness in the distribution of r the confidence limits for ρ are not symmetrically placed about the observed r, but $.80 - .65 = .15$, and $.65 - .43 = .22$.

Test of the Hypothesis That Two Independent Populations Have the Same Correlation.   If a sample is drawn from each of two independent, normal bivariate populations the difference $r_1 - r_2$ should fluctuate around $\rho_1 - \rho_2$. If each r is transformed to z_r, the difference $z_{r_1} - z_{r_2}$ will also fluctuate around $\zeta_1 - \zeta_2$ as mean and will have as standard deviation

$$\sqrt{ \frac{1}{n_1 - 3} + \frac{1}{n_2 - 3} }$$

Then

(13-24)
$$z = \frac{(z_{r_1} - z_{r_2}) - (\zeta_1 - \zeta_2)}{\sqrt{ \dfrac{1}{n_1 - 3} + \dfrac{1}{n_2 - 3} }}$$

may be treated as a normal deviate. If the hypothesis that $\zeta_1 = \zeta_2$ is rejected, the hypothesis $\rho_1 = \rho_2$ must also be rejected. If the hypothesis $\zeta_1 = \zeta_2$ is sustained, the hypothesis $\rho_1 = \rho_2$ must also be sustained.

In his research concerning *Children's Collecting Activity*[1] Durost studied the collections made by 50 boys and 50 girls between the ages of 10 and 14.

---

[1] Walter N. Durost, *Children's Collecting Activity Related to Social Factors*, New York. Teachers College, Columbia University, Bureau of Publications, 1932.

Among the correlations obtained was a correlation between mental age and the average rating of the child's collections (each of the collections of each child being rated as to quality and the average taken for the child). For boys this correlation was .31 and for girls, .06. Do these figures provide evidence that the relation between mental maturity and quality of collections made is higher for boys than for girls?

	r		z_r	$n-3$	$\dfrac{1}{n-3}$
Boys	.31		.32	47	$.021277 = \sigma^2{}_{z_{rb}}$
Girls	.06		.06	47	$.021277 = \sigma^2{}_{z_{rg}}$
		$z_{rb} - z_{rg} = .26$			$.042554 = \sigma^2{}_{z_{rb}} + \sigma^2{}_{z_{rg}}$

$$z = \frac{z_{rb} - z_{rg}}{\sigma_{z_{rb}-z_{rg}}} = \frac{.26}{\sqrt{.0426}} = \frac{.26}{.206} = 1.26$$

With so small a value of z, it would be inappropriate to assume that the relationship is higher for boys than for girls.

The Size of a Correlation Coefficient.   Several different bases exist for interpreting the size of a coefficient of correlation, and much confusion has arisen because what might be termed "large" in one connection would be "small" in another. To draw up a numerical scale and to say that coefficients in one particular range are large and in another negligible only adds to this confusion. A value of r may be "large" because it is too large to be consistent with the null hypothesis $\rho = 0$; or because it is large enough to be useful for predictive purposes; or because it is larger than values commonly obtained in similar circumstances. To use a single word "large" to cover these three quite disparate meanings is sheer verbal laziness. We shall now look at each in turn.

1.   *Interpretation in terms of statistical significance.* To say that a sample value of r is significant at, for instance, the .01 level means that, if a large number of samples of the given size are drawn from an uncorrelated population (that is, a population in which $\rho = 0$), 1 percent of such samples would be expected to produce a value of r at least as large numerically as the observed value. This is the concept which has been under discussion in the preceding pages. A very small r may be significant if the sample is very large. Thus, for 500 cases, $r = .09$ is significant but still too small to be useful for most purposes. It must always be kept in mind that the terms "significant" and "nonsignificant" are really comments on the relation of a sample to the hypothesis which is being tested.

2. *Interpretation in terms of usefulness of the regression equation for predictive purposes.* Chapter 12 explained that the usefulness of a regression equation depends on how closely the estimated and actual values agree. The correlation coefficient provides the clue to this agreement. Formula (13-1a) can be rewritten as

$$(13\text{-}25) \qquad 1 - r^2 = \frac{\Sigma(y - y_x)^2}{\Sigma(y - \bar{y})^2}$$

Here we see that $1 - r^2$ expresses the proportion which the sum of squares of deviations from the regression line is to the sum of squares of deviations from the mean. The smaller this proportion, the better the prediction.

3. *Interpretation in terms of correlations commonly obtained.* Sometimes, when a research worker comments that he obtained a high correlation between two variables, he means only that r was higher than previous experience had led him to anticipate. So many factors affect the size of r that it is well nigh impossible to draw up a list of typical values which would not be misleading. To build up a sense of what is reasonable to expect, a researcher requires experience, familiarity with the research literature in his own field, and careful attention to factors which make one situation different from others with which it is being compared.

Reliability of Measurement. A very important use of the correlation coefficient is to evaluate the reliability of measurements. Suppose a physical education teacher, wanting to measure the physical fitness of a group of boys at the beginning of a term, records the number of push-ups each boy can do. How dependable is such a single record? If exactly the same task is repeated a week later, there will be slight differences in individual performance. The correlation coefficient between the two records is called a *reliability coefficient* and in this situation indicates the similarity of performance of individuals on the repetition of a task. If this reliability coefficient appears to be undesirably low, the teacher may decide to use the average of several scores made on different days rather than a single day's score.

If the task is an intellectual or psychological one, repetition of precisely the same task will seldom serve a useful purpose. Two applications of precisely the same history test would be likely to yield an almost perfect correlation because each person would miss nearly the same questions on both occasions, even though the scores were not reliable indications of the students' knowledge of the subject. In such situations it is customary to develop two equivalent forms of the test and to correlate scores on the two forms. Each form contains only a sample out of all the many questions which might be asked and so is an imperfect measuring instrument. The correlation between scores on the two forms gives information about the similarity of those forms and measures the *reliability of the test*. If the two forms are administered on the same day, variation in the physical or emotional state of the pupils has little effect on the outcome, and so this reliability coefficient is really descriptive of the test. If the two forms are administered on different days, the correlation between the scores on the two forms is affected both by the differences between the items in the two forms and by day-to-day changes in the persons who take the test.

Sometimes only one form is available, and the items are divided into two half tests called "split halves." The correlation between scores on the two halves is, of course, lower than the correlation between two complete test forms, and it is the latter which is meant by "the reliability of the test." The reliability of the entire test can, however, be estimated from the correlation between the half tests.

Let R = correlation between two comparable test forms
$\quad r$ = correlation between two half tests.

(13-26)   Then $$R = \frac{2r}{1 + r}$$

This formula is known as the Spearman-Brown Prophecy formula because it was published by Carl Spearman and by William Brown in two papers which appeared simultaneously.

The reliability coefficient is very easily affected by the range of scores. A reliability coefficient computed for pupils in grades seven to nine would be considerably higher than one computed for pupils in a single grade only.

EXERCISE 13-3

1. Suppose that a 40-item test has been divided into two half tests of 20 items each and that the correlation between scores on the half tests, administered to 60 pupils, is $r = .52$. What would the correlation probably be between scores on two similar tests of 40 items each? Would this value necessarily be achieved? What may this correlation be expected to be between the two 20-item tests if given to 120 pupils instead of 60?

2. Suppose the correlation between two forms of a test has been found to be .45, and the test user feels this is too low a reliability coefficient to serve his purposes. What reliability coefficient might he hope to achieve if he could add similar items to make the test twice as long? Do you think this result would be achieved if he lengthened the test by adding poorer items than those in the original test?

Validity.   A common use of measurement is to predict scores on one variable from scores on another. Thus college entrance examinations, or high school grades, may be used to predict performance in college. Attempts have been made to predict delinquency from factors in home environment. The accuracy with which a prediction of this sort can be made is called the *validity* of a measure in making a particular prediction. A validity, or validity coefficient, is usually expressed as a correlation coefficient between the measure used in making the prediction and the measure which is predicted.

Validity differs from reliability in that reliability is an evaluation of the consistency in repeated measures of the same sort, whereas validity is an evaluation of the relationship between measures on different variables. Reliability is an inherent characteristic of a measure, but validity is a relationship between two measures. Thus in reports on a mental test, the reliability will be reported as an aspect of the tests, but validities are reported separately for each of the variables the test may be used to predict. Thus a test in mathematics might have a validity of .50 in predicting success in engineering school and a validity near zero in predicting success as a salesman.

Validity itself is influenced by reliability, and a test with low reliability will necessarily have low validity in all its predictions.

Personnel Selection.   A very important application of the correlation coefficient is the standardization and evaluation of procedures used in selection of persons for employment or for admission to school. The selection procedure provides a measure by which subsequent performance on the job or in the school can be estimated.

The user of the test has two problems. He must select a score or standard such that all candidates who attain or surpass this score are selected, and the remaining candidates are rejected. He must also check to ascertain whether the test is actually effective in selecting the more desirable candidates and in rejecting the less desirable.

The solution of both of these problems is accomplished by relating the scores of the selection test to those of a measure of achievement in the occupation or school for which applicants are selected. This new measure is commonly called the *criterion*, because it is the criterion against which the selection procedure is evaluated. Thus the selection procedure might be an entrance examination to a college, and the criterion might be college grades. Another example is provided by a test which leads to employment in a position and ratings of job performances given by supervisors.

The correlation between a selection test and a related criterion is the *validity* of the test or the *validity coefficient* because this correlation measures the validity or correctness of the test in predicting criterion scores.

Table 13-3 has been prepared to demonstrate the effectiveness of several validity coefficients in a selection procedure. The table consists of six subtables, each of which represents a distribution of 100 individuals according to several levels of the test score and of the criterion. Consider, for example, the first subtable at the left. Suppose that the standard for this group has been set to select the higher scoring half of the candidate

Table 13-3   Effectiveness of Selection Procedures for Three Different Validity Coefficients (100 candidates)

		Number of Candidates with Test Score in				
Validity Coefficient	Criterion Score	Lower Half	Upper Half*	Lowest Third	Middle Third	Top Third†
r = 0	Top third	17	17	12	10	12
	Middle third	16	16	10	12	10
	Lowest third	17	17	12	10	12
r = .50	Top third	9	25	5	10	19
	Middle third	16	16	10	12	10
	Lowest third	25	9	19	10	5
r = .80	Top third	4	30	1	8	25
	Middle third	16	16	8	16	8
	Lowest third	30	4	25	8	1

* Upper half of candidates selected
† Upper third of candidates selected

group. Of the 50 so selected, 17 are in the top third of the criterion group, 16 are in the middle third, and 17 are in the lowest third. We notice also that the rejected candidates have exactly the same distribution according to the criterion. We see, therefore, that when the validity coefficient is zero, choice of the higher group on test score provides no advantage. In fact, one is just as well off without using the test. When the validity is 0.50, if we choose the upper half of the group on test score, we find that fully 25 are in the top third of the criterion group, but only 9 are in the lowest third. It is clear, therefore, that here there is a distinct gain in using the test over using no test at all. A similar reading of the lowest subtable in the left column shows an even greater advantage in using the test when the validity coefficient is 0.80.

In the right-hand column of Table 13-3 it is assumed that we select only the top third of the group on the basis of test score. The first sub-table in this column shows at once that the test is not helpful in selection when validity is zero. The second subtable shows that 19 of the 34 se-lected subjects are in the upper third of the criterion group. This is 56 percent of the selected group, whereas only 50 percent were in the upper third when half were chosen. Thus a choice of upper third rather than upper half provides a distinct improvement in selection when the validity coefficient is 0.50. A similar gain is demonstrated for the validity coef-ficient of 0.80.

The conclusion from this analysis of Table 13-3 is that a test is in-creasingly useful in selection as its validity coefficient increases. When the validity coefficient is positive, selection is improved if the standard for selection is set higher.

The situation is analogous when the validity coefficient is negative, but in that case we choose from the lower level of test score rather than from the upper level.

Phi Coefficient. Sometimes the scatter diagram consists of only two rows and two columns. If the frequencies in the four cells of such a table are indicated by the letters a, b, c, and d and placed as in the adjacent sketch, the correlation coefficient can be obtained by a very convenient short cut, and the result is usually called ϕ (phi) to inform the reader that it was obtained from a fourfold table.

	0	1	
1	b	a	$a + b$
0	d	c	$d + c$
	$b + d$	$a + c$	

$$(13\text{-}27) \qquad \phi = \frac{ad - bc}{\sqrt{(a + b)(d + c)(b + d)(a + c)}}$$

Suppose that out of 60 men, 23 approve a particular proposition, whereas out of 80 women, 27 approve it. Is there a relation between sex and approval of the proposition?

	Men	Women	
Approve	23	27	50
Disapprove	37	53	90
	60	80	140

$$\phi = \frac{37(27) - 53(23)}{\sqrt{50(90)(60)(80)}} = -.047$$

The relationship is practically zero, certainly negligible.

Correlation among Ranks.   Sometimes where there is no satisfactory device for scoring a trait, individuals can be placed in a rank order in respect to the degree of the trait they exhibit. In such a situation it is possible to say for any two individuals which one is higher on the scale for the trait but not possible to say how much higher. The reader will readily think of many situations of this kind, as, for example, the attempt to place in order of merit the performances of contestants for an award in some contest or to rank persons on some intangible such as "friendliness" or "conscientiousness." If the group is large, even determining the order of individuals becomes difficult, so a correlation coefficient based on rank order is obviously most useful when samples are not large.

The correlation between two sets of ranks is called *rank order correlation* or *Spearman's rank order coefficient* (because it was originally proposed by Carl Spearman) or sometimes *rho* (because he denoted it by the Greek letter ρ). The formula is

(13-28)
$$R = 1 - \frac{6\Sigma d^2}{n(n^2 - 1)}$$

where n is the number of individuals ranked and d is the difference in the ranks assigned to the same individual. In the computation, a useful check is provided by the fact that $\Sigma d = 0$. This formula is derived by applying the usual product moment formula to the ranks.

Suppose 12 cakes submitted in a competition at a country fair have been ranked by two judges with results as shown in Table 13-4. Then

$$R = 1 - \frac{6(40)}{12(144 - 1)} = .86$$

Now let the ranks assigned by Judge I be denoted x and those assigned

Table 13-4   Computation of Coefficient of Correlation among Ranks Assigned to 12 Cakes by 2 Judges

Cake	Rank Assigned by		d	d^2
	Judge I	*Judge II*		
A	7	6	1	1
B	8	4	4	16
C	2	1	1	1
D	1	3	−2	4
E	9	11	−2	4
F	3	2	1	1
G	12	12	0	0
H	11	10	1	1
I	4	5	−1	1
J	10	9	1	1
K	6	7	−1	1
L	5	8	−3	9
			0	$\Sigma d^2 = 40$

$$R = 1 - \frac{6(40)}{12(143)} = 1 - \frac{20}{143} = \frac{123}{143} = .86$$

by Judge II be denoted y. Then

$$\Sigma x = 78 \qquad\qquad \Sigma y = 78$$
$$\Sigma x^2 = 650 \qquad\qquad \Sigma y^2 = 650$$
$$\Sigma(x - \bar{x})^2 = 650 - \frac{78^2}{12} = 143 \qquad\qquad \Sigma(y - \bar{y})^2 = 650 - \frac{78^2}{12} = 143$$
$$\Sigma xy = 630 \qquad\qquad \Sigma(x - \bar{x})(y - \bar{y}) = 630$$
$$-\frac{(78)(78)}{12} = 123$$

and

$$r = \frac{123}{\sqrt{(143)(143)}} = \frac{123}{143} = .86$$

This computation illustrates the fact that the rank order coefficient and the product moment coefficient applied to the ranks are identical. However, if scores are transformed to ranks, the product moment correlation of the ranks (which is the rank order coefficient) is almost certain to be different from the product moment coefficient of the original scores.

14

CIRCUMSTANCES
AFFECTING THE SIZE
OF CORRELATION
AND REGRESSION
COEFFICIENTS

The outcome of a correlational study is often difficult to interpret because the correlation between two variables not only reflects the strength of the intrinsic relation between those variables but may also be affected by other circumstances, such as the reliability with which the variables have been measured and the selection of subjects for the study. You can be more realistic about the meaning of coefficients you obtain if you are aware of such possible effects.

Effect of Changing Units. Some changes in data leave the correlation coefficient entirely unchanged.

1. Multiplying (or dividing) all the scores of either variable by the same positive number has no effect at all on r, but if the scores on one of the variables are multiplied by a negative number and those on the other are not, then the sign of the coefficient is changed. However, if all the values of x are multiplied by the number a and all values of y are multiplied by the number c, the regression coefficient $\hat{b}_{yx}$ will be multiplied by c/a, and the regression coefficient $\hat{b}_{xy}$ will be multiplied by a/c.

Thus the correlation between height and weight will have identically the same value (except for rounding errors) whether measurements are taken in inches and pounds or in centimeters and kilograms. The equations to predict weight from height will be different in the two situations, but the result in pounds of predicting the weight of a particular person from his height in inches will be equivalent to the result in kilograms of predicting his weight from his height in centimeters.

2.   Adding the same number to all values of one variable has no effect on either r or $\hat{b}$ but will affect the constants in the regression equations.

Statements (1) and (2) call attention to the fact that, in computing the sums of products and of squares which are used in obtaining a correlation coefficient, no use whatever is made of the scales of the variables. However, the scales are needed for obtaining the regression equations since the step interval is required for finding the regression coefficient, and both means are required for finding the constant term in the equation.

Effect of Grouping Scores.   Grouping scores into class intervals will not greatly affect the value of r or $\hat{b}$ provided the intervals are not made very large. Although the most accurate results are achieved by use of ungrouped data, computations made by subdivision of the scales into 10 or 15 intervals give satisfactory results for many purposes. The extreme procedure of subdividing the scales into two or three intervals and then computing the correlation and regression coefficients should be avoided since the distortion would then be considerable.

Variable Errors of Measurement.   By now you are aware that nothing can be measured with complete precision. Every good research worker eliminates all possible bias from his measures and reduces accidental errors of observation and measurement to a minimum, but he can never completely free his observations from accidental error. If these variable errors which remain are purely chance errors, they are as likely to be positive as to be negative and to have only a negligible effect on measures of central tendency. They tend to increase measures of variability, and they tend to *decrease measures of relationship*. This is an important matter. It means that an observed correlation is almost always lower than the correlation between true measures of the same traits would be. It means that if measurement is made with unreliable instruments, the observed correlation coefficient may be very low even when the traits concerned are closely related.

The reduction in the size of r because of variable errors of measurement is called *attenuation*. To illustrate its effect, an experiment was performed in which purely random errors were attached to the language scores and the intelligence quotients of the 109 subjects in Appendix Table IX.

A variable error was attached to each score in this way. Two decks of cards were shuffled together, and a card was drawn. The number on the first card drawn was taken as the error to be added to the score of the first child, the number on the second card drawn as the error for the second child, and so forth. Red cards were considered to indicate positive errors and black cards, negative. After every tenth drawing the cards were reshuffled. For case 1, the original language score was 17, the first card drawn was a six of hearts, and so the score plus error is 17 + 6 = 23. (It must not be supposed that the original scores were free from measurement error!)

Results of computing from the original scores and from the scores with random errors attached were as follows:

	Original Scores	Scores with Errors
Mean language score	21.98	22.08
Mean IQ	110.04	110.02
Standard deviation, language scores	6.5	10.3
Standard deviation, IQ	17.5	19.6
Correlation coefficient	.59	.20

The means have been changed so little by the variable errors that results have to be carried to four or five places before any difference appears. One mean has been very slightly increased, the other very slightly decreased. Both standard deviations have been noticeably increased. The correlation coefficient has been greatly decreased.

Effect of Number of Cases.   The number of cases on which r or $\hat{b}$ is calculated does not influence the values of these coefficients in the sense of making them consistently larger or consistently smaller. The number of cases does, however, affect the accuracy with which a relationship between two variables is determined. The matter of accuracy of the measurement of relationship is discussed more fully on the basis of statistical inference in Chapter 13.

Linearity of Regression.   Sometimes the means of the arrays of one variable (or both) lie on a line which is not straight. (Either a row or a column is called an array. A straight line is called *linear* or *rectilinear*. A line that is not straight is called *nonlinear* or *curvilinear*.) Thus the regression on age for almost any physical measure of a person is nonlinear if taken over the entire life span. Several such curved regression lines

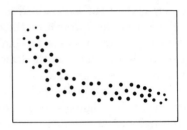

have been shown in Figure 12-6 on page 208. When a measure of overhead expense is related to the number of individuals involved, the scatter diagram often resembles the adjacent sketch. In some psychological studies there are variables for which a rating at either extreme tends to be associated with low performance or poor adjustment, and a rating in the middle range to be associated with better achievement or more healthy adjustment. An extreme situation is represented in the sketch below where column means are represented by small crosses and row means by small circles. A curved regression line fitted to the small crosses would represent the data rather well, so that it would be possible to make a fairly good estimate of y from x, but a line fitted to the row means would have little meaning.

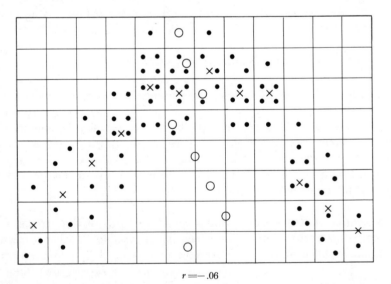

$$r = -.06$$

Since the correlation coefficient is based on the assumption of linearity of regression, it would be practically meaningless in data such as are sketched here in which there is obviously a relation between the variables, although the coefficient of correlation is only -0.06. There is a measure of relation which is applicable in such situations, called the *correlation ratio*. It is equal to r when regression is precisely linear, and greater than r when regression is not linear. The computation of the correlation ratio will not be described in this book.

Selection of Cases.   The effect on correlation of some particular selectivity in the data can be very perplexing. For example, if you were asked what is the correlation between age and intelligence quotient, you might say promptly that it is zero. Yet if you should compute that correlation for the 109 fourth-grade pupils of Appendix Table IX, you would discover that $r \equiv -.72$, as in Figure 13-3 on page 223. Why? Because of selectivity of subjects. It has already been noted that within any single grade (here the fourth) the younger children tend to be brighter than the older ones, producing a negative correlation between age and a measure of intelligence. If all children in a school system are measured, that correlation would almost certainly be near zero.

It has already been noted that mere increase or decrease in the number of individuals, without any change in the nature of those individuals, has little effect on the size of r. However, if the form of distribution is considerably changed, r may also be considerably changed. Some of the ways in which this can occur will be discussed in the next four sections.

Eliminating Cases near the Mean.   Sometimes a research worker gathers data on a large number of cases and then eliminates from consideration a block of cases near the center of the distribution. If he computes a coefficient of correlation from the two extreme groups, he is likely to get a value numerically larger than if he had used the full distribution.

For example, consider Figure 13-2 on page 222 in which, for all 109 cases, the correlation between intelligence quotient and language score has been found to be $r = .58$. If you should eliminate the 48 cases with language score between 19 and 27, leaving the distribution of Figure 14-1, the correlation computed for the remaining 61 cases would be $r = .69$. It is possible to make a good estimate of the correlation in an entire distribution from certain statistics computed from cases at the two ends of that distribution, but special tables are required for that purpose.[1]

---

[1] These tables were computed and privately distributed by John C. Flanagan. They have been reproduced by permission in R. L. Thorndike, *Personal Selection, Test and Measurement Techniques*, John Wiley & Sons, Inc., 1949, and in H. M. Walker and J. Lev, *Statistical Inference*, Holt, Rinehart and Winston, Inc., 1953.

Language score

Intelligence quotient	7–9	10–12	13–15	16–18	19–21	22–24	25–27	28–30	31–33	34–36	37–39	$f\,y$
147–153									1			1
140–146								2	1			3
133–139								1	1	1		3
126–132								4	1	1		6
119–125			1	4				3	1	1		10
112–118			2	1				2	1		1	7
105–111		2		4								6
98–104			2	6				2				10
91–97	1	1		1				1				4
84–90		1	2	1								4
77–83		1	2									3
70–76				1								1
63–69	1	1		1								3
$f\,x$	2	6	9	19				15	6	3	1	61

$r = .69$

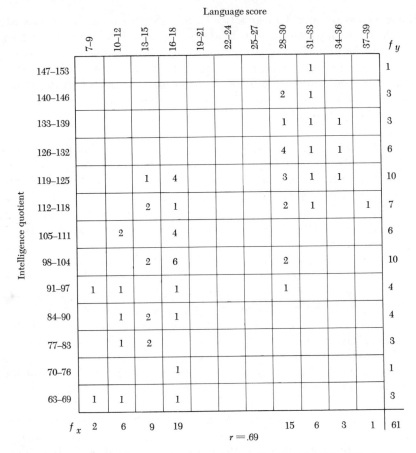

Figure 14-1   Distribution of Figure 13-2 on Page 222 After the Elimination of 48 Cases with Language Score between 19 and 27

Correlation in a Composite Group when the Subgroups Have Unequal Means.   Sometimes two groups with widely separated means are thrown together to form one bivariate distribution. The correlations obtained from a composite group of this sort are spurious and meaningless.

This phenomenon was dramatically illustrated in one of the earliest educational studies making use of correlation, made before people had become sensitive to the distortion which might occur. Simpson[2] studied two groups: *A*, consisting of professors and advanced students in Colum-

[2] *Correlations of Mental Abilities,* New York, Teachers College Bureau of Publications, 1912.

bia University; and B, consisting of unemployed men in New York missions, none of whom had ever held a position demanding a high grade of intelligence. Among the scores he secured were these:

x_1 — Auditory memory for words
x_2 — Auditory memory for a connected passage
x_3 — Visual recognition of forms

The following correlations were obtained:

	Group A	Group B	Combined Groups
r_{12}	.23	.73	.82
r_{13}	.37	.15	.71
r_{23}	.14	.06	.66

If two or more groups have the same means on both variables, the correlation obtained from the combined groups is not distorted in any serious fashion. When the groups have different means on one or both variables, the effect of combining them may be to produce a correlation coefficient larger than the correlation in the separate groups, as in the example just cited or smaller or even different in sign. Some such effects are schematically suggested in Figure 14-2.

Two or more groups with unequal means may be pooled in order to obtain a common coefficient of correlation which will reflect the relation of the two variables within the groups and will not be affected by the differences between the groups. To do this formula (14-1) may be used

$$14\text{-}1) \quad r = \frac{\Sigma x_1 y_1 + \Sigma x_2 y_2 + \Sigma x_3 y_3 - \dfrac{(\Sigma x_1)(\Sigma y_1)}{n_1} - \dfrac{(\Sigma x_2)(\Sigma y_2)}{n_2} - \dfrac{(\Sigma x_3)(\Sigma y_3)}{n_3}}{\sqrt{\Sigma x_1^2 + \Sigma x_2^2 + \Sigma x_3^2 - \dfrac{(\Sigma x_1)^2}{n_1} - \dfrac{(\Sigma x_2)^2}{n_2} - \dfrac{(\Sigma x_3)^2}{n_3}} \sqrt{\Sigma y_1^2 + \Sigma y_2^2 + \Sigma y_3^2 - \dfrac{(\Sigma y_1)^2}{n_1} - \dfrac{(\Sigma y_2)^2}{n_2} - \dfrac{(\Sigma y_3)^2}{n_3}}}$$

The formula as presented here is for a composite of three subgroups, but can easily be adapted to a different number of groups. Such expressions as Σx_1 and Σx_2 indicate summation for the subgroup only.

Often the presence of a single extreme case presents the research worker with a dilemma. Thus Rowell found a 14-year-old child with an IQ of 45 in a third-grade class. Either to include him or to exclude him from her computations would obviously misrepresent the situation. Under such circumstances the best procedure may be to state the situation so the reader can understand it and show how the extreme case affects results.

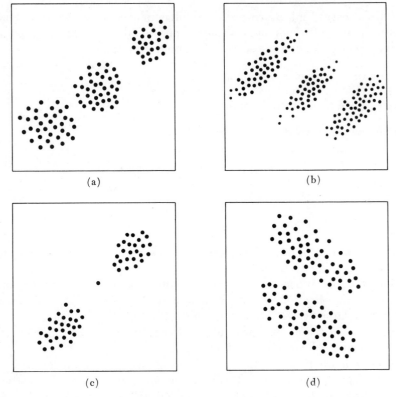

Figure 14-2   Sketches to Illustrate How the Value of the Correlation Coefficient May Be Affected by the Combination of Two or More Groups with Unequal Means

  (a) In each group, r is near zero. In composite group, it is high and positive.

  (b) In each group, r is positive. In composite group, it is negative.

  (c) In each group, r is positive but low. In composite group, it is positive and much higher.

  (d) In each group, r is negative. In composite group, it is near zero.

Rowell (unpublished study) found correlations as follows:

Traits	With Extreme Case	Without Extreme Case
CA and IQ	$-.500$	$-.430$
CA and MA	$-.171$	$-.078$
MA and IQ	$-.917$	$-.919$

Effect of Range.   One of the chief troubles in using the correlation coefficient is the effect of the range of scores on the value of this coefficient.

If the group is restricted in range on one of the variables, it may be so restricted on the others, and the correlation may be greatly reduced in value. If the group has a great range of values on one variable and if non-zero correlation exists, then the range on the other variable is likely to be relatively great, and the correlation may be large in value.

This effect is so disconcerting to some statisticians that they prefer not to use the coefficient of correlation at all. Instead, they use the regression formula and the related standard error of estimate as a description of the relation between two variables. The correlation coefficient is, nevertheless, generally used as a measure of relationship. The precaution must be taken, however, to keep the range of scores in mind when interpreting a correlation coefficient.

As an example of the effect of changing the range of scores, consider the correlation between IQ and language for the children listed in Figure 13-2 on page 222. For the entire group this correlation is .58, but when the IQ range is restricted by considering only children with IQ of 126 or higher from this group, the correlation is reduced to .20. When children with IQ less than 91 or IQ greater than 125 are excluded, the correlation is reduced to .24. See Exercise 14-1.

The matter of range in test scores is very important in evaluating validity and reliability coefficients. Such a coefficient based on a group which is highly homogeneous in ability is likely to be low for that reason alone, in comparison with such a coefficient for a group very diverse in ability.

A problem of restriction of range arises when only those who achieve high test scores are selected for employment, and the later success of these people is used as a criterion for judging the validity of the selection process. Since the validity coefficient, the correlation between selection score and measure of success, is computed for a group which had attained or exceeded the selection standard, this group is restricted on the range of scores in the selection test. The validity coefficient is, therefore, much lower than if the entire candidate group could have been used in computing the coefficient. In other words, the accomplishment of the test in eliminating a large number of undesirable candidates is not made evident by the computation.

One way to recapture the information about the validity of the test for the entire candidate group merely by using data for the selected group is to estimate a correlation coefficient by extension. Suppose that, for the selected group, the standard deviation of test scores is s_x and the correlation with the criterion is r_{xy}. Since the entire candidate group was given the selection test, the standard deviation for this group is available. Call this S_x. Then the validity coefficient for the entire group R_{xy} can be estimated by the formula

$$(14\text{-}2) \qquad R_{xy} = \frac{r_{xy}S_x}{\sqrt{r_{xy}{}^2S_x{}^2 + s_x{}^2(1 - r_{xy}{}^2)}}$$

EXERCISE 14-1

1. Verify the values of r, $\hat{b}_{yx}$, $\hat{b}_{xy}$, s_x, and s_y listed below which would be obtained from the data of Figure 13-2 on page 222 if certain parts of the distribution were eliminated and computations were made from the cases indicated:

	Cases retained	n	s_x	s_y	r	$\hat{b}_{yx}$	$\hat{b}_{xy}$
(a)	Entire group	109	6.5	17.8	.58	1.60	.21
(b)	Language score between 19 and 27 inclusive	48	2.4	14.4	.16	.93	.026
(c)	IQ between 91 and 125 inclusive	75	5.8	9.6	.24	.40	.15
(d)	Language score below 19	36	2.8	16.6	.39	2.28	.07
(e)	Language score above 27	25	2.6	13.9	.13	.70	.025
(f)	Language score below 19 or above 27	61	8.4	20.2	.69	1.65	.29
(g)	IQ below 91	15	4.5	7.9	.26	.45	.15
(h)	IQ above 125	19	4.0	6.8	.20	.34	.11
(i)	IQ below 91 or above 125	34	7.8	28.9	.85	3.14	.23

2. In situation (b), the extreme values of x were removed. What did this removal do to s_x? s_y? r? $\hat{b}_{yx}$? $\hat{b}_{xy}$? Which was affected more, s_x or s_y?

3. In situation (c), the extreme values of y were removed. What did this removal do to s_x? s_y? r? $\hat{b}_{yx}$? $\hat{b}_{xy}$? Which was affected more, s_x or s_y?

4. In situation (f), the middle of the distribution of x was removed. What did this do to s_x? s_y? r? $\hat{b}_{yx}$? $\hat{b}_{xy}$? Which was affected more, s_x or s_y?

5. In situation (i), the y scores near the mean were removed. What did this removal do to s_x? s_y? r? $\hat{b}_{yx}$? $\hat{b}_{xy}$? Which was affected more, s_x or s_y?

Variation in a Third Trait. Sometimes part of the difficulty in interpreting a coefficient of correlation comes from the fact that both of the correlated variables appear to be related to a third variable in such a way that the intrinsic relation between the two correlated variables is either exaggerated or suppressed by variation in the third. This topic will be treated at greater length in Chapter 18.

15

INFERENCES ABOUT
STANDARD DEVIATIONS
AND VARIANCES

Although the primary purpose of this chapter is to develop methods of making inferences about population variances σ^2 and population standard deviations σ, a second purpose is to introduce two theoretical distributions: χ^2 (chi-square) and F. These distributions are fundamental in the theory and practice of statistical inference in a wide range of situations apart from those considered in the present chapter.

Relation of the Variance to the Chi-Square Statistic.   The sample variance was defined in formula (6-4) on page 77 as $s^2 = \Sigma(x - \bar{x})^2/(n - 1)$. To make inferences about s^2 we need to know its sampling distribution. This distribution cannot be given directly in terms of s^2 itself because the value of s^2 depends on the units in which the original scores are expressed. This limitation can be overcome by the very natural device of expressing each score in units of the standard deviation of the population, σ. The resulting statistic, called chi-square, may be written as

$$(15\text{-}1) \qquad\qquad \chi^2 = \frac{\Sigma(x - \bar{x})^2}{\sigma^2}$$

or

$$(15\text{-}2) \qquad\qquad \chi^2 = \frac{(n - 1)s^2}{\sigma^2}$$

χ is the lower case form of the Greek letter chi, pronounced ki to rhyme with sky.

A Class Project to Develop Familiarity with Chi-Square.   To help his students acquire a feeling for the statistic and for its sampling distribution, a teacher of statistics gave his class the data of Appendix Table X, for which the mean is $\mu = 121.6$, the variance is $\sigma^2 = 1380.4$, the standard deviation is $\sigma = 37.15$, and the distribution is approximately normal. He asked each student to draw at random—by means of the table of random numbers—a sample of 5 cases and from their scores to compute s^2 and χ^2. The sample was to be drawn with replacement, so if the same number was drawn twice it would be used twice. Drawing with replacement has the effect of transforming this finite population of 447 cases into an infinitely large population with the same proportional frequencies. Among the samples reported by students were the five shown here:

Sample	Scores Reported	$\Sigma(x - \bar{x})^2$	s^2	χ^2
A	169, 137, 123, 104, 98	3246.8	811.7	2.35
B	201, 184, 86, 84, 28	21515.2	5378.8	15.59
C	120, 142, 100, 136, 162	2184.0	546.0	1.58
D	136, 112, 102, 98, 96	1076.8	269.2	.78
E	175, 86, 102, 109, 55	7785.2	1946.3	5.64

As soon as the teacher looked at the results, he knew that something was wrong with sample B. (How he knew this will be explained shortly.) Since the computation was not at fault, he asked the student to re-examine his data. The student discovered that he had copied a score of 128 as 28. When he corrected this error, he found $\Sigma(x - \bar{x})^2 = 11795$, $s^2 = 2948.8$, and $\chi^2 = 8.54$ instead of 15.59. The teacher considered this value reasonable. Let us now see how the teacher was able to detect the error.

Degrees of Freedom for the Variance.   In order to interpret a value of χ^2 it is necessary to be able to read a probability table such as Appendix Table IV on page 357. As in the tables for "Student's" distribution, with which you are already familiar, each row of this table is in reality a separate table for a particular number of degrees of freedom; therefore before consulting the table one must decide how many degrees of freedom the situation provides.

On page 152 it was stated that the number of degrees of freedom for t as defined in formula (9-5) is $\nu = n - 1$. This statement holds also for χ^2 as defined in formula (15-1) or (15-2). In both cases it is the sample variance which has $n - 1$ degrees of freedom.

It is not difficult to see why the variance or the standard deviation of a sample of n cases has $n - 1$ degrees of freedom. If you are asked to write down n numbers to represent the scores for such a sample, you have n free choices, one for each case in the sample. However, if the mean is specified, you have only $n - 1$ free choices because the nth score is fixed, and it must be a number which can be added to the $n - 1$ numbers already selected to produce $\Sigma x = n\bar{x}$.

In later chapters we shall encounter situations in which the number of degrees of freedom is something other than $n - 1$. These situations will be discussed as they arise.

The Chi-Square Distribution.   The values of χ^2 computed for 125 samples of 5 cases each drawn by students as described earlier in this chapter, have been tabulated and their frequency distribution is presented in the histogram of Figure 15-1. The smooth curve superimposed on the histogram is the theoretical chi-square curve for four degrees of freedom. ($n = 5$, so $\nu = 5 - 1$.)

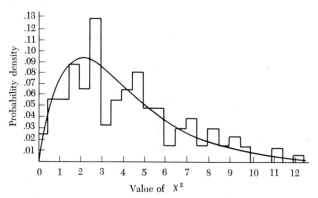

Figure 15-1   Distribution of $\Sigma(x - \bar{x})^2/\sigma^2$ in 125 Samples of 5 Cases Each and Theoretical χ^2 Distribution with 4 Degrees of Freedom

The empirical distribution of χ^2 obtained from data is always discrete, represented by a step curve like this histogram. The theoretical χ^2 curve is always smooth. The larger the number of samples under consideration, the more closely will the histogram approximate the smooth curve.

As the number of degrees of freedom changes, so does the shape of the smooth chi-square curve. Figure 15-2 presents five of the smooth curves for which probabilities are given in Appendix Table IV. Each curve corresponds to one row of that table.

It must be noted that χ^2 is never negative. As ν increases, the curves become less skewed, more nearly symmetrical. When ν is larger than 1, the highest point on the curve occurs at $\chi^2 = \nu - 2$.

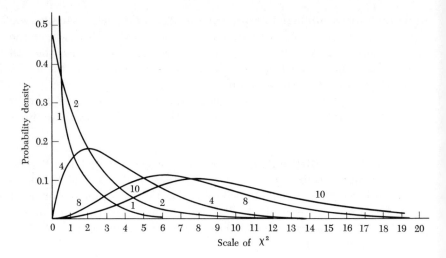

Figure 15-2 χ^2 Curves for 1, 2, 4, 8, and 10 Degrees of Freedom

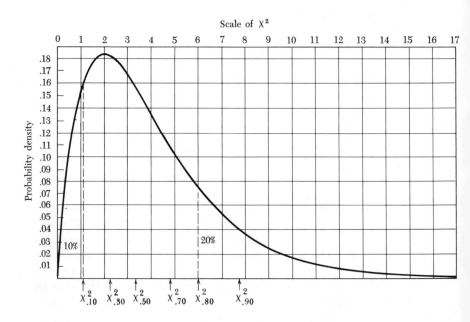

Figure 15-3 Smooth χ^2 Curve for 4 Degrees of Freedom with Selected Percentiles Indicated on the Base Line

Reading the Chi-Square Table.   Now let us examine in detail one of these curves, the one for $\nu = 4$ shown in Figure 15-3 and the corresponding row in Appendix Table IV. There in the row for $\nu = 4$ and the column headed $\chi_{.10}^2$ the cell entry is 1.1. This entry means

1. If $\nu = 4$, $\chi_{.10}^2 = 1.1$.
2. The 10th percentile for the χ^2 distribution for 4 degrees of freedom is 1.1.
3. When a true hypothesis is tested in a situation affording 4 degrees of freedom, the probability is .10 that a random sample will yield a value of χ^2 smaller than 1.1, and the probability is .90 that such a sample will yield a value larger than 1.1.
4. The area to the left of the point 1.1 in Figure 15-3 is 10 percent of the total area under the curve.

In the same row of Table IV we read that $\chi_{.90}^2 = 7.8$, and this point is marked on the base line of Figure 15-3. How would you interpret this value?

Beyond (that is, to the right of) what point would the upper 5 percent of the area lie? Locate this point. Do the same for $\chi_{.99}^2$.

Some problems call for a one-tail test, some for a two-tail test. The logic governing the choice will be considered as applications are described; at this point we are concerned with reading the probability table. When a one-tail test is required, the region of significance is almost invariably in the upper tail.

EXERCISE 15-1

By examination of Appendix Table IV, verify enough of the following statements to be sure you are reading the table correctly.

1. The 95th percentile of the χ^2 distribution is
    (a) 6.0 when $\nu = 2$.   (c) 31.4 when $\nu = 20$.
    (b) 22.4 when $\nu = 13$.   (d) 43.8 when $\nu = 30$.

2. The 2nd percentile of the χ^2 distribution is
    (a) 6.61 when $\nu = 16$.   (c) 11.3 when $\nu = 23$.
    (b) 3.06 when $\nu = 10$.   (d) 4.18 when $\nu = 12$.

3. The median (50th percentile) of the χ^2 distribution is
    (a) 2.4 when $\nu = 3$.   (d) 19.3 when $\nu = 20$.
    (b) 5.4 when $\nu = 6$.   (e) 22.3 when $\nu = 23$.
    (c) 9.3 when $\nu = 10$.   (f) 25.3 when $\nu = 26$.

4. In the χ^2 distribution when $\nu = 11$
   (a)   1 percent of the area lies to the left of 3.05.
   (b)   10 percent of the area lies below $\chi^2 = 5.6$.
   (c)   50 percent of the area lies between $\chi^2 = 7.6$ and $\chi^2 = 13.7$.
   (d)   10 percent of the area lies to the right of $\chi^2 = 17.3$.
   (e)   0.1 percent of the area lies to the right of $\chi^2 = 31.3$.

5. In the χ^2 distribution for $\nu = 20$, the proportion of area
   (a)   to the right of $\chi^2 = 30$ is between .05 and .10.
   (b)   to the left of $\chi^2 = 16$ is between .25 and .50.
   (c)   to the left of $\chi^2 = 5$ is less than .005.
   (d)   to the right of $\chi^2 = .50$ is less than .001.

6. In testing a hypothesis by means of a χ^2 test, when α and ν are as indicated, the critical region would be as shown below:

	α	ν	One-Tail Region	Two-Tail Region
(a)	.05	7	$\chi^2 > 14.1$	$\chi^2 < 1.7$ and $\chi^2 > 16.0$
(b)	.01	3	$\chi^2 > 11.3$	$\chi^2 < .07$ and $\chi^2 > 12.8$
(c)	.02	15	$\chi^2 > 28.3$	$\chi^2 < 5.23$ and $\chi^2 > 30.6$
(d)	.10	8	$\chi^2 > 13.4$	$\chi^2 < 2.7$ and $\chi^2 > 18.2$
(e)	.01	16	$\chi^2 > 32.0$	$\chi^2 < 5.1$ and $\chi^2 > 34.3$
(f)	.05	6	$\chi^2 > 12.6$	$\chi^2 < 1.2$ and $\chi^2 > 14.4$

7. For the specified degrees of freedom, the probability of obtaining a value of χ^2 larger than the one listed is as indicated.

	ν	χ^2	Probability
(a)	2	7.1	$.025 < \text{Prob} < .05$
(b)	11	25.1	$.005 < \text{Prob} < .01$
(c)	26	8.0	$\text{Prob} > .995$
(d)	21	16.3	$\text{Prob} = .75$
(e)	30	60.0	$\text{Prob} < .001$

Testing an Observed Variance against a Theoretical Variance.   Now let us return to the problem described at the beginning of this chapter, in which students were asked to draw samples of 5 cases each at random from a population for which the variance was known to be $\sigma^2 = 1380.4$. If a given sample had either a suspiciously large or a suspiciously small value of χ^2, the hypothesis that it was a random sample from this population would be rejected. This might come about through failure to follow directions, to copy results correctly, or for various other reasons. (One student actually used code numbers instead of scores!) Thus the situation calls for a two-tail test.

As each sample contains 5 cases, there are 4 degrees of freedom. Therefore the teacher knows that in the long run 95 percent of all samples should produce a value of χ^2 such that $.48 < \chi^2 < 11.1$. Locate the points .48 and 11.1 on the base line of Figure 15-3. The range between these points is a *region of acceptance* for the hypothesis $\sigma^2 = 1380.4$. Now locate the point $\chi^2 = 2.35$ reported in sample A, and mark its position A. You will note that it falls very close to the median of the distribution. Do the same for the other 4 points. Only the result for sample B falls outside the region of acceptance, and when the value for B is corrected, the new value is well within the region of acceptance.

Thus if the hypothesis $\sigma^2 = 1380.4$ were tested against the results from any one of these samples, it would be held to be a reasonable hypothesis. Does this prove that the sample actually was correctly drawn? Does it prove that computation was correct?

If $.48 < \chi^2 < 11.1$ is a region of acceptance, and $(n - 1)s^2/\sigma^2 = \chi^2$ or $s^2 = \chi^2(1380.4)/4$, then $165.6 < s^2 < 3830.6$ is also a region of acceptance.

The following is a generalization of this procedure. Suppose the hypothesis has been formulated that σ^2 has a specified value, say σ_0^2. To test this hypothesis, calculate the statistic $\Sigma(x - \bar{x})^2/\sigma_0^2$ or $(n - 1)s^2/\sigma_0^2$. With $\nu = n - 1$, refer to Appendix Table IV and determine a critical region corresponding to whatever level of significance has been selected. If the calculated statistic falls in this critical region, reject the hypothesis; otherwise accept it.

Interval Estimate for the Variance.   The method of making an interval estimate for the population variance σ^2 is similar to that for the mean. Most readers will find it helpful at this point to review the concept of confidence interval for the mean which was developed in Chapter 9.

The steps in the procedure of making an interval estimate for σ^2 are outlined in Table 15-1. Note that one has complete freedom of choice for the confidence coefficient c, but that customarily some convenient number slightly smaller than 1.00 is chosen, say .95 or .90 or .80. Formulas for the two limits are

(15-3a)   Upper limit: $\dfrac{\Sigma(x - \bar{x})^2}{\chi^2_{(1-c)/2}} = \dfrac{\Sigma x^2 - (\Sigma x)^2/n}{\chi^2_{(1-c)/2}}$

(15-3b)   Lower limit: $\dfrac{\Sigma(x - \bar{x})^2}{\chi^2_{(1+c)/2}} = \dfrac{\Sigma x^2 - (\Sigma x)^2/n}{\chi^2_{(1+c)/2}}$

The interval limit for σ^2 made with confidence coefficient c is therefore

(15-4a) $\dfrac{\Sigma(x - \bar{x})^2}{\chi^2_{(1+c)/2}} < \sigma^2 < \dfrac{\Sigma(x - \bar{x})^2}{\chi^2_{(1-c)/2}}$

Table 15-1   Steps in the Computation of an Interval Estimate for σ^2

Step	Procedure
1.	Select a value of the confidence coefficient c.
2.	Draw a random sample of n cases.
3.	Calculate $\Sigma(x - \bar{x})^2$ or its more easily obtained equivalent $\Sigma x^2 - (\Sigma x)^2/n$.
4.	From the χ^2 table with $\nu = n - 1$ read the values of $\chi^2_{(1-c)/2}$ and $\chi^2_{(1+c)/2}$.
5.	Calculate: Upper limit: $\dfrac{\Sigma(x - \bar{x})^2}{\chi^2_{(1-c)/2}}$ Lower limit: $\dfrac{\Sigma(x - \bar{x})^2}{\chi^2_{(1+c)/2}}$
6.	Set up the interval estimate according to formula (15-4).

or equivalently

(15-4b) $$\frac{(n - 1)s^2}{\chi^2_{(1+c)/2}} < \sigma^2 < \frac{(n - 1)s^2}{\chi^2_{(1-c)/2}}$$

and the corresponding interval for σ is

(15-5) $$\sqrt{\Sigma(x - \bar{x})^2/\chi^2_{(1+c)/2}} < \sigma < \sqrt{\Sigma(x - \bar{x})^2/\chi^2_{(1-c)/2}}$$

To understand the considerations which may influence the choice of confidence coefficient, examine the adjacent list of intervals based on the data of sample E described at the beginning of this chapter:

Confidence Coefficient	Interval Estimate Based on Data of Sample E
.50	$1442 < \sigma^2 < 4055$
.80	$998 < \sigma^2 < 7077$
.90	$819 < \sigma^2 < 10965$
.95	$701 < \sigma^2 < 16219$
.98	$585 < \sigma^2 < 25951$
.99	$522 < \sigma^2 < 37072$

It is obvious that the interval becomes wider as the confidence coefficient becomes greater. In other words, if it seems necessary to have great confidence that an interval actually contains the parameter, that confidence is purchased at the expense of a wide interval; if it seems necessary to make a statement that the parameter lies within narrow limits, that appearance of precision is purchased at the expense of reducing confidence that the parameter actually lies in the interval. In these six intervals only the first, which is the narrowest, fails to contain the parameter.

If it is important to have a narrow interval and also to have considerable confidence that the interval encloses the parameter, this result can be obtained by increasing sample size. The following intervals have been computed with $c = .95$ and $s^2 = 50$. Note how they shrank as n increases.

n	Interval
5	$18 < \sigma^2 < 417$
10	$24 < \sigma^2 < 167$
20	$29 < \sigma^2 < 107$
30	$32 < \sigma^2 < 91$
61	$36 < \sigma^2 < 74$
101	$39 < \sigma^2 < 67$

Tests when the Number of Degrees of Freedom is Large. When the number of degrees of freedom is larger than those shown in Table IV, the standard deviation has a sampling distribution which is nearly normal. (Its distribution is not normal in general but only for large samples.) Then the test of significance may be carried out by computing

$$(15\text{-}6) \qquad z = \frac{s}{\sigma} \sqrt{2n - 2} - \sqrt{2n - 2}$$

and referring it to the normal probability table. Rejection of the hypothesis $\sigma = A$ implies rejection of the hypothesis $\sigma^2 = A^2$ and vice versa, at the same level of significance.

To obtain an interval estimate for σ with confidence coefficient c using the relation of formula (15-6), a first step is to read the values of $z_{(1-c)/2}$ and $z_{(1+c)/2}$ from a table of normal probability. Then

$$(15\text{-}7a) \qquad \text{Lower limit for } \sigma = \frac{s\sqrt{2n - 2}}{z_{(1+c)/2} + \sqrt{2n - 2}}$$

$$(15\text{-}7b) \qquad \text{Upper limit for } \sigma = \frac{s\sqrt{2n - 2}}{z_{(1-c)/2} + \sqrt{2n - 2}}$$

also

$$(15\text{-}8) \qquad \frac{s^2(2n - 2)}{(z_{(1+c)/2} + \sqrt{2n - 2})^2} < \sigma^2 < \frac{s^2(2n - 2)}{(z_{(1-c)/2} + \sqrt{2n - 2})^2}$$

provides an approximate interval estimate for σ^2 with confidence coefficient c.

How good is this approximation? To answer this question let us make an interval estimate using the exact formula (15-4) and also for the same data an estimate using the approximation in formula (15-8). Since $\nu = 100$ is the largest entry available in our χ^2 table, we shall use a sample of 101 cases. Assume $s^2 = 25$ and $c = .95$. Then by formula (15-4)

$$\frac{100(25)}{129.6} < \sigma^2 < \frac{100(25)}{74.2}$$

or

$$19.3 < \sigma^2 < 33.7$$

and by formula (15-8)

$$\frac{25(200)}{(1.96 + \sqrt{200})^2} < \sigma^2 < \frac{25(200)}{(-1.96 + \sqrt{200})^2}$$

or

$$19.3 < \sigma^2 < 33.7$$

Both ends of the interval obtained from formula (15-8) agree with the correct values from (15-4). However it can be seen from Table 15-2 that the approximation is less satisfactory when n is small and less satisfactory when $c = .99$ than when $c = .95$.

Table 15-2 Interval Estimates for σ^2 Made with Formulas (15-4b) and (15-8) for Samples of Different Size and Different Confidence Level

	n	s^2	c	*Formula*	*Interval Estimate*
A	101	25	.95	(15-4b)	$19.3 < \sigma^2 < 33.7$
A'	101	25	.95	(15-8)	$19.3 < \sigma^2 < 33.7$
B	101	25	.99	(15-4b)	$17.8 < \sigma^2 < 37.1$
B'	101	25	.99	(15-8)	$17.8 < \sigma^2 < 37.3$
C	21	25	.95	(15-4b)	$14.6 < \sigma^2 < 52.1$
C'	21	25	.95	(15-8)	$14.6 < \sigma^2 < 52.6$
D	21	25	.99	(15-4b)	$12.5 < \sigma^2 < 67.6$
D'	21	25	.99	(15-8)	$12.6 < \sigma^2 < 71.0$
E	9	25	.99	(15-4b)	$9.1 < \sigma^2 < 154$
E'	9	25	.99	(15-8)	$9.2 < \sigma^2 < 197$

EXERCISE 15-2

1. Some of the following statements describe a region of acceptance and some
are interval estimates of a parameter. They have a superficial similarity of
appearance which confuses some people. Place the letter R in front of each
statement you think is a region of acceptance and E in front of each one which
you think is an interval estimate:

_____ (a) $17.5 < \bar{x} < 24.2$.   _____ (e) $42.3 < \mu < 51.5$.

_____ (b) $3.4 < \sigma^2 < 6.2$.   _____ (f) $.41 < r < .59$.

_____ (c) $\bar{x} > 42.3$.   _____ (g) $\chi^2 < 9.5$.

_____ (d) $9.4 < s^2 < 15.6$.   _____ (h) $.63 < \rho < .75$.

2. Some people confuse the terms "confidence coefficient" and "significance
level." Place the letter C in front of each expression which you think correctly
applies to a confidence coefficient, the letter S in front of each which you
think correctly applies to a level of significance, and the letter X in front of
each which applies to neither.

A number which represents:

_____ (a) The proportion of random samples which will lead to the rejection
of a hypothesis if it is true.

_____ (b) The proportion of random samples such that an interval estimate
based on them contains the population parameter.

_____ (c) The proportion of random samples leading to the acceptance of a
hypothesis which is false.

_____ (d) The proportion of random samples leading to the acceptance of
a true hypothesis.

_____ (e) The probability that a statistic computed from a random sample
will fall in the critical region if the hypothesis tested is true.

3. If you are told that each of the following numbers is either a level of significance or a confidence coefficient, how would you classify them. Mark each
one either S or C as in the preceding question.

_____ (a) .05.   _____ (c) .98.   _____ (e) .01.

_____ (b) .95.   _____ (d) .02.   _____ (f) .99.

4. To test the hypothesis that $\sigma^2 = 35$, the following sample is available:
10, 19, 16, 13, 21, 14, 15, 27, 22, 24, 17, 18. Choose a significance level, complete the test, and formulate a conclusion.

5. From the data of Question 4, construct an interval estimate for σ^2 with
$c = .95$; construct an interval estimate for σ.

6. Using $c = .50$ construct an interval estimate for σ^2 from each of the five
samples displayed at the beginning of this chapter. (In Sample B make the
correction of changing 28 to 128.) How many of these intervals contain the
population value $\sigma^2 = 1380.4$? Does your answer incline you to think $c = .50$
is a reasonable choice? Why?

7. In a sample of 181 cases the standard deviation is 7.51. Using $c = .90$ make an interval estimate for σ. For σ^2.

8. In a sample of 196 cases, $s^2 = 137$. Test the hypothesis that the population variance is 100.

9. Verify enough of the interval estimates in Table 15-2 to satisfy yourself that you know how they were calculated. Does the approximation seem to be better when c is larger or when it is smaller?

Comparison of the Variances from Two Independent Samples. Earlier sections of this chapter dealt with the comparison of a sample variance s^2 and a theoretical value σ^2. A natural next step is the comparison of the variances from two independent samples.

If σ_1^2 and σ_2^2 are the variances of two normal populations, it is often of interest to test the hypothesis

$$H : \sigma_1^2 = \sigma_2^2 = \sigma^2$$

The statistic used to test this hypothesis is the ratio of the two sample variances, a statistic which the well-known Indian statistician P. C. Mahalanobis named F in honor of Sir Ronald Fisher, who developed the concept but used a somewhat different form of the statistic.

$$(15\text{-}9) \qquad\qquad F = \frac{s_1^2}{s_2^2}$$

Note that in the null hypothesis there is no assumption concerning the means of these populations. Also note that in the F-statistic there is no requirement about sample size.

Obviously the hypothesis $H : \sigma_1^2 = \sigma_2^2$ would be rejected if s_1^2 is either much larger or much smaller than s_2^2, so a two-tail test is required. However the tables of the F distribution, about to be described, are set up to show only values of F larger than 1.00. Therefore in problems of the type now under consideration, the variance ratio is written with the larger variance in the numerator, and the level of significance given in the table is doubled. Chapter 17 will deal with problems calling for a one-tail test and variance ratios in which the larger variance is not always in the numerator.

In using the F table, the numbers of cases n_1 and n_2 (corresponding to s_1^2 and s_2^2) are converted to degrees of freedom by the relations $\nu_1 = n_1 - 1$ and $\nu_2 = n_2 - 1$. The calculated value of F is then referred to the F table with ν_1 and ν_2 degrees of freedom. The way in which the F table is used will be described in the following sections.

Reading Tables of the F Distribution.   In the tables for the t and χ^2 distributions each horizontal row corresponds to a specific number of degrees of freedom and presents a unique probability distribution. The F-statistic has ν_1 degrees of freedom for the numerator variance and ν_2 for the denominator; therefore for every combination of ν_1 and ν_2 there is a unique probability distribution. Publishing these distributions in detail comparable to that of the t or χ^2 tables would require many pages. As an economy measure, only a few percentile values are published for each ν_1, ν_2 pair. Snedecor, who published the first F table, gave only the 95th and 99th percentile values in each cell, and his table is still widely used. It is the one published as Appendix Table V in this book. More recently other percentile values have been completed and are available in advanced texts or in books of tables.

The 95th percentile of the F distribution with 5 degrees of freedom in the numerator and 8 in the denominator will be denoted $F_{.95}(5,8)$. This value is found in Appendix Table V as the upper value in the cell which is in the column headed 5 and the row headed 8. In this cell we find the two numbers 3.69 and 6.63. The upper number is the 95th percentile; the lower number the 99th.

Verify these values by reading from Table V, or at least enough values to satisfy yourself that you understand how to find them.

$$F_{.95}(3,20) = 3.10 \qquad F_{.99}(50,10) = 4.12$$
$$F_{.99}(16,2) = 99.44 \qquad F_{.99}(10,50) = 2.70$$
$$F_{.99}(1,30) = 7.56 \qquad F_{.95}(6,100) = 2.19$$
$$F_{.99}(12,3) = 27.05 \qquad F_{.95}(100,6) = 3.71$$
$$F_{.95}(8,25) = 2.34 \qquad F_{.95}(50,50) = 1.60$$
$$F_{.95}(3,40) = 2.84 \qquad F_{.99}(1,\infty) = 6.64$$
$$F_{.95}(20,34) = 1.89 \qquad F_{.99}(10,\infty) = 2.32$$

Illustrations

1.   In sample A, $n_A = 15$ and $s_A{}^2 = 36$. In sample B, $n_B = 22$ and $s_B{}^2 = 15$. Is the hypothesis $\sigma_A{}^2 = \sigma_B{}^2$ tenable? $F = \frac{36}{15} = 2.4$. In Table V in the cell which lies in column 14 and row 21 we find the two numbers 2.20 and 3.07. As 2.20 is the 95th percentile of the probability distribution, it would represent a significance level of .05 if this were a one-tail test but a level of $2(.05) = .10$ for a two-tail test. Similarly 3.07, which is the 99th percentile, represents a significance level of $2(.01) = .02$. Since the computed $F = 2.4$ is between 2.20 and 3.07, the probability of obtaining $F > 2.4$ if the hypothesis is true is between .02 and .10. If a more precise statement of significance is required, it would be necessary to consult other tables.

2. Given $n_A = 13$, $s_A{}^2 = 9.2$, $n_B = 17$, and $s_B{}^2 = 16.8$. Then $F = 16.8/9.2 = 1.83$. For $\nu_1 = 16$ and $\nu_2 = 12$, $F_{.95} = 2.60$. Since the observed $F < F_{.95}$, the level of significance must be greater than $2(.05) = .10$, but how much greater we cannot say.

3. Given $n_A = 76$, $s_A{}^2 = 48.3$, $n_B = 16$, and $s_B{}^2 = 14.2$. Then $F = 48.3/14.2 = 3.40$. For $\nu_1 = 75$ and $\nu_2 = 15$, $F_{.99} = 3.00$. Since $F > F_{.99}$, its level of significance is less than $2(.01) = .02$.

4. Given $n_A = 80$, $s_A{}^2 = 16.8$, $n_B = 41$, and $s_B{}^2 = 35.2$. Then $F = 35.2/16.8 = 2.1$. In the F table there is no row for $\nu_2 = 79$. In the cell for $\nu_1 = 40$ and $\nu_2 = 70$ we find $F_{.99} = 1.88$. In the cell for $\nu_1 = 40$ and $\nu_2 = 80$ we find $F_{.99} = 1.84$. Since the observed F is larger than either of these it is unquestionably larger than $F_{.99}(40,79)$, which does not appear in the table. Hence it is significant at the .02 level.

5. Given $n_A = 230$, $s_A{}^2 = 78.4$, $n_B = 306$, and $s_B{}^2 = 74.5$. Then $F = 78.4/74.5 = 1.05$. The table contains no column for $\nu_1 = 229$ and no row for $\nu_2 = 305$. However we may look at the four cells nearest to the missing cell. Their entries are

	$\nu_1 = 200$	$\nu_1 = 500$
$\nu_2 = 200$	1.26	1.22
	1.39	1.33
$\nu_2 = 400$	1.22	1.16
	1.32	1.19

Since the observed F is smaller than $F_{.95}$ in any of these cells, it is correct to conclude that $F = 1.05 < F_{.95}(229,305)$, and therefore F is not significant even at the .10 level.

EXERCISE 15-3

Compute F, indicate the values of ν_1 and ν_2, and give the two-tail significance level α for each of the following situations:

	n_A	n_B	$s_A{}^2$	$s_B{}^2$	F	ν_1	ν_2	α
1.	13	21	45.6	12.4	—	—	—	—
2.	13	21	12.4	45.6	—	—	—	—
3.	25	15	21.6	15.2	—	—	—	—
4.	76	41	37.6	33.8	—	—	—	—
5.	120	90	24.2	31.6	—	—	—	—
6.	35	75	29.4	16.8	—	—	—	—
7.	52	27	15.6	36.3	—	—	—	—
8.	31	31	34.0	16.2	—	—	—	—
9.	1	20	38.2	7.5	—	—	—	—
10.	10	∞	13.3	2.1	—	—	—	—

Special Case of $\nu_1 = 1$**.** It is often very helpful to remember that whenever $\nu_1 = 1$,

(15-10) $$F_{1-\alpha} = t_{1-\frac{1}{2}\alpha}^2$$

In formula (15-10) the symbol α refers to the level of significance. Hence, $1 - \alpha$ is the complement of α and $F_{1-\alpha}$ is the $(1 - \alpha)$th percentile of the F distribution. For example, if $\alpha = .05$, then $1 - \alpha = .95$ and $F_{1-\alpha}$ is the 95th percentile of F. In this case, $t_{1-\frac{1}{2}\alpha}$ is the 2.5th and $t_{1+\frac{1}{2}\alpha}$ the 97.5th percentile of t. This relationship makes it possible to ascertain the significance level for certain values of F not listed in Appendix Table V. First, however, you should be sure you understand the relation between F and t^2 by verifying the following values:

$F_{.95}(1,13) = 4.67$	$\sqrt{4.67} = 2.161$	$t_{.975}(13) = 2.16$
$F_{.95}(1,26) = 4.22$	$\sqrt{4.22} = 2.054$	$t_{.975}(26) = 2.06$
$F_{.99}(1,4) = 21.20$	$\sqrt{21.20} = 4.604$	$t_{.995}(4) = 4.60$
$F_{.95}(1,60) = 4.00$	$\sqrt{4.00} = 2.00$	$t_{.975}(60) = 2.00$
$F_{.95}(1,\infty) = 3.84$	$\sqrt{3.84} = 1.9596$	$t_{.975}(\infty) = 1.96$
$F_{.99}(1,\infty) = 6.64$	$\sqrt{6.64} = 2.5768$	$t_{.995}(\infty) = 2.576$

The last two items above show that when $\nu_1 = 1$ and $\nu_2 = \infty$,

$$\sqrt{F_{.95}} = z_{.975} \quad \text{and} \quad \sqrt{F_{.99}} = z_{.995}$$

so that in such cases one may refer $\sqrt{F}$ to a table of normal probability.

Now suppose that for $\nu_1 = 1$ and $\nu_2 = 26$ we have obtained $F = 6.03$. A direct approach to the F table shows only that $F_{.95} < 6.03 < F_{.99}$ so

$$.02 < \alpha < .10$$

because in taking s_1^2/s_2^2 the tabled level of significance is doubled.

However $\sqrt{6.03} = 2.456$ and for 26 degrees of freedom $t_{.975} = 2.06$ and $t_{.99} = 2.48$. Therefore

$$t_{.975} = 2.06 < \sqrt{6.03} < t_{.99} = 2.48$$

and

$$t_{.01} = -2.48 < -\sqrt{6.03} < t_{.025} = -2.06$$

Hence

$$.02 < \alpha < .05$$

Again, suppose $\nu_1 = 1$ and $\nu_2 = 1000$ and $F = 5.02$. Although $\nu_2 \neq \infty$, it is so large that we may treat $\sqrt{F}$ as normally distributed. Then $\sqrt{F} = \sqrt{5.02} = 2.24 = z$. Reference to the normal probability table shows that $2.24 = z_{.9875}$. Then $\alpha = 2(1 - .9875) = .025$ and $5.02 = F_{.975}$.

Comparison of Two Variances Based on Related Scores.   Formula (15-9) can be used only to compare two variances from independent samples. Sometimes it is as important to compare the variances of scores obtained from the same individuals under two different circumstances as to compare the means.

Suppose the same test is given twice to several classes with a four-month period between the initial and final tests. In one class the teacher puts great effort on stimulating the brightest pupils to advance to the limit of their ability and pays little attention to the pupils with low scores on the first test. The result is likely to be an increase in the variance and also a positive correlation between gain and initial score.

In another class the teacher concentrates effort on bringing up the low-scoring pupils with little attention to those whose initial scores are high. The result is likely to be a decrease in variance and also a negative correlation between gain and initial score.

In a third class all the pupils make similar gains, and the result is that the variance does not change much in either direction, and the correlation between gain and initial score is negligible.

The discovery of a significant change in variability may be as important as the discovery of a significant change in mean.

The statistic used to test the hypothesis that the population variance is the same on the two occasions,

$$H : \sigma_1{}^2 = \sigma_2{}^2 = \sigma^2,$$

is

(15-11) $$t = \frac{(s_2{}^2 - s_1{}^2)\sqrt{n-2}}{2s_1 s_2 \sqrt{1 - r_{12}{}^2}}$$

This statistic has "Student's" distribution with $n - 2$ degrees of freedom. An equivalent form which may be easier to compute is

(15-12) $$t = \frac{\{\Sigma(x_1 - \bar{x}_1)^2 - \Sigma(x_2 - \bar{x}_2)^2\}\sqrt{n-2}}{2\sqrt{\Sigma(x_1 - \bar{x}_1)^2 \Sigma(x_2 - \bar{x}_2)^2 - \{\Sigma(x_1 - \bar{x}_1)(x_2 - \bar{x}_2)\}^2}}$$

Comparison of the Variances of Several Populations.   In Chapter 9 we tested the hypothesis that the means of several populations are equal. We can also test the hypothesis that the variances of several populations are equal and methods for doing so are included in most advanced texts. A reader who needs to make such a test may find it indexed under such a heading as "Homogeneity of variance," "Heterogeneity of variance," "Bartlett's Test," "Cochran's test for equality of variance," or "Hartley's test for equality of variance."

16

INFERENCES ABOUT POPULATIONS ON CATEGORICAL VARIABLES

Methods will now be presented for testing hypotheses about categorical populations, namely those in which a po ulation is indicated by several classes and by the proportion of individuals in each class. The methods to be developed are an extension of those appearing in Chapter 11, where the discussion was limited to populations consisting of only two classes. The tests will involve a comparison of the frequencies in the various classes of an observed sample with the frequencies expected on the basis of the hypothesis being tested. The statistic used for these tests provides a measure of the discrepancy between the two sets of frequencies. It is called chi-square because its sampling distribution is approximated by the chi-square distribution discussed in Chapter 15, although the formula for this chi-square statistic is very different from formula (15-1).

Two main types of application will be discussed: (I) situations in which the expected frequencies are based on a theoretical distribution, and (II) situations in which the expected frequencies are based on the hypothesis that two variables are independent of each other.

I. COMPARING AN OBSERVED FREQUENCY DISTRIBUTION WITH ONE BASED ON A THEORETICAL DISTRIBUTION

Heuristic Development. A very simple illustration will help the reader develop an understanding of problems of this type. The subjects of an experiment are given three small cubes marked A, B, and C and are told that one of them is slightly heavier than the other two. As a test of weight discrimination they are to select the heaviest of the three cubes. Actually the cubes have been constructed as nearly equal in weight as possible, so that it should be a matter of chance which cube any given subject may select. The hypothetical distribution of subjects in the population, classified according to decision as to which cube is heaviest, would therefore be that shown in Table 16-1.

Table 16-1   Population Distribution in Three Classes

Class	Proportion in Class
A	$\frac{1}{3}$
B	$\frac{1}{3}$
C	$\frac{1}{3}$
All classes	1

Suppose now that 30 subjects are selected and each is asked to state which cube he considers heaviest. A subject is given no information about the choices of other persons and no clue except what he can get by lifting the cubes as often as he wishes. Suppose that on the basis of their decisions these 30 subjects are classified as indicated in Table 16-2.

Table 16-2   Sample Distribution in Three Classes

Class	Number in Class
A	9
B	7
C	14
All classes	30

We shall now consider a test of the hypothesis that the distribution of Table 16-2 is that of a random sample from the population described in Table 16-1.

In order to relate the sample to the population, one calculates the frequency expected in each class on the basis of the hypothetical population distribution, multiplying the total number of cases in the sample ($n = 30$) by the proportion of the population in that class. The observed and expected frequencies are juxtaposed in Table 16-3.

Table 16-3   Observed and Expected Frequencies in Three
Classes

Class	Observed Frequency	Expected Frequency	Difference
A	9	10	−1
B	7	10	−3
C	14	10	4
All classes	30	30	0

The differences in the final column of Table 16-3 show the extent to which the observed frequencies depart from theory. What is needed now is a summary value which describes the total discrepancy for the three classes combined. Clearly the sum of the discrepancies is no help because that sum is always zero. One might then consider using the sum of the squares of the discrepancies, but this is also unsatisfactory because it depends on the total number of cases in the sample.

A measure of discrepancy which does work is obtained by calculating the square of each difference, dividing it by the corresponding expected frequency, and finding the sum of the ratios thus formed. The resulting statistic is called chi-square, for reasons which will be discussed on page 269. For the problem now under discussion

$$\chi^2 = \frac{(-1)^2}{10} + \frac{(-3)^2}{10} + \frac{(4)^2}{10} = 2.6$$

To test the hypothesis that the population from which this sample was drawn has equal proportions in the three classes, this value of 2.6 is referred to the χ^2 table with degrees of freedom equal to the number of classes minus 1—that is, $\nu = 3 - 1 = 2$. Since the computed χ^2 is small, even smaller than $\chi^2_{.75}$, the hypothesis is accepted.

An absolute requirement for the use of the χ^2 test in problems of the sort we are now considering is that the classification of one individual must not in any way be affected by the prior classification of some other individual. This criterion was presumably met in the situation just described; we shall now describe two ways in which it might have been violated.

Suppose that instead of a sample of 30 persons the experimenter had used 10 subjects each of whom went through the experiment three times, on different days. The 30 records thus produced cannot be considered as those of independent individuals. Many subjects would remember their first choice; some would consciously or unconsciously repeat it; some would consciously change it. In any case the category in which a subject is placed on his second or third trial cannot be held to be independent of his category on the first.

Suppose again that the experimenter has 30 independent subjects but all are in the same room throughout the judging and hear each other's decisions. Suppose the 7th subject has heard his predecessors say $A - B - A - A - C - A$. If he is at all suggestive, does he approach the experiment with a completely open mind? Can it be asserted that his choice—which, of course, means his classification—is truly independent of the choices of his six predecessors?

Now let us consider a situation in which the criterion of independent classification is presumably met but in which the observed distribution may be that of a sample from some population other than the one hypothesized in Table 16-1. Suppose the investigator has 30 subjects who make their decisions independently and report them secretly. Again he uses three cubes marked A, B, and C, all having as near the same weight as possible, but now A is three inches on a side, B two inches, and C one inch. His psychological hypothesis is that such differences in size do not influence the perception of weight; in statistical terms this hypothesis becomes $P_A = P_B = P_C = \frac{1}{3}$. He will reject this hypothesis if χ^2 proves to be significantly large. The χ^2 test is entirely appropriate to use.

Symbolic Formulation.   To obtain adequate generality, symbolism is needed. For the ith class, the hypothetical proportion will be denoted p_i, the observed frequency f_i, the hypothetical frequency F_i; with $\Sigma P = 1.00$, $\Sigma f_i = n$, $\Sigma F_i = n$, and $F_i = P_i n$. The symbolism for four classes summarized in Table 16-4 can easily be extended to any number of classes.

Table 16-4   Symbolism for Four Classes

Class	Hypothetical Proportion	Observed Frequency	Hypothetical Frequency	Discrepancy	$\dfrac{(f_i - F_i)^2}{F_i}$
A_1	P_1	f_1	F_1	$f_1 - F_1$	$(f_1 - F_1)^2/F_1$
A_2	P_2	f_2	F_2	$f_2 - F_2$	$(f_2 - F_2)^2/F_2$
A_3	P_3	f_3	F_3	$f_3 - F_3$	$(f_3 - F_3)^2/F_3$
A_4	P_4	f_4	F_4	$f_4 - F_4$	$(f_4 - F_4)^2/F_4$
Sum	1	n	n	0	χ^2

In the type of problem we are now discussing, χ^2 may be defined as

(16-1)
$$\chi^2 = \sum \frac{(f_i - F_i)^2}{F_i}$$

An algebraically equivalent formula which provides a much more economical routine is

(16-2)
$$\chi^2 = \sum \frac{f_i^2}{F_i} - n$$

Degrees of Freedom.   In problems of the type described in this section, the population is considered to have k classes on a single variable. The sampling variability is assumed to apply to samples all of which have n cases, n being constant from sample to sample. Thus frequencies could be arbitrarily assigned to $k - 1$ of the k classes, but the frequency in the kth class would have to be whatever is required to bring the total to n. Thus the number of degrees of freedom is $k - 1$.

The number of degrees of freedom for problems of the type described in Section II is obtained by a different method which will be described in that section.

Sampling Distribution.   In Chapter 15 there was a discussion of the statistic $\chi^2 = \Sigma(x - \bar{x})^2/\sigma^2$, and its sampling distribution was described as a smooth curve with a form depending upon the number of degrees of freedom. The new statistic described above by formula (16-1) and the one to be introduced on page 274 in formula (16-4) have sampling distributions which are discrete, not smooth and continuous. These statistics are also called chi-square because the continuous chi-square distribution approximates the exact discrete distribution in much the same way that the continuous normal distribution approximates the histogram shown in Figure 8-1 on page 108. This approximation tends to be closer when n is large, poorer when n is small. The form of the smooth chi-square curve does not depend at all on n but only on the degrees of freedom.

Goodness of Fit.   The procedure discussed in the preceding paragraphs is customarily referred to as a test of *goodness of fit*. One wishes to ascertain whether the distribution of n individuals into classes observed in a sample can be reasonably regarded as fitting a hypothetical proportionate distribution of individuals in a population.

Thus, in the problem of choosing the cube which seems heaviest of the three presented, the 30 subjects were distributed into three classes in accordance with their choice of cube. In the hypothetical population the individuals were distributed with proportions $\frac{1}{3}$, $\frac{1}{3}$, $\frac{1}{3}$. The chi-square test was then carried out to ascertain whether the distribution of individuals in the sample represented a good fit to the hypothetical population distribution apart from sampling variability.

The reader should notice that this situation differs from the one discussed in Chapter 12 where a line was fitted to data. There the line was adapted through sample estimates of parameters to fit the data. In the present situation the proportions are held rigid and the question is simply whether the sample does or does not fit them. If one were fitting proportions to the data these would be 9/30, 7/30, 14/30 as can be observed from Table 16-2.

Applications. Let us examine the distribution of the digits in some portion of the Table of Random Numbers on pages 372 to 374. In the first 5 rows of that table there are 350 digits, the frequency distribution of which is given in Table 16-5. If a truly random process was used to develop this table, the expected proportion for each digit is 1/10 and the expected frequency 35.

Table 16-5   Computation of χ^2 from the Frequency Distribution of the Digits in the First Five Rows of Appendix Table XI ($H: P_0 = P_1 = \cdots = P_9 = 1/10$)

Digit	f_i	F_i	$f_i - F_i$	$(f_i - F_i)^2/F_i$
0	42	35	7	49/35
1	41	35	6	36/35
2	27	35	-8	64/35
3	36	35	1	1/35
4	29	35	-6	36/35
5	33	35	-2	4/35
6	43	35	8	64/35
7	30	35	-5	25/35
8	25	35	-10	100/35
9	44	35	9	81/35
Sum	350	350	0	$460/35 = 13.14 = \chi^2$

For 9 degrees of freedom, $\chi^2 = 13.14$ is not significant and so throws no suspicion on the randomness of the digits in this portion of the table. Note however, that randomness has not been established. The digits might conceivably have been arranged in a systematic order and yet the observed frequency for each digit might be close to the expected frequency.

An application in a different context is found in a study by Williams on the relation of season of birth and mental retardation. Previous studies had suggested to him the possibility that the seasonal pattern of births might not be quite the same for mentally retarded as for normal children.

From the *Registrar General* of England, Williams obtained the proportion of all births occurring in each of the three seasons named in the stub of Table 16-6. These proportions, which specify the population distribution, are given in the column headed p_i of that table.

Table 16-6 Computation of χ^2 to Compare the Seasonal Distribution of Birth Dates of Retarded Children with the Nationwide Distribution for each 4-Month Period*

Season	p_i	F_i	f_i	$f_i - F_i$	$(f_i - F_i)^2/F_i$
Jan. to Apr.	.343	91	88	−3	.0989
May to Aug.	.343	91	111	20	4.3956
Sept. to Dec.	.314	83	66	−17	3.4819
Total	1.000	265	265	0	$7.98 = \chi^2$

* SOURCE: Adapted from "Date of Birth, Backwardness and Educational Organization," *British Journal of Educational Psychology*, 34 (Nov. 1964), 247–255.

He obtained a sample of 265 children enrolled in schools for the educationally subnormal, and tabulated their birthdays, obtaining the frequencies in the column headed f_i.

For two degrees of freedom, $\chi^2 = 7.98$ is significant at the .05 level. In fact $\chi_{.98}^2 < 7.98 < \chi_{.99}^2$. For $\nu = 2$, $\chi^2 = 7.98$ is significant at the .02 level. Therefore the hypothesis that retarded children have the same seasonal distribution of births as normal children must be rejected.

EXERCISE 16-1

1. A student in a class of probability theory threw 4 pennies repeatedly until he had completed 80 throws. At each throw he counted the number of heads he observed, this number ranging from 0 to 4. The following table shows the

number of observed throws associated with each possible number of heads. The table also shows the proportion of heads in the population of throws if the experiment had been conducted fairly.

Number of heads on one throw	0	1	2	3	4
Observed number of throws	9	25	33	11	2
Proportion of throws in population	1/16	1/4	3/8	1/4	1/16
Expected number of throws	5	20	30	20	5

Would you be willing to agree that the student has apparently been using unbiased coins and throwing them properly? Answer this question by carrying out a chi-square test.

II. TESTS OF INDEPENDENCE IN CONTINGENCY TABLES

Nature of Problem.   In a contingency table the data are laid out in a two-way distribution which resembles a correlation table. In a correlation table the two variables are always scaled; in a contingency table they are categorical, and the categories do not necessarily have any order.

In Section I we dealt with population distributions on a single variate; now we shall deal with bivariate distributions. In Section I we tested a hypothesis about the proportion of cases in the population classes; now we shall test the hypothesis that in the population the two variables are independent.

The word "independent" is stronger than the word "uncorrelated." If two variables are independent they are necessarily uncorrelated, but they may be uncorrelated without being independent. The diagram on page 242 illustrates such a situation. Since $r = -.06$, the two variables are practically uncorrelated. But because information about the value of x provides considerable information about the value of y, they are not independent.

Symbolism.   In its most general form a contingency table may be described as having r rows and c columns. For any given cell the observed frequency is denoted by f and the theoretical frequency by F with two subscripts, of which the first names the row and the second names the column. This order of subscripts is a convention which is almost always observed. Thus f_{23} indicates the observed frequency in a cell in row 2 and column 3, and f_{ij} is a general expression for the frequency in the cell in the ith row and jth column; f_{ij} is understood to mean the frequency in an unspecified cell.

It is convenient to have a simple symbol for the sum of the frequencies in a row or in a column. As all the frequencies in any one row, say row 3, have the same first subscript and a variable second subscript (f_{31}, f_{32}, f_{33} . . .), any one of them is designated as f_{3j}. The sum for j going from 1 to 4 is

$$f_{31} + f_{32} + f_{33} + f_{34} = \sum_{j=1}^{4} f_{3j} = f_3.$$

Similarly all the frequencies in one column have a variable first subscript and a common second subscript, so any one of them in column 4 is denoted f_{i4}, and the sum for the frequencies with i going from 1 to 5 is

$$f_{14} + f_{24} + f_{34} + f_{44} + f_{54} = \sum_{1}^{5} f_{i4} = f_{.4}$$

Replacing the variable symbol i or j with a dot gives a symbol easy to recognize. It avoids confusion, making it clear that $f_1.$, $f_3.$, and $f_7.$ are marginal values for rows whereas $f_{.1}$, $f_{.3}$, and $f_{.7}$ are marginal values for columns.

A contingency table with r rows and c columns is represented symbolically in Table 16-7 and the corresponding theoretical frequencies in Table 16-8. Note that marginal frequencies are the same for the two tables.

Table 16-7 Symbols for a Contingency Table
with r Rows and c Columns

	x_1	x_2	x_j	x_c	
y_1	f_{11}	f_{12} $\cdots$ f_{1j} $\cdots$	f_{1c}	$f_1.$
y_2	f_{21}	f_{22} $\cdots$ f_{2j} $\cdots$	f_{2c}	$f_2.$
	.	.	.	.	.
	.	.	.	.	.
y_i	f_{i1}	f_{i2} $\cdots$ f_{ij} $\cdots$	f_{ic}	$f_i.$
	.	.	.	.	.
	.	.	.	.	.
y_r	f_{r1}	f_{r2} $\cdots$ f_{rj} $\cdots$	f_{rc}	$f_r.$
	$f_{.1}$	$f_{.2}$ $\cdots$ $f_{.j}$ $\cdots$	$f_{.c}$	$f.. = n$

Now we want to test the hypothesis that the variables x and y are independent. If this is strictly true, then all the theoretical entries in any row are proportional to the frequencies in the horizontal margin, and

Table 16-8   Theoretical Values in a Contingency Table
with r Rows and c Columns

	x_1	x_2		x_j		x_c	
y_1	F_{11}	F_{12}	$\cdots$	F_{1j}	$\cdots$	F_{1c}	$f_{1.}$
y_2	F_{21}	F_{22}	$\cdots$	F_{2j}	$\cdots$	F_{2c}	$f_{2.}$
	.	.		.		.	
	.	.		.		.	
y_i	F_{i1}	F_{i2}	$\cdots$	F_{ij}	$\cdots$	F_{ic}	$f_{i.}$
	.	.		.		.	
	.	.		.		.	
y_r	F_{r1}	F_{r2}	$\cdots$	F_{rj}	$\cdots$	F_{rc}	$f_{r.}$
	$f_{.1}$	$f_{.2}$		$f_{.j}$		$f_{.c}$	$f_{..} = n$

all the theoretical frequencies in any column are proportional to the frequencies in the vertical margin. In symbols

$$\frac{F_{i1}}{f_{.1}} = \frac{F_{i2}}{f_{.2}} = \cdots = \frac{F_{ij}}{f_{.j}} = \cdots = \frac{f_{i.}}{n}$$

and

$$\frac{F_{1j}}{f_{1.}} = \frac{F_{2j}}{f_{2.}} = \cdots = \frac{F_{ij}}{f_{i.}} = \cdots = \frac{f_{.j}}{n}$$

Therefore the expected frequency is

$$(16\text{-}3) \qquad F_{ij} = \frac{(f_{i.})(f_{.j})}{n}$$

The computation of χ^2 may now be completed by a formula similar to formula (16-1)

$$(16\text{-}4) \qquad \chi^2 = \sum_i \sum_j \frac{(f_{ij} - F_{ij})^2}{F_{ij}}$$

or by an algebraically equivalent formula similar to (16-2)

$$(16\text{-}5) \qquad \chi^2 = \sum \sum \frac{f_{ij}^2}{F_{ij}} - n$$

or by another which is also algebraically equivalent but which usually involves less arithmetic

$$(16\text{-}6) \qquad \chi^2 = n \left\{ \sum \sum \frac{f_{ij}^2}{(f_{i.})(f_{.j})} - 1 \right\}$$

EXERCISE 16-2

1. What symbol would be used for the observed frequency in cells a, b, c, d, and e?

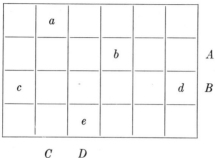

$$C \quad D$$

2. What symbol would be used for the marginal totals A, B, C, and D?

3. What symbol would represent
   (a) $f_{11} + f_{12} + f_{13} + f_{14} + f_{15} + f_{16}$?
   (b) $f_{51} + f_{52} + f_{53} + f_{54} + f_{55} + f_{56}$?
   (c) $f_{15} + f_{25} + f_{35} + f_{45}$?

4. For the diagram in Question 1, what sum is represented by the symbol $f_{.3}$? By $f_{3.}$? By $f_{.2}$?

5. Expand $\displaystyle\sum_{i=1}^{3} f_{i4}$; $\displaystyle\sum_{j=1}^{3} f_{2j}$.

6. What symbol would represent the frequency expected in each of the cells a, b, c, d, and e? On the hypothesis that the two variables are independent, by what formula would those expected frequencies be computed?

7. In a contingency table with marginal frequencies as indicated here, what are the cell values of the F_{ij}'s?

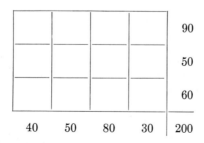

8. Try to assign cell values to these 12 cells, using first one order of assignment and then another. For how many cells is a free choice possible?

9. Draw a contingency table with 5 rows and 3 columns, and write in a set of marginal frequencies. To how many of the 15 cells of the table can cell frequencies be freely assigned? Does this number of free choices appear to be related to the size of n? To the size of the marginal frequencies? To the values of r and c?

Degrees of Freedom for an $r \times c$ Contingency Table. The student who has answered Questions 8 and 9 in Exercise 16-2 has probably recognized that in a table with r rows and c columns, only $(r-1)(c-1)$ cell frequencies can be freely assigned. If he says that in such a table χ^2 has $(r-1)(c-1)$ degrees of freedom, he will be correct.

A more general formulation which is not dependent on the intuitive approach of counting the number of values which can be assigned at will is provided by the rule

For any problem the number of degrees of freedom is the number of variables reduced by the number of independent restrictions on those variables.

In the present context the variables are the cell frequencies which are rc in number. The marginal totals are assumed to be constant for a population of samples of which the one under consideration is a member. This assumption imposes $r + c$ restrictions; but since the sum of the row totals is the same as the sum of the column totals, both being n, the total number of restrictions is reduced by 1 to give $r + c - 1$ independent restrictions. Hence the number of degrees of freedom is $rc - (r + c - 1) = (r-1)(c-1)$.

Illustration of a 3×3 Contingency Table. As an example of such a situation consider data taken from Rope's study of *Opinion Conflict and School Support*. Each of 1464 adult residents of Pittsburgh, Pa., was interviewed on a variety of issues related to tax support. One of the questions asked was, "Do you think tax money should, or should not, be spent on nursery schools for children less than four and a half years old?" Responses were classified as (1) favorable, (2) no opinion, and (3) unfavorable. The problem was to determine whether there is a relationship between type of response and age of respondent. Each individual was then classified both according to type of response and according to age. The resulting distribution appears in Table 16-9. It may be noted that age, though actually a scaled variable, has here been treated as categorical.

The values of F_{ij}, computed by formula (16-3) are shown in Table 16-10. The computation of χ^2 is carried out by formula (16-5) as follows:

$$\chi^2 = \frac{(153)^2}{154.4} + \frac{(35)^2}{42.5} + \cdots + \frac{(25)^2}{18.8} + \frac{(160)^2}{162.9} - 1464 = 4.1$$

Table 16-9   Frequency of Observed Response to the Question of Spending Tax Money for Nursery Schools, Classified According to Nature of Response and Age of Respondent*

	Group Aged 20–34	Group Aged 35–54	Group over 54	Total
Favorable response	153	182	65	400
No opinion	35	50	25	110
Unfavorable response	377	417	160	954
Total	565	649	250	1464

* SOURCE: Taken from Rope, *Opinion Conflict and School Support.*

Table 16-10   Expected Frequency of Response to the Question of Spending Tax Money for Nursery Schools, on the Hypothesis that Age and Nature of Response are Independent (Data from Table 16-9)

	Group Aged 20–34	Group Aged 35–54	Group over 54	Total
Favorable response	154.4	177.3	68.3	400
No opinion	42.5	48.8	18.8	110
Unfavorable response	368.2	422.9	162.9	954
Total	565.0	649.0	250.0	1464

The same result can be obtained a little more easily by using formula (16-6), taking one column at a time and factoring out $1/f_{ij}$ for that column as follows:

$$\frac{1}{565}\left\{\frac{(153)^2}{400}+\frac{(35)^2}{110}+\frac{(377)^2}{954}\right\}=\frac{218.6411}{565}=.3870$$

$$\frac{1}{649}\left\{\frac{(182)^2}{400}+\frac{(50)^2}{110}+\frac{(417)^2}{954}\right\}=\frac{287.8108}{649}=.4435$$

$$\frac{1}{250}\left\{\frac{(65)^2}{400}+\frac{(25)^2}{110}+\frac{(160)^2}{954}\right\}=\frac{40.0787}{250}=\frac{.1723}{1.0028}$$

$$\chi^2 = 1464(1.0028 - 1) = 4.099 \text{ or } 4.1$$

It will be of interest to the reader to note that in Table 16-10 the equalities of proportions such as

$$\frac{F_{11}}{f_{.1}} = \frac{F_{12}}{f_{.2}} = \frac{F_{13}}{f_{.3}} = \frac{f_{1.}}{n}$$

hold true for all rows and all columns.

Contingency Table in which One Trait Is Dichotomous. When one or both variables have only two classes, formulas (16-4), (16-5), and (16-6) can be reduced to special formulas which involve considerably less arithmetic. We shall consider first the special method applicable when one variable has two classes and the other has more than two.

Let us suppose that 1000 university students have been interviewed and asked which of 4 possible candidates they prefer as the next president of the United States. These persons are then classified by sex and by preference as in Table 16-11. These data could be used to test the hypothe-

Table 16-11 Preferences for the Next President of the United States Expressed by 600 Men and 400 Women Students in University X

	Number Expressing Preference		
Preference	Men	Women	Both
A	86	66	152
B	175	120	295
C	202	144	346
D	34	34	68
Undecided	103	36	139
Total	600	400	1000

sis that among university students preference for presidential candidates is independent of sex. Computing expected frequencies by formula (16-3) and then computing χ^2 by the method already described will yield $\chi^2 = .16$ which is less than $\chi^2_{.005} = .21$ for $\nu = 4$. Clearly the data give no justification for rejecting the hypothesis that opinion on this issue is independent of sex.

To facilitate a simpler routine of computation we may use the symbolism shown here for two columns and five rows but easily extended to any number of rows.

a_1	b_1	n_1
a_2	b_2	n_2
a_3	b_3	n_3
a_4	b_4	n_4
a_5	b_5	n_5
A	B	n

It is to be noted that

$$a_i + b_i = n_i$$
$$\Sigma a_i = A$$
$$\Sigma b_i = B$$
$$\Sigma n_i = n$$
$$A + B = n$$

Then

(16-7a)
$$\chi^2 = \left\{ \sum \frac{a_i^2}{n_i} - \frac{A^2}{n} \right\} \frac{n^2}{AB}$$

or

(16-7b)
$$\chi^2 = \left\{ \sum \frac{b_i^2}{n_i} - \frac{B^2}{n} \right\} \frac{n^2}{AB}$$

For the data of Table 16-11, the application of formula (16-7a) yields

$$\chi^2 = \left\{ \frac{(86)^2}{152} + \frac{(175)^2}{295} + \frac{(202)^2}{346} + \frac{(34)^2}{68} + \frac{(103)^2}{139} - \frac{(600)^2}{1000} \right\} \frac{(1000)^2}{(600)(400)}$$
$$= 15.5$$

The student may find it profitable to apply formula (16-7b) to the same data as a check.

Contingency Table in which Both Traits are Dichotomous. In this situation, formulas (16-4), (16-5), (16-6), or (16-7) can be used but a special formula (16-8) provides an easier routine. It is derived by straightforward algebra from any of the preceding formulas if the frequencies in the four cells are denoted a, b, c, and d and $n = a + b + c + d$.

b	a	$a + b$
d	c	$c + d$
$b + d$	$a + c$	

(16-8)
$$\chi^2 = \frac{(ad - bc)^2 n}{(a + b)(c + d)(b + d)(a + c)}$$

Suppose a polling organization has asked 200 adults living in cities and 300 living in rural districts their opinions about a proposal before the state legislature, with results as follows:

	Urban	Rural	Total
Approve	132	175	307
Disapprove	68	125	193
Total	200	300	500

Then

$$\chi^2 = \frac{[(132)(125) - (175)(68)]^2(500)}{(200)(300)(307)(193)}$$
$$= \frac{(4600)^2(500)}{(200)(300)(307)(193)} = 3.0$$

The number of degrees of freedom in a 2×2 contingency table is $(2 - 1)(2 - 1) = 1$. For 1 degree of freedom $\chi^2_{.95} = 3.8$, hence $\chi^2 = 3.0$ is not significant, and it cannot be concluded that opinion on this issue is different for urban and rural areas.

Relation of χ^2 to z.   Because the normal probability table is available in much greater detail than the χ^2 table, it is advantageous to make use of the fact that when $\nu = 1$,

(16-9)
$$z_{1-\alpha/2} = \sqrt{\chi^2_{1-\alpha}}$$

Suppose that with $\nu = 1$ χ^2 is found to be 8.6. All the χ^2 table can tell us is that $\chi^2_{.995} < 8.6 < \chi^2_{.999}$. The relationship of (16-9) allows us to say that $z = \sqrt{8.6} = 2.93$ and $2.93 = z_{.9983}$.

Verify the following values when $\nu = 1$:

$$\chi^2_{.50} = .46 \qquad \sqrt{.46} = .68 \qquad z_{.75} = .6745$$
$$\chi^2_{.90} = 2.7 \qquad \sqrt{2.7} = 1.64 \qquad z_{.95} = 1.645$$
$$\chi^2_{.99} = 6.6 \qquad \sqrt{6.6} = 2.57 \qquad z_{.995} = 2.576$$

Relation of the ϕ Coefficient to χ^2 with 1 Degree of Freedom. The reader may have noticed that formula (16-8) is very similar to formula (13-27) on page 236. In fact

$$(16\text{-}10) \qquad \phi = \sqrt{\chi^2/n}$$

The ϕ coefficient measures the relationship between the two dichotomous variables; χ^2 may be used to test the significance of that relationship. The ϕ coefficient does not depend in any way on the size of n; χ^2 increases with sample size, if each of the four cell frequencies is multiplied by a, then χ^2 is multiplied by a. The statistic ϕ is a measure of correlation, of relationship, but does not of itself give any indication as to whether that relationship is significant or not. The statistic χ^2 provides a test of the significance of the relationship but does not of itself say anything about the extent of the relationship.

The 2×2 Contingency Table with Small Frequencies. The theoretical chi-square distribution is a smooth distribution for a continuous variable. The exact probability distribution for χ^2 is discrete. When all marginal frequencies in a contingency distribution are large, the smooth theoretical distribution gives a good approximation to the exact discrete distribution. If the marginal frequencies are not large but $\nu > 1$, the approximation may be fairly good. But when one or more of the marginal frequencies is small and $\nu = 1$ as in a 2×2 table, the probabilities obtained from Appendix Table IV may be very inaccurate. To see why this is so, imagine a sample of 300 cases, which is far from small, with marginal frequencies as shown. There are only 7 values which the cell frequency a can take; 0, 1, 2, 3, 4, 5, or 6; only 7 possible 2×2 tables. Thus χ^2 can have only 7 values, and the exact probability distribution has only 7 discrete steps. The correspondence between this step distribution and the smooth curve would be poor.

b	a	6
d	c	294
240	60	300

There is a method for computing the exact probability for each of these 7 tables but it is outside the scope of this text. Most advanced texts present it under the heading "Fisher's exact test."

To bring the probabilities read from Appendix Table IV into closer correspondence with these exact probabilities, the value of χ^2 is reduced slightly by what is known as Yates' correction for continuity:

$$(16\text{-}11) \qquad \chi_y{}^2 = \frac{(|ad - bc| - n/2)^2 n}{(a + b)(a + c)(b + d)(d + c)}$$

This correction should always reduce the value of χ^2. Thus if $a = 12$, $d = 4$, $b = 10$, and $c = 8$, we have $ad - bc = -32$, so the numerator is $(|-32| - 17)^2(34) = (32 - 17)^2(34)$, because $n/2$ is subtracted from the absolute value of $ad - bc$, that is from a positive number.

Suppose a thematic apperception test has been given to 15 boys and 15 girls aged 14 to 16 and that each has been classified as to whether he wrote at least one story involving a theme of violence.

	Boys	Girls	Both
Used theme of violence	6	1	7
Did not use theme of violence	9	14	23
Total	15	15	30

Six boys and one girl did write such a story. If χ^2 is incorrectly computed by formula (16-8) the result would be $\chi^2 = 4.66 > \chi_{95}{}^2$. When Yates' correction is applied, $\chi_y{}^2 = 2.98 < \chi_{.95}{}^2$.

Comparison of Two Proportions Based on the Same Individuals. An interesting type of problem arises when the same individuals are measured twice, and it is required to compare the proportions showing a given characteristic on the two occasions. Suppose for example that the same 200 pupils in the ninth grade were tested with an instrument designed to detect race prejudice and were classified as showing high or low prejudice and subsequently took the test a second time. The results of the two tests are shown in Table 16-12.

This table looks superficially like the 2×2 contingency tables previously discussed, but there is a very important difference. The entries in the four cells of this table cannot be considered independent because each individual was measured twice and thus one of the fundamental assumptions underlying the probability distribution for χ^2 is not met.

The number showing high prejudice has decreased from $a + b = 80$ on the first test to $a + c = 54$ on the second, so the change is $(a + b) -$

Table 16-12   Results of Testing the Same Individuals for
Degree of Prejudice, on Two Different Occasions

First Test	Second Test		
	Low	High	
High	$b = 54$	$a = 26$	$a + b = 80$
Low	$d = 92$	$c = 28$	$c + d = 120$

$$b + d = 146 \quad a + c = 54$$

$(a + c) = b - c = 26$. (Similarly the number showing low prejudice has
increased from $c + d = 120$ on the first test to $b + d = 146$ on the second, the change being $(c + d) - (b + d) = c - b = -26$.) Clearly we
are concerned only with b and c, the numbers of individuals changing, and
can ignore a and d, the numbers not changing.

If, among those individuals who change, the direction of change is a
matter of chance only, the expected number changing in each direction is
$(b + c)/2$. For these data, we now have

f_i	F_i	$(f_i - F_i)^2/F_i$
54	41	$(13)^2/41$
28	41	$(-13)^2/41$

and $\chi^2 = 2(169)/41 = 8.24$. Or in general

(16-12)
$$\chi^2 = \frac{(b - c)^2}{b + c}$$

By formula (16-12), $\chi^2 = (26)^2/82 = 8.24$. Then $z = \sqrt{8.24} = 2.87$,
which is significant at the .004 level.

However if formula (16-8) had been used, we should have supposed
χ^2 to be only 2.04 and should have mistakenly held it to be nonsignificant.

Misuses of the Chi-Square Test.   A number of mistakes in using or in
reporting χ^2 are so common that they justify a word of caution.

**1.   Failure to describe precisely the situation for which the computa-
tion was made.** All too often a published study contains the statement
that "chi-square was computed and found significant," but the author

fails to say to which of various possible arrangements of his data the computation relates.

2. The use of classes which are not mutually exclusive. Suppose the teachers in a school want to make a comparative study of the problem behavior of boys and girls and decide to classify each child as to sex and as to the type of problem he typically presents, using the following categories: tardiness, lying, disturbing class routine, inattention in class, cheating on tests, teasing younger children, and no serious problem. Psychologically the classification scheme is very poor, but at the moment we are particularly concerned with its statistical shortcomings. For one thing it is not exhaustive, some children may not belong in any of these classes. For another thing, the classes are not mutually exclusive; some children may belong in more than one class. When the data are gathered nothing can be done with them. The plan was faulty and cannot be saved after the data are in. This illustration may sound a little silly, but problems nearly as bad crop up in published papers all too often.

3. Choosing a particular category for comparison after examining the data. It is a truism in statistics that a hypothesis cannot be tested with the data which suggested it. When a chi-square test with more than 1 degree of freedom is significant, the research worker often has a keen desire to pinpoint the discrepancy. Looking at his frequency distribution he may notice that there is a large discrepancy between observed and expected values in a particular category, so he combines the frequencies in all the other categories. If he started with a $k \times 2$ table he now has a 2×2 table, for which he computes chi-square with 1 degree of freedom. This procedure is not to be recommended. A $k \times 2$ table had $k - 1$ degrees of freedom only. The number of 2×2 tables which could be produced by combining categories in different ways is large, much larger than $k - 1$, and the resulting values of χ^2 are not independent. A method exists for partitioning χ^2 into independent 2×2 tables each with 1 degree of freedom, but this method is not appropriate for an elementary text. Note that if the reason for studying the frequencies in a particular category arises from the logic of the problem and is recognized before the data are seen, such special analysis is justified.

4. Comparison of percentages without regard to sample size. Suppose a newspaper reports that a poll shows 68 percent of persons in Group A in favor of a proposed amendment to the state constitution and 32 percent opposed; 52 percent of Group B in favor and 48 percent opposed. It concludes that there is a great deal more approval for the amendment in Group A than in Group B. No conclusion is possible without knowledge of the sample size. Note these distributions each of which has the proportions specified.

	+	−	
A	204	96	300
B	156	144	300
	360	240	600

$\chi^2 = 16.00$

	+	−	
A	68	32	100
B	52	48	100
	120	80	200

$\chi^2 = 5.33$

	+	−	
A	51	24	75
B	39	36	75
	90	60	150

$\chi^2 = 4.00$

	+	−	
A	17	8	25
B	13	12	25
	30	20	50

$\chi^2 = 1.33$

5.   Indiscriminate comparison of frequencies. Suppose two classes of 20 children are taught by two different methods A and B. Each child is then given a 15-question test, with results as indicated here.

	A	B
Number of right answers	252	206
Number of wrong answers	48	94

If χ^2 is computed from these figures it will be devoid of all meaning. It is children not answers which are independent individuals, and the analysis must be made in terms of children. Suppose the score for each child is available and a frequency distribution can be set up:

Score	A	B	Both
15	3	1	4
14	3	1	4
13	5	1	6
12	4	2	6
11	3	5	8
10	1	3	4
9	1	3	4
8		2	2
7		1	1
6			
5		1	1
	20	20	40

The means of the two groups could be compared by the method described on page 162. Another legitimate method would be to divide the distribution of scores as near the median of the combined group as convenient, and obtain a 2 × 2 table from which χ^2 can be computed:

	A	B	
Scoring 12 or more	15	5	20
Scoring 11 or less	5	15	20
	20	20	

6. Failure to apply Yates' correction when frequencies are small. This situation has already been discussed on page 281.

EXERCISE 16-3

For each of the following sets of data, set up the table appropriate for a computation of χ^2, decide what is the appropriate formula to use, perform the computation, state the number of degrees of freedom, and interpret the results.

1. Maurice Chavan examined children enrolled in Schools for the Educationally Subnormal in Wales and children in ordinary schools. Out of 110 boys in ordinary schools he considered 101 to be in good health and out of 110 in ESN schools, only 69.

2. In the same study, Chavan reports that out of 110 ESN boys 69 were in good health whereas out of 51 ESN girls 21 were in good health.

3. A test writer constructing a test of information about agriculture tries out a preliminary form with 10 boys who grew up on farms and 8 who grew up in a city. He wants to be fairly sure that boys with a rural background make higher scores than boys with an urban background. The scores of these boys were:

Urban: 36, 52, 39, 48, 41, 46, 42, 48
Rural: 53, 61, 42, 67, 51, 48, 53, 67, 58, 56

Can you test in more than one way the hypothesis that urban and rural boys perform equally well on this test?
Divide the scores into two classes "above 50" and "50 or below," and set up a 2 × 2 table. Compute χ^2. Do you need to apply Yates' correction?
Also test the hypothesis that the means of urban and rural boys are the same.

4. A psychology class had been concerned as to whether a statistically significant relation exists between the Rorschach and other personality tests. They decided to obtain for each of 8 persons an MMPI and a Rorschach protocol, to eliminate all extraneous clues and to ask a group of psychologists to try to match protocols with MMPI. A student who had studied the theory of probability worked out the probability distribution for the number of correct matchings to be expected if matching was by pure chance. The probabilities and the record of 60 psychologists are as follows:

Number of Correct Matchings	Proportion Expected by Chance	Number of Psychologists
4–8	.0191	5
3	.0611	9
2	.1840	16
1	.3679	18
0	.3679	12
	1.0000	60

Was the number of correct matchings too great to be ascribed to chance?

5. In a certain high school, one group of 30 pupils received instruction in automobile driving, and 25 of them passed a driver's examination similar to the state examination for a driver's license. Out of 120 pupils of the same age who did not receive the instruction, 71 passed the examination.

17

COMPARISON
OF SEVERAL
INDEPENDENT
POPULATIONS

Methods for testing hypotheses about the means of three or more populations will be presented in this chapter. Chapter 10 dealt with hypotheses about the means of two samples, and so the methods described there are a special case of the methods described in this chapter. These methods involve the subdivision or analysis into components of the total variation inherent in a set of observations and are therefore collectively spoken of as the analysis of variance.

The analysis of variance is a very large topic about which extensive and scholarly books are written. It can be used to solve a great variety of problems, some of them quite complex. Because it is useful in so many situations and occupies such a prominent place in modern statistical work, even an elementary text such as this one should introduce the topic to its readers. Although we can present here only its simplest aspects, we shall try to lay a foundation of understanding which will enable the reader to go on to more advanced treatments when he needs to do so.

Testing hypotheses about the means of several independent populations on a single variable is probably the simplest problem in the field. A sample is drawn at random from each population, and these samples are not matched in any way.

Before a numerical example is taken up, several basic matters must be considered. The assumptions which will define the populations must be stated; in other words, a mathematical model must be constructed.

Symbolism will be needed which will facilitate describing the necessary calculations with clarity.

Mathematical Model. Each of the populations sampled is assumed to have a normal distribution; all populations are assumed to have the same variance but not necessarily the same mean. This model is illustrated for three populations in the diagram below; the illustration can easily be extended to any number of populations.

Population	Mean	Variance
A	μ_A	σ^2
B	μ_B	σ^2
C	μ_C	σ^2

Most of the advanced texts present statistical tests for determining whether in a particular situation the assumptions of normality and equality of variance are reasonable. In most actual situations visual inspection of the frequency distributions and computed variances is adequate. Recent empirical studies have shown that the test for equality of means about to be described is remarkably *robust* against non-normality unless the latter is extreme, and extreme non-normality has high visibility. Inequality of variance is more serious when sample sizes are small and unequal. Its effects can be greatly reduced by the simple expedient of making all n's equal and not too small.

In terms of the preceding diagram the hypothesis to be tested is

$$H : \mu_A = \mu_B = \mu_C = \mu$$

Outline of Test Procedure. In order to carry out the test procedure for the hypothesis just stated, formulas are required which may appear complex at first reading. Because of this factor, the mathematics entering into the formulas will be developed in some detail.

It may be helpful at this point to give an overview of the test procedure so as to give meaning to the discussion which follows. The reader will note that the hypothesis to be discussed in this chapter is an extension of the hypothesis that two means are equal, which was presented in Chapter 10, page 162. To test this earlier hypothesis one uses the expression for t in formula (10-2). The value of t so obtained is then referred to the table of the t distribution.

To test the hypothesis that two means are equal, it is sufficient to use the difference $\bar{x}_1 - \bar{x}_2$ between sample means. However, when more than two means are to be compared, a difference is not sufficient. One then turns to an alternate formulation, namely a variance among means.

It is of interest to note the relation between t^2 and the variance among the means of 2 groups. If $\mu_1 - \mu_2 = 0$, t^2 from formula (10-6) on page 165 can by algebraic manipulation be converted into the equivalent expression

$$(17\text{-}1) \qquad t^2 = \frac{n_1(\bar{x}_1 - \bar{x})^2 + n_2(\bar{x}_2 - \bar{x})^2}{[\Sigma(x_1 - \bar{x}_1)^2 + \Sigma(x_2 - \bar{x}_2)^2]/(n_1 + n_2 - 2)}$$

Here $\bar{x}$ is the mean of all $n_1 + n_2$ cases in the sample.

In this formula the numerator is a variance among the weighted means. The denominator involves deviations from means. The procedure to be described for dealing with several groups is an extension of the procedure for two groups.

The extension to several groups will be developed in a series of steps as follows:

1. Use of double summations as needed for computations to follow.
2. Conversion of formulas expressed in deviation form into formulas expressed as sums of scores and of their squares, to simplify computation.
3. Partition of sums of squares into two parts, one of which is the variation of means of groups about the mean of all cases and the other the variation within groups.
4. Discussion of degrees of freedom appropriate to analysis of variance.
5. Concept of the mean squares. These mean squares correspond to the numerator and denominator of the expression for t^2 in formula (17-1).
6. Testing the hypothesis of equality of means by an F ratio. The reader will recall that t^2 also has the F distribution. See page 263.

Symbolism Using Double Subscripts. In Chapter 16, page 272, double subscripts were used to designate frequencies in a contingency table. They will now be used to identify a score in terms of the population from which it is drawn and of the individual to whom it belongs.

Suppose a variable is denoted X. Then we might denote the score of

individual 3 in a sample from population 1 as x_{13}
individual 4 in a sample from population 5 as x_{54}
an unspecified individual in a sample from population 2 as $x_{2\alpha}$
individual 2 in a sample from an unspecified population as x_{i2}
an unspecified individual in a sample from an unspecified population as $x_{i\alpha}$

This notation is summarized in Table 17-1.

The symbol α here has nothing whatever to do with its previous use

Table 17-1   Notation for Scores in Samples from Different Populations

Population Sampled	Scores in Sample
1	$x_{11}, x_{12}, x_{13}, \cdots x_{1\alpha} \cdots x_{1n_1}$
2	$x_{21}, x_{22}, x_{23}, \cdots x_{2\alpha} \cdots x_{2n_2}$
i	$x_{i1}, x_{i2}, x_{i3}, \cdots x_{i\alpha} \cdots x_{in_i}$
k	$x_{k1}, x_{k2}, x_{k3}, \cdots x_{k\alpha} \cdots x_{kn_k}$

as level of significance. In the symbolism shown above and in Table 17-1, it is a generic subscript relating to *any individual*, and i is a generic subscript relating to *any population*. To emphasize in the notation the distinction between a subscript relating to a population or a sample and a subscript denoting an individual in a sample, a Latin letter is used for the former and a Greek letter for the latter.

We shall need symbolism to represent such sums as $x_{11} + x_{12} + x_{13} + x_{14} + x_{15}$. Any one of these scores may be denoted as $x_{1\alpha}$, so their sum is obviously $\Sigma x_{1\alpha}$. To show very clearly that α is taking values from 1 to 5 we could write $\sum_{\alpha=1}^{5} x_{1\alpha}$ or $\Sigma_{\alpha=1}^{5} x_{1\alpha}$, meaning "the sum of $x_{1\alpha}$ with α going from 1 to 5." The symbols $\sum_{\alpha} x_{1\alpha}$ and $\Sigma_{\alpha} x_{1\alpha}$ make it clear that the summation is over values of α but do not say which values.

Now look at the scores for sample 1 listed in Table 17-2. Any one of these scores is $x_{1\alpha}$. Consequently

$$\Sigma x_{1\alpha} = 11 + 21 + 15 + 46 + 19 = 112$$
$$\Sigma x_{1\alpha}^2 = 11^2 + 21^2 + 15^2 + 46^2 + 19^2 = 3264$$
$$\sum \frac{x_{1\alpha}}{n_1} = \frac{1}{5}(11 + 21 + 15 + 46 + 19) = 22.4$$
$$(\Sigma x_{1\alpha})^2 = 112^2 = 12544$$

In the last column of Table 17-2, the sums are $\Sigma x_{1\alpha} = 112$, $\Sigma x_{2\alpha} = 41$,

Table 17-2   Scores in Samples from Four Populations

Sample	List of Scores	Sum
1	11, 21, 15, 46, 19	112
2	5, 7, 19, 10	41
3	10, 25, 25, 10, 15, 17	102
4	21, 21, 10, 19, 11	82

$\Sigma x_{3\alpha} = 102$, and $\Sigma x_{4\alpha} = 82$. We now need a symbol to represent the sum of these sums. Any one of the sums could be called $\Sigma_{\alpha} x_{i\alpha}$, so obviously and simply their sum is $\Sigma_i(\Sigma_{\alpha} x_{i\alpha})$ or more explicitly $\Sigma_{i=1}^{4}\Sigma_{\alpha=1}^{n_i} x_{i\alpha}$.

EXERCISE 17-1

Answer these questions by referring to Table 17-2.

1. What are the values of n_1, n_2, n_3, and n_4?

2. What are the values of x_{23}, x_{41}, and x_{34}?

3. What symbol would describe the score whose value is 46? 7?

4. What is the value of $\Sigma x_{2\alpha}^2$?

5. What is the value of $\Sigma_i \Sigma_\alpha x_{i\alpha}$?

6. What scores are to be added in $\Sigma_{\alpha=3}^{5} x_{3\alpha}$? What is their sum?

7. What is the value of $\Sigma_{\alpha=2}^{4} x_{4\alpha}$?

Use an appropriate summation sign with indication of the limits of summation to express each of the following:

8. $x_{51} + x_{52} + x_{53} + x_{54} + x_{55} + x_{56}$.

9. $x_{31}^2 + x_{32}^2 + x_{33}^2 + x_{34}^2$.

10. $x_{1\alpha} + x_{2\alpha} + x_{3\alpha} + x_{4\alpha}$.

11. $\Sigma_{\alpha=1}^{4} x_{1\alpha} + \Sigma_{\alpha=1}^{7} x_{2\alpha} + \Sigma_{\alpha=1}^{2} x_{3\alpha}$.

12. $\Sigma_{\alpha=1}^{4} x_{1\alpha}^2 + \Sigma_{\alpha=1}^{4} x_{2\alpha}^2 + \Sigma_{\alpha=1}^{4} x_{3\alpha}^2$.

Write out in detail the sum denoted by these symbols:

13. $\Sigma_{\alpha=1}^{3} x_{4\alpha}$.

14. $\Sigma_{\alpha=1}^{4} x_{7\alpha}$.

15. $\Sigma_{i=1}^{2}(\Sigma_{\alpha=1}^{3} x_{i\alpha})$.

Study Table 17-3 to see if you can find the symbol which represents each of these expressions:

16. The entire number of cases in all three samples.

17. The mean of sample 3.

18. The sum of all the scores in all the samples.

19. The sum of the squares of the deviation of the scores in sample 2 from the mean of that sample.

Table 17-3 Symbols and Formulas for Computations on Three Independent Samples

Sample Number	Number of Cases	Sum of Scores	Sum of Squares of Scores	Mean	Sum of Squares of Deviations from Mean
1	n_1	$\sum_\alpha x_{1\alpha}$	$\sum_\alpha x_{1\alpha}^2$	$\bar{x}_1 = \dfrac{\sum_\alpha x_{1\alpha}}{n_1}$	$\sum_\alpha (x_{1\alpha} - \bar{x}_1)^2 = \sum_\alpha x_{1\alpha}^2 - \dfrac{\left(\sum_\alpha x_{1\alpha}\right)^2}{n_1}$
2	n_2	$\sum_\alpha x_{2\alpha}$	$\sum_\alpha x_{2\alpha}^2$	$\bar{x}_2 = \dfrac{\sum_\alpha x_{2\alpha}}{n_2}$	$\sum_\alpha (x_{2\alpha} - \bar{x}_2)^2 = \sum_\alpha x_{2\alpha}^2 - \dfrac{\left(\sum_\alpha x_{2\alpha}\right)^2}{n_2}$
3	n_3	$\sum_\alpha x_{3\alpha}$	$\sum_\alpha x_{3\alpha}^2$	$\bar{x}_3 = \dfrac{\sum_\alpha x_{3\alpha}}{n_3}$	$\sum_\alpha (x_{3\alpha} - \bar{x}_3)^2 = \sum_\alpha x_{3\alpha}^2 - \dfrac{\left(\sum_\alpha x_{3\alpha}\right)^2}{n_3}$
Combined samples	$n = \sum_i n_i$	$\sum_i \sum_\alpha x_{i\alpha}$	$\sum_i \sum_\alpha x_{i\alpha}^2$	$\bar{x} = \dfrac{\sum_i \sum_\alpha x_{i\alpha}}{n}$	$\sum_i \sum_\alpha (x_{i\alpha} - \bar{x})^2 = \sum_i \sum_\alpha x_{i\alpha}^2 - \dfrac{\left(\sum_i \sum_\alpha x_{i\alpha}\right)^2}{n}$
Within samples					$\sum_i \sum_\alpha (x_{i\alpha} - \bar{x}_i)^2 = \sum_i \sum_\alpha x_{i\alpha}^2 - \sum_i \dfrac{\left(\sum_\alpha x_{i\alpha}\right)^2}{n_i}$

20. The square of the sum of the scores in sample 1.

21. The sum of the squares of the scores in sample 3.

22. The square of the sum of all the scores in all the samples.

23. The sum of the squares of all the scores in all the samples.

Formulas for Computation. In Chapter 6 you met formula (6-14) $\Sigma(x - \bar{x})^2 = \Sigma x^2 - (\Sigma x)^2/n$, which could be put into words as "The sum of the squares of the deviations of scores from their mean is equal to the sum of the squares of those scores reduced by a fraction whose numerator is the square of the sum of those scores and whose denominator is n." Compare the formulas in the first 3 rows of the last column of Table 17-3 with formula (6-14), and note that they differ from it only because subscripts have been inserted. In the final row the same formula appears with two summation signs because $(x_{i\alpha} - \bar{x}_i)^2$ has been summed first for all the individuals in each sample and then $\Sigma(x_{i\alpha} - \bar{x}_i)^2$ has been summed for all samples. In Chapter 6 you had to distinguish $(\Sigma x)^2$, which is the square of the sum of the scores, from Σx^2, which is the sum of the squares of the scores; now you must distinguish $(\Sigma\Sigma x_{i\alpha})^2$ and $\Sigma\Sigma x_{i\alpha}^2$. If you were not previously able to answer Questions 19 to 23 in Exercise 17-1, probably you can do so after reading this paragraph.

It is important to note the difference between the lines headed "Combined samples" and "Within samples." In the former the scores of all samples are combined and treated as a single sample of $n = \Sigma n_i$ cases. In the latter the sums of the squared deviations for the separate samples are added.

In Table 17-4 the formulas of Table 17-3 have been applied to the first two samples shown in Table 17-2.

Table 17-4   Application of Formulas of Table 17-3 to the First Two Samples in Table 17-2

Sample	n	Sum of Scores	Sum of Squared Scores	Mean	Sum of Squared Deviations
1	5	112	3264	22.40	755.20
2	4	41	535	10.25	114.75
Combined Samples	9	153	3799	17	1198
Within Samples					869.95

EXERCISE 17-2

Apply the formulas of Table 17-3 to all four samples in Table 17-2, and show results arranged as in Table 17-4.

Partition of the Sum of Squares into Components. The sum of the squares of the deviations of all the scores from the means of all the scores, $\Sigma_i \Sigma_\alpha (x_{i\alpha} - \bar{x})^2$ can be partitioned into two components as shown in formula 17-2.

$$(17\text{-}2) \quad \Sigma_i \Sigma_\alpha (x_{i\alpha} - \bar{x})^2 = \{\Sigma_i n_i (\bar{x}_i - \bar{x})^2\} + \{\Sigma_i [\Sigma_\alpha (x_{i\alpha} - \bar{x}_i)^2]\}$$

Note that the first component relates to the variation of the means of different samples around the combined mean. It would be zero if all samples had exactly the same mean. The second component relates to the variation within samples. It would be zero if in each sample all the scores were exactly equal. Substituting the numerical data of Table 17-4 into formula (17-2) may clarify its meaning:

$$1198 = \{5(22.40 - 17)^2 + 4(10.25 - 17)^2\} + \{755.20 + 114.75\}$$
$$= \{145.8 + 182.25\} + \{869.95\}$$
$$= 328.05 + 869.95$$

The sum of the squares of the deviations of all the scores from the mean of all the scores

equals

the sum of the squares of the deviations of the sample means from the mean of all the scores, each square being multiplied by the number of cases in the sample

plus

the total of all the sums of the squares of deviations of scores from the mean of their own sample

---

or

> Total sum of squares
>
> equals
>
> sum of squares between means
>
> plus
>
> sum of squares within samples

These statements can be further condensed to read:

Total SS = SS for means + SS within groups.

As computing formulas we have

$$(17\text{-}3) \qquad \sum_i \sum_\alpha (x_{i\alpha} - \bar{x})^2 = \sum_i \sum_\alpha x_{i\alpha}{}^2 - \frac{\left(\sum_i \sum_\alpha x_{i\alpha}\right)^2}{n}$$

$$(17\text{-}4) \qquad \sum_i n_i(\bar{x}_i - \bar{x})^2 = \sum_i \frac{\left(\sum_\alpha x_{i\alpha}\right)^2}{n_i} - \frac{\left(\sum_i \sum_\alpha x_{i\alpha}\right)^2}{n}$$

$$(17\text{-}5) \qquad \sum_i \sum_\alpha (x_{i\alpha} - \bar{x}_i)^2 = \sum_i \sum_\alpha x_{i\alpha}{}^2 - \sum_i \frac{\left(\sum_\alpha x_{i\alpha}\right)^2}{n_i}$$

If the right-hand members of these formulas are substituted for the left-hand members in formula (17-2), formula (17-6) is obtained, and in it the identity is clear.

$$(17\text{-}6) \quad \sum_i \sum_\alpha x_{i\alpha}{}^2 - \frac{\left(\sum_i \sum_\alpha x_{i\alpha}\right)^2}{n} = \left\{ \sum_i \frac{\left(\sum_\alpha x_{i\alpha}\right)^2}{n_i} - \frac{\left(\sum_i \sum_\alpha x_{i\alpha}\right)^2}{n} \right\}$$

$$+ \left\{ \sum_i \sum_\alpha x_{i\alpha}{}^2 - \sum_i \frac{\left(\sum_\alpha x_{i\alpha}\right)^2}{n_i} \right\}$$

Degrees of Freedom. This phrase has come up several times in earlier chapters. An arbitrary statement made with no attempt at justification occurs on page 152 of Chapter 9 and on page 164 of Chapter 10 in relation to the t-test for means; another occurs on page 227 of Chapter 13 in relation to a t-test for r. On page 250 of Chapter 15 an intuitive explanation is given as to why the variance or the standard deviation is said to have $n - 1$ degrees of freedom and on page 276 of Chapter 16 a similar explanation is given for the degrees of freedom in a contingency table. These explanations are based on the idea of the number of values of a variable which under particular circumstances can be assigned arbitrarily.

An explanation which is consistent with the foregoing but rather more enlightening in the context of the analysis of variance is to say that in choosing a sample of n cases there are altogether n degrees of freedom of which one belongs to the mean and $n - 1$ to the variation of scores about that mean. Statisticians commonly speak of *partitioning* the n degrees of freedom and of *allocating* one degree of freedom to the mean and $n - 1$ to the variation around the mean. (It makes no difference whether variation is expressed in terms of the standard deviation, the variance, the sum of the squares of deviations usually called the sum of squares or SS, or the mean square called MS.)

We are now concerned with a situation involving k populations from each of which a sample of n_i cases is drawn, $n = \Sigma n_i$. In each sample there is 1 degree of freedom for the sample mean and $n_i - 1$ for variation around that mean. But the k means function like a sample of k cases varying around the grand mean $\bar{x}$, the mean of all n cases. These k cases then have 1 degree of freedom for the grand mean $\bar{x}$ and $k - 1$ for the variation of $\bar{x}_i$ around $\bar{x}$. Since each sample has $n_i - 1$ degrees of freedom for the variation of scores around its mean $\bar{x}_i$, the k samples have altogether $\Sigma_{i=1}^{k} (n_i - 1) = n - k$ degrees of freedom for variation within samples. If all n cases are combined and treated as one sample, there is 1 degree of freedom for the grand mean $\bar{x}$ and $n - 1$ for variation around $\bar{x}$.

Formulas (17-2) and (17-6) showed a partitioning of the sum of squares $\Sigma_i \Sigma_\alpha (x_{i\alpha} - \bar{x})^2$ into two components. Now we can partition its degrees of freedom $n - 1$ into two corresponding parts:

For total sum of squares . . . $n - 1$ degrees of freedom
For sum of squares between means . . . $k - 1$ degrees of freedom
For sum of squares within groups . . . $\Sigma(n_i - 1) = n - k$ degrees of freedom

Mean Square.   The variance defined by formula (6-4) on page 77 has the number of degrees of freedom as its denominator; so has the conditional standard deviation defined in formula (12-5) on page 203.

Similarly we shall now divide each of the two component sums of squares (SS) by its degrees of freedom to obtain the mean square (MS), as shown in Table 17-5. For ease of reference these mean squares will also be set down as formulas (17-7), and (17-8) and (17-9).

Table 17-5 Mean Square between Means and Mean Square within Samples

Formula for SS	Common Descriptive Terms	d.f.	Mean Square MS
$\displaystyle\sum_{i=1}^{k} n_i(\bar{x}_i - \bar{x})^2$ or $\displaystyle\sum_{i=1}^{k} \frac{\left(\sum_{\alpha} x_{i\alpha}\right)^2}{n_i} - \frac{\left(\sum_{i}\sum_{\alpha} x_{i\alpha}\right)^2}{n}$	SS for means or SS between groups or Treatment sum of squares or SS_B	$k - 1$	$\dfrac{\sum_{i} n_i(\bar{x}_i - \bar{x})^2}{k - 1}$
$\displaystyle\sum_{i}\sum_{\alpha} (x_{i\alpha} - \bar{x}_i)^2$ or $\displaystyle\sum_{i}\sum_{\alpha} x_{i\alpha}^2 - \sum_{i} \frac{\left(\sum_{\alpha} x_{i\alpha}\right)^2}{n_i}$	SS within samples or SS within groups or SS among individuals within groups or Error SS or SS_W	$n - k$	$\dfrac{\sum_{i}\sum_{\alpha} (x_{i\alpha} - \bar{x}_i)^2}{n - k}$
$\displaystyle\sum_{i}\sum_{\alpha} (x_{i\alpha} - \bar{x})^2$ or $\displaystyle\sum_{i}\sum_{\alpha} x_{i\alpha}^2 - \frac{\left(\sum_{i}\sum_{\alpha} x_{i\alpha}\right)^2}{n}$	Total SS	$n - 1$	

The mean square between means is

$$(17\text{-}7) \qquad \mathrm{MS}_B = \frac{\sum_{i} n_i(\bar{x}_i - \bar{x})^2}{k - 1} = \frac{\sum_{i} \dfrac{\left(\sum_{\alpha} x_{i\alpha}\right)^2}{n_i} - \dfrac{\left(\sum_{i}\sum_{\alpha} x_{i\alpha}\right)^2}{n}}{k - 1}$$

The mean square within samples is

$$(17\text{-}8) \quad \mathrm{MS}_W = \frac{\sum_i \left\{ \sum_\alpha (x_{i\alpha} - \bar{x}_i)^2 \right\}}{n - k} = \frac{\sum_i \sum_\alpha x_{i\alpha}^2 - \sum_i \dfrac{\left(\sum_\alpha x_{i\alpha} \right)^2}{n_i}}{n - k}$$

Also formula (17-9) is convenient to use:

$$(17\text{-}9) \quad \mathrm{MS}_B = \frac{\text{Total SS} - \text{SS}_W}{k - 1}$$

It can be proved by mathematics (it is too advanced to show here) that when the hypothesis $\mu_1 = \mu_2 \cdots \mu_k = \mu$ is true, each of these mean squares is an *unbiased estimate* of the common variance σ^2. An unbiased estimate was defined on page 146. When this hypothesis is not true, the mean square within samples is still an unbiased estimate of the variance σ^2, but the mean square between means tends to be larger than σ^2. This is the reason for using the term "mean square" rather than "variance."

Independence of the Mean Squares. If the assumptions stated in the mathematical model hold, then the two mean squares are independent, even though computed from the same data. Imagine that after computing the MS_W and the MS_B for one set of k samples, you draw a second set of k samples of the same size, and a third, and so on, each time computing the mean squares, until you have, let us say, 100 pairs of mean squares. Then you compute the correlation coefficient between these paired mean squares. That correlation coefficient would be very close to zero. This is the meaning of the statement that the mean squares are independent.

If the population distributions are so skewed that μ and σ^2 are almost equal, the mean squares will be correlated, and the F test about to be described will be inappropriate without some special adjustment outside the scope of this book.

It would also be possible to separate the total SS into two pieces in some other arbitrary fashion, but if the two pieces were not independent they could not be used to test the hypothesis of equal population means.

Test of Hypothesis. The availability of two independent estimates of σ^2 reminds us of formula (15-9) on page 260, in which the ratio of two variances from independent samples was used to test the hypothesis that $\sigma_1^2 = \sigma_2^2$. Accordingly we shall form the ratio

$$(17\text{-}10) \quad F = \frac{\mathrm{MS}_B}{\mathrm{MS}_W} = \frac{\mathrm{SS}_B}{\mathrm{SS}_W} \cdot \frac{n - k}{k - 1}$$

In this ratio the MS_B is always placed in the numerator even if it happens to be smaller than MS_W.

If the hypothesis is true that population means are equal, then MS_B and MS_W should be about the same size, and F should be near unity, sometimes larger, sometimes smaller. Differences among population means tend to make MS_B larger than MS_W; consequently a large value of F tends to put suspicion on the null hypothesis and may, if large enough, lead to its rejection.

As in Chapter 15, the computed value of F is referred to Appendix Table V. However, in the problems now under consideration we reject the null hypothesis only when F is significantly large, hence we have a one-tail test with region of significance in the upper tail only, that is, we reject when $F > F_{1-\alpha}$.

Summary of Computations.   As an illustration in miniature, the data in the first two samples of Table 17-2 may be used. To compute F it is necessary to obtain MS_B and MS_W and to obtain these the various sums, and sums of squares found in formulas (17-7) and (17-8) are needed. Actually these can be obtained from Table 17-4.

$$MS_W = \frac{869.95}{9-2} = 124.28$$

$$\text{Total SS} = 1198.0$$

$$MS_B = \frac{1198.0 - 869.95}{2-1} = 328.05$$

$$F = \frac{328.05}{124.28} = 2.64$$

If Table 17-4 had not already been developed, the following computations might have been made:

$$\sum_i \sum_\alpha x_{i\alpha} = 153$$

$$\left(\sum_i \sum_\alpha x_{i\alpha}\right)^2 / n = \frac{(153)^2}{9} = 2601$$

$$\sum_i \sum_\alpha x_{i\alpha}{}^2 = 3799$$

$$\text{Total SS} = 3799 - 2601 = 1198$$

$$SS_W = 3799 - \frac{112^2}{5} - \frac{41^2}{4} = 869.95$$

$$SS_B = 1198.0 - 869.95 = 328.05$$

Also

$$SS_B = 5(22.40 - 17)^2 + 4(10.25 - 17)^2 = 328.05$$

or
$$SS_B = \frac{112^2}{5} + \frac{41^2}{4} - 2601 = 328.05$$

These results are customarily presented in the format of Table 17-6. Since $F < F_{.95}$, there is no reason to question the hypothesis that $\mu_1 = \mu_2$. When the research worker dealing with actual data reaches the point of deciding whether to accept or reject the hypothesis tested, he should express his final interpretation in terms of the concrete problem under consideration.

Table 17-6 Analysis of Variance of Data of Samples 1 and 2 in Table 17-2

Source of Variation	SS	d.f.	MS	F	$F_{.95}$
Between means	328.05	1	328.05	2.64	5.59
Within samples	869.95	7	124.28		
Total	1198.0	8			

Computations When Original Scores Are Not Available. Sometimes one wishes to compute F from published data for which the original scores are not available. It would then be impossible to use formulas (17-8) and (17-7). If the mean and standard deviation are given for each sample and the number of cases, the formulas can be adapted as follows:

(17-11) $$\text{Total sum of scores} = \sum_i n_i \bar{x}_i$$

(17-12) $$\text{SS within groups} = \sum_i (n_i - 1)s_i^2$$

(17-13) $$\text{SS between means} = \sum_i n_i \bar{x}_i^2 - \frac{\left(\sum_i n_i \bar{x}_i\right)^2}{\sum_i n_i}$$

A difficulty often encountered in trying to use published data in this way is that the original computer may not have retained enough digits to have any significant digits in the SS between means. Another difficulty is that sometimes one cannot be sure whether n or $n - 1$ was used as denominator for the variance. Formula (17-12) is based on the assumption that $n - 1$ was used.

EXERCISE 17-3

1. For the four samples of Table 17-2 make the necessary computations, and present them in a table with the format of Table 17-6. What hypothesis is being tested? What is your conclusion about the hypothesis if you have set .05 as the value of α?

2. The three sections of a statistics class whose midterm scores had the distributions shown in Table 5-3 on page 55, may be thought of as samples from these populations:

   Section I: Students who anticipate no difficulties with statistics either because they like mathematics or have high scores on a placement test and who enroll in a class meeting once a week for 2 hours taught by Teacher A.

   Section II: Students not clearly qualifying for either Section I or Section III who enroll in a class meeting twice a week for $1\frac{1}{2}$ hours each time taught by Teacher B.

   Section III: Students who anticipate real difficulties with statistics because they had difficulties with mathematics or have low scores on a placement test and who enroll in a class meeting twice a week for two hours each time with extra sessions as needed taught by Teacher C.

   The adjustment in the amount of class time and in speed of instruction was an attempt to overcome poor preparation by special teaching effort, an attempt to make $\bar{x}_1 = \bar{x}_2 = \bar{x}_3$ in spite of initial disadvantages on the part of some students. The fact that three different teachers taught the three sections makes it impossible to make a clear interpretation of the results, but for the present purposes we shall disregard that factor. (This was designed as a teaching plan, not as a scientific experiment.)

   Table 6-3 on page 90 gives the derived data which can be used in formulas (17-12) and (17-13). Complete the analysis, and draw up a table with the format of Table 17-6, and write a statement interpreting the results.

3. In an experiment designed to find out if one method of instruction secures better results than another, pupils are assigned at random to three classes: C, a control group; E_1, a group using experimental method 1; E_2, a group using experimental method 2. (The reader is free to define C, E_1, and E_2 in terms of his particular field of interest. In general it would be better to make $n_1 = n_2 = n_C$, but here we have used unequal numbers in order to give the reader the experience of working with such.) A criterion test measure of

progress (again the reader may define such in his particular situation) shows results as follows:

Population Sampled	n	Sum of Scores	Sum of Squared Scores
E_1	28	1519	84,759
E_2	33	1610	81,966
C	27	941	37,139

Complete the analysis of variance.

All Comparisons among Means. When an analysis of variance has been completed and the F ratio has been found significant so the null hypothesis that all k population means are equal is rejected, important questions are still unanswered. What is the nature of the differences among the means? Is one population mean different from all the others? Are there several pairs of populations for which $\mu_i - \mu_j$ is not zero?

Recognizing these questions for the first time, some persons are tempted to make a t test comparing every mean with every other mean. This seemingly reasonable approach is actually wrong because, if carried out with α as significance level, the probability of false rejection of the hypothesis $\mu_i - \mu_j = 0$ is greater than α. When k is the number of populations, there are only $k - 1$ degrees of freedom among the means, but the number of pairwise comparisons among the means is $k(k - 1)/2$.

k	$k - 1$	$\dfrac{k(k - 1)}{2}$
2	1	1
3	2	3
4	3	6
8	7	28
12	11	66

To see why it would be misleading to perform $k(k - 1)/2$ t tests, let us consider what would happen if $k = 12$ and the null hypothesis $\mu_1 = \mu_2 = \cdots = \mu_{12} = \mu$ is actually true. If we set $\alpha = .05$, there is probability .05 of obtaining an F which would cause us falsely to reject that hypothesis. But since $.05(66) = 3.3$, we should expect about 3 of the 66 pairs of sample means by chance to produce a value of $t > t_{.95}$. Since all the population means are really equal, it is a matter of chance which pairs these would be, but the data would seem to select a few pairs and lead us to the incorrect conclusion that for them $\mu_i \neq \mu_j$. This incorrect procedure is sometimes spoken of as multiple t tests.

Obviously a method is needed for answering the questions suggested in the first paragraph of this section and several have been proposed. The one presented here was proposed by Scheffé.[1] It has the particular merit that it can be used when the numbers of cases in the samples are unequal, in contrast to other methods which require equality of sample size.

The Scheffé method leads to a confidence interval for a contrast among population means. Consider first the simplest contrast which is a difference between two means, $\mu_i - \mu_j$. Thus if $k = 3$, there are 3 contrasts of this simple type: $\mu_1 - \mu_2$, $\mu_1 - \mu_3$, and $\mu_2 - \mu_3$.

Making an interval estimate with confidence coefficient c for all differences among k means or for any particular differences requires knowledge of

$n_1, n_2, n_3 \cdot \cdot \cdot n_k$

$\bar{x}_1, \bar{x}_2, \bar{x}_3 \cdot \cdot \cdot \bar{x}_k$

MS_W or the error mean square obtained from the prior analysis of variance

F_c, read from Appendix Table V as the cth percentile for $k - 1$ and $n - 1$ degrees of freedom

Then the confidence limits for the contrast $\mu_i - \mu_j$ are

$$(17\text{-}14) \quad \text{Upper limit: } (\bar{x}_i - \bar{x}_j) + \sqrt{(k - 1)F_c(\text{MS}_W)\left(\frac{1}{n_i} + \frac{1}{n_j}\right)}$$

$$\text{Lower limit: } (\bar{x}_i - \bar{x}_j) - \sqrt{(k - 1)F_c(\text{MS}_W)\left(\frac{1}{n_i} + \frac{1}{n_j}\right)}$$

It is interesting to note that if $t^2_{(1+c)/2}$ were substituted for $(k - 1)F_c$, the resulting formulas would describe the limits estimated by the incorrect multiple t-test procedure. This fact illustrates how that procedure would lead to deceptive conclusions.

In applying these formulas, if one limit is found to be positive and the other negative, 0 is a possible value of $\mu_i - \mu_j$, and the observed difference $\bar{x}_i - \bar{x}_j$ is nonsignificant. If both are positive, the hypothesis is rejected, and it is conceded that $\mu_i > \mu_j$; if both are negative the hypothesis is rejected, and it is conceded that $\mu_i < \mu_j$.

Now returning to the data of Question 3 in Exercise 17-3, we have

$n_1 = 28 \qquad \bar{x}_1 = 54.25 \qquad \text{MS}_W \doteq 118.99$

$n_2 = 33 \qquad \bar{x}_2 = 48.79 \qquad F_{.95}(2,85)$ is not shown

$n_c = 27 \qquad \bar{x}_c = 34.85 \qquad$ in the table, but 3.10 is

$\qquad\qquad\qquad\qquad\qquad\qquad\qquad$ a close approximation

[1] Henry Scheffé, *The Analysis of Variance*, John Wiley & Sons, 1959, pp. 66–72.

Hence for $\mu_1 - \mu_2$ the lower limit is

$$54.25 - 48.79 - \sqrt{2(3.10)(118.99)(\tfrac{1}{28} + \tfrac{1}{33})} = 5.46 - 6.98 = -1.52$$

and the upper limit is $5.46 + 6.98 = 12.44$.

The limits for the other two contrasts are computed in similar manner, and for all three we have

$$-1.52 < \mu_1 - \mu_2 < 12.44$$
$$12.07 < \mu_1 - \mu_c < 26.73$$
$$6.89 < \mu_2 - \mu_c < 20.99$$

It must be concluded that the sample means of the two experimental groups do not differ significantly from each other but that each of them is significantly higher than the mean of the control group.

More complex comparisons can be made in similar manner. For example, it might seem desirable to compare the control group with the combination of the two experimental groups, the contrast being $\mu_1 + \mu_2 - 2\mu_c$. Here the weights assigned to the three means are 1, 1, and -2. It is a requirement of the method that the sum of the weights shall be 0, and $1 + 1 - 2 = 0$. Then the lower limit would be

$$\bar{x}_1 + \bar{x}_2 - 2\bar{x}_3 - \sqrt{(k-1)F_c(\mathrm{MS}_W)\left(\frac{1}{n_1} + \frac{1}{n_2} + \frac{4}{n_c}\right)} = 33.34 - 12.57$$
$$= 20.77$$

and the upper limit would be

$$33.34 + 12.57 = 45.91$$

or

$$20.77 < \mu_1 + \mu_2 - 2\mu_c < 45.91$$

We conclude that the mean score for the experimental groups considered jointly exceeds the mean score for the control group.

The method may be generalized to k populations with weights w_1, w_2, $\cdots$ w_k, with the requirement that for any selected contrast $\Sigma w = 0$ and that not every w shall be 0. Thus

$3\mu_1 - 2\mu_4 - \mu_5$ is a contrast because $3 - 2 - 1 = 0$

$\mu_1 + \mu_3 - \mu_5 - \mu_6$ is a contrast because $1 + 1 - 1 - 1 = 0$

$4\mu_2 - \mu_3 - 2\mu_4 - \mu_6$ is a contrast because $4 - 1 - 2 - 1 = 0$

But $3\mu_4 - \mu_1 - \mu_5$ is not a contrast because $3 - 1 - 1 \neq 0$

In each of the foregoing it is understood that any μ_i not named in the contrast has a weight $w_i = 0$.

When a contrast is thus defined, its confidence limits are

$$(17\text{-}15) \qquad \sum_{i=1}^{k} w_i \bar{x}_i \pm \sqrt{(k-1)F_c(\text{MS}_W) \sum_{i-1}^{k} \frac{w_i^2}{n_i}}$$

where F has $k-1$ and $n-k$ degrees of freedom as before.

Analysis of Variance from Ranks.   Sometimes it is impossible to give subjects a numerical score on a variable X and yet possible to place them in a meaningful order on that variable. The question may then arise as to whether several populations have exactly the same distribution on X. The method of analysis of variance previously discussed cannot be used because there are no scores. Sometimes it seems obvious that the mathematical model described on page 289 is not appropriate either because the sample variances are too disparate or because the distributions are too far from the normal or both. In either of these situations a method devised by Kruskal and Wallis[2] is useful.

The scores in the several samples are first thrown into one combined distribution and converted to ranks, as illustrated in the right-hand section of Table 17-7. As there are $n = \Sigma n_i$ cases in the combined distribution, the mean of all n ranks is $(n+1)/2$ and variance $(n^2 - 1)/12$. (These formulas are familiar to mathematicians. The nonmathematical reader can easily convince himself of their correctness by choosing some small value of n, setting down the digits 1, 2, 3, . . . n, and computing their mean and variance.) Because of the difficulty of ranking any large number of individuals, the method cannot easily be used if n is large. For each sample the sum of the ranks assigned to its subjects is computed and denoted R_i.

If the samples actually come from populations with the same distribution, the mean ranks in those samples should differ only by chance. This argument suggests the use of a statistic similar to that of formula (15-1) on page 249, $\chi^2 = \Sigma(x - \bar{x})^2/\sigma^2$, with mean rank R_i/n_i taking the place of x, $(n+1)/2$ taking the place of $\bar{x}$, and $(n^2 - 1)/12$ taking the place of σ^2. Aside from a factor $(n-1)/n$, which is negligible unless n is very small, too small to make any analysis worthwhile, this substitution produces the statistic

$$(17\text{-}16) \qquad H = \frac{12}{n(n+1)} \sum \frac{R_i^2}{n_i} - 3(n+1)$$

[2] Kruskal, Wm. H. and Wallis, W. A., "Use of Ranks in One-Criterion Variance Analysis," *Journal of the American Statistical Association*, 47 (1952), 583–621.

Table 17-7   Illustration of Computation for Analysis of Variance from Ranks

Sample Data in Score Form			Transformation to Ranks					
A	B	C	Score	Rank	Group	Score	Rank	Group
59	40	11	60	1	B	19	13	C
0	28	7	59	2	A	17	14	B
42	60	8	42	3	A	16	15	C
12	38	29	40	4	B	15	16	C
37	23	27	38	5	B	12	17	A
9	17	19	37	6	A	11	18	C
3	24	16	30	7	C	9	19	A
4		15	29	8	C	8	20	C
		30	28	9	B	7	21	C
n_i 8	7	9	27	10	C	4	22	A
R_i 116	56	128	24	11	B	3	23	A
$\dfrac{R_i{}^2}{n_i}$ 1682	448	1820.44	23	12	B	0	24	A

$$\Sigma R_i = 300 = \frac{24(25)}{2} \qquad \sum \frac{R_i{}^2}{n_i} = 3950.44$$

$$n = \Sigma n_i = 24$$

$$H = \frac{12}{24(25)} (3950.44) - 3(25) = 4.01$$

$$4.01 < \chi_{.90}{}^2 = 4.6$$

which has approximately the χ^2 distribution with $k - 1$ degrees of freedom, k being as usual the number of populations compared.

Often several subjects have the same score, causing ties in the ranking. Then each of the tied ranks is assigned the mean score for the tie, as in Table 17-8. For each of the ties $(t - 1)t(t + 1)$ is computed, where t is the number of tied scores. See Table 17-8. Then the correction for ties is

(17-17)
$$C = 1 - \frac{\Sigma(t - 1)t(t + 1)}{n(n^2 - 1)}$$

and the previously computed H is divided by C to obtain a corrected value H_c.

Table 17-8 Illustration of the Computation for Analysis of Variance from Ranks with Ties (Data from Table 17-2 on Page 291)

Score	f	Rank		Group			$(t-1)t(t+1)$
46	1	1	A				0
25	2	2.5			C, C		6
21	3	5	A			D, D	24
19	3	8	A	B		D	24
17	1	10			C		0
15	2	11.5	A		C		6
11	2	13.5	A			D	6
10	4	16.5		B	C,C	D	60
7	1	19		B			0
5	1	20		B			0
							$\overline{126}$

	R_i	n_i	$R_i{}^2/n_i$
A	39	5	304.2
B	63.5	4	1008.06
C	59.5	6	590.04
D	$\underline{48}$	$\underline{5}$	$\underline{460.8}$
	210	20	2363.10

$$H = \frac{12}{20(21)}(2363.1) - 3(21)$$

$$= 4.517$$

Correction for ties:

$$1 - \frac{126}{20(399)} = .9842$$

Check: $\dfrac{20(21)}{2} = 210$

$$H_c = \frac{4.517}{.9842} = 4.59$$

$$4.59 < \chi_{.90}{}^2$$

18

MULTIPLE REGRESSION AND CORRELATION

In Chapter 12 it was demonstrated how knowledge of the scores on one variable helps to provide information about scores on another variable. By the use of a regression formula and the related standard error of estimate, a frequency distribution is provided for the dependent or criterion variable for each score on the independent or predictor variable. This distribution is called a *conditional distribution* because it is conditioned on the particular value of the independent variable to which it is related.

To make this discussion more concrete, it will be helpful to turn to the illustration relating to the course in statistics which was used in Chapter 12. In the prediction of a midterm score y from a prognostic test score in arithmetic x the following were obtained by calculation: the regression formula of y on x given by the expression $y_x = 31.62 + .664x$ and the standard error of estimate which has the value 7.98. (See page 202.) For a particular value of x, say $x = 20$, the conditional distribution has a mean provided by the regression formula which is $y_{20} = 31.62 + .664(20) = 44.90$ and standard deviation = 7.98.

It may be expected that additional precision in estimation of values on the dependent variable can be obtained by using more than one independent variable. For example, in the study of the data in the course of statistics appearing in Appendix Table VIII, one might use not only arithmetic test scores but also scores on the test of artificial language to predict scores on the midterm test.

In this chapter, the use of two predictors will be studied in detail. Indications of the way in which problems with more than two predictors are dealt with will also be provided.

The Multiple Regression Model. This model will be expressed in population terms corresponding to the model in Chapter 12. Consequently, symbols in the expression to be described here will appear as parameters. Sample estimates of the parameters will be developed in a later section.

Let the *predictor* or *independent* variables be denoted as X_1 and X_2 and the *criterion* or *dependent* variable as Y. Then in parametric form the regression formula may be written as

(18-1) $$\mu_{y.12} = a + b_1 x_1 + b_2 x_2$$

The expression at the left of the equality sign indicates that the formula represents a mean of a distribution. The term $\mu_{y.12}$ is a mean of Y scores which depends on the choice of values of the X_1 and X_2 variables.

The expression to the right of the sign of equality indicates the form of relationship between this mean and the predictors. In this expression a, b_1, and b_2 are constant for all values of X_1 and X_2. The expression is called linear because the variables X_1 and X_2 appear only in the first degree. Geometrically, the expression in formula (18-1) represents a plane in the three dimensional space with coordinates Y, X_1, and X_2.

There is also a standard deviation of Y scores about the regression value $\mu_{y.12}$, and that standard deviation may be denoted as $\sigma_{y.12}$. Although $\sigma_{y.12}$ may depend on the particular choice of values of X_1 and X_2, it is customary to assume that it is constant for all values of X_1 and X_2.

The mean and standard deviation jointly specify a frequency distribution in Y which is conditional on the choice of values of X_1 and X_2. There is usually a different conditional distribution for each choice of a pair of values of X_1 and X_2, and these distributions have the same mean only if b_1 and b_2 are both zero.

One should distinguish between σ_y and $\sigma_{y.12}$ since the former relates to the distribution of all Y scores taken together. It is important to note that $\sigma_{y.12}$ cannot exceed σ_y and is commonly smaller.

For convenience the assumption is often made that the Y scores corresponding to each conditional distribution are distributed normally. This assumption is important in developing tests of significance.

Formula (18-1) can, of course, be expanded to any number of predictors. Thus for three predictors, the formula is

$$\mu_{y.123} = a + b_1 x_1 + b_2 x_2 + b_3 x_3$$

Expressions for Regression Estimates from Data. The next problem to be considered is finding expressions for sample estimates of the regression coefficients. In terms of statistics, the following expression for the regression formula will be adopted:

(18-2) $$y_{12} = \hat{a} + \hat{b}_1 x_1 + \hat{b}_2 x_2$$

where x_1 and x_2 are specific values of the variables X_1 and X_2; $\hat{a}$, $\hat{b}_1$, and $\hat{b}_2$ (called a circumflex, b one circumflex, and so forth) are numerical constants calculated from data; and y_{12} is a sample estimate of $\mu_{y.12}$. The method of calculating these constants is described in a later section.

The reader will note that the population values or parameters in formula (18-1) are written in Latin rather than in Greek letters as parameters are customarily written in other parts of this book. The reason is that the Greek letters β_1, β_2, and so on are used commonly to describe another formula for the regression equation to be discussed in a later section. The coefficients in the two formulas are distinguished in common usage as the b and beta coefficients.

The conditional standard deviation which is an estimate of $\sigma_{y.12}$ will be denoted as $s_{y.12}$. It is also called the standard error of estimate.

Sums, Sums of Squares, and Sums of Crossproducts. Prior to the actual calculation of $\hat{a}$, $\hat{b}_1$, and $\hat{b}_2$ it is necessary to set up expressions which combine the scores into sums, sums of squares for the variables separately, and into crossproducts for pairs of variables.

The sums are written in the familiar fashion as Σy, Σx_1, and Σx_2. The sums of squares and crossproducts are written in deviation form as shown in Table 18-1.

Table 18-1 Sums of Squares and Crossproducts for Three Variables

Symbolic Expression	In Terms of Deviations from Means	As Used for Calculation
S_{yy}	$\Sigma(y - \bar{y})^2$	$\Sigma y^2 - (\Sigma y)^2/n$
S_{11}	$\Sigma(x_1 - \bar{x}_1)^2$	$\Sigma x_1{}^2 - (\Sigma x_1)^2/n$
S_{22}	$\Sigma(x_2 - \bar{x}_2)^2$	$\Sigma x_2{}^2 - (\Sigma x_2)^2/n$
S_{y1}	$\Sigma(x_1 - \bar{x}_1)(y - \bar{y})$	$\Sigma x_1 y - (\Sigma x_1)(\Sigma y)/n$
S_{y2}	$\Sigma(x_2 - \bar{x}_2)(y - \bar{y})$	$\Sigma x_2 y - (\Sigma x_2)(\Sigma y)/n$
S_{12}	$\Sigma(x_1 - \bar{x}_1)(x_2 - \bar{x}_2)$	$\Sigma x_1 x_2 - (\Sigma x_1)(\Sigma x_2)/n$

The Normal Equations. The actual estimates of the coefficients in the regression equation can now be obtained as solutions of a set of simultaneous equations known as the normal equations. The reader should notice that the use of the word normal here is unrelated to the normal distribution.

The normal equations expressed in terms of the symbols of Table 18-1 have the following form:

$$
\begin{aligned}
S_{11}\hat{b}_1 + S_{12}\hat{b}_2 &= S_{y1} \\
S_{12}\hat{b}_1 + S_{22}\hat{b}_2 &= S_{y2}
\end{aligned}
$$

(18-3)

These equations can be solved directly for $\hat{b}_1$ and $\hat{b}_2$, and when there are three or more predictors the solution is obtained in this way. However, when there are only two predictors, expressions for the $\hat{b}$'s can be written readily as

(18-4)
$$
\begin{aligned}
\hat{b}_1 &= \frac{S_{y1}S_{22} - S_{y2}S_{12}}{S_{11}S_{22} - S_{12}{}^2} \\[2mm]
\hat{b}_2 &= \frac{S_{y2}S_{11} - S_{y1}S_{12}}{S_{11}S_{22} - S_{12}{}^2}
\end{aligned}
$$

The constant $\hat{a}$ of the regression formula can be written as

(18-5)
$$
\hat{a} = \bar{y} - \hat{b}_1\bar{x}_1 - \hat{b}_2\bar{x}_2
$$

Estimate of Conditional Variance. The sample estimate of the conditional variance will be denoted $s_{y.12}{}^2$ and defined as

(18-6)
$$
s_{y.12}{}^2 = \frac{\Sigma(y - y_{12})^2}{n - 3}
$$

This is analogous to formula (12-5) on page 203. In the formula $s_y{}^2 = \Sigma(y - \bar{y})^2/(n - 1)$, the denominator takes account of the fact that the n deviations from $\bar{y}$ have $n - 1$ degrees of freedom. In the formula $s_{y.x}{}^2 = \Sigma(y - y_x)^2/(n - 2)$, the denominator takes account of the fact that the n deviations from a regression line have $n - 2$ degrees of freedom. Formula (18-6) relates to a situation in which the n deviations from a regression plane have $n - 3$ degrees of freedom. If instead of 2 predictor variables we were using k predictors, the denominator would be $n - k - 1$.

Formula (18-6) does not provide a convenient routine for computation, but by algebraic manipulation it can be transformed into an equivalent formula which is computationally easier to use:

(18-7)
$$
s_{y.12}{}^2 = \frac{S_{yy} - \hat{b}_1 S_{y1} - \hat{b}_2 S_{y2}}{n - 3}
$$

The Coefficient of Multiple Correlation. The correlation between actual values of y and the values of y_{12} obtained from a regression equation involving multiple predictor variables is known as the *coefficient of*

multiple correlation. For two predictor variables the population value is denoted as $\rho_{y.12}$ and the sample estimate as $r_{y.12}$.

A second definition, analogous to formula (13-1a) on page 213 is

$$(18\text{-}8) \qquad r_{y.12}{}^2 = 1 - \frac{\Sigma(y - y_{12})^2}{\Sigma(y - \bar{y})^2}$$

As in the case of one predictor, when estimation is good y_{12} tends to approximate y closely, the numerator of the fraction in the formula is small and $r_{y.12}{}^2$ is close to 1. When estimation is poor, the numerator is not much smaller than the denominator and $r_{y.12}{}^2$ is close to $1 - 1$ or 0.

For the determination of $r_{y.12}$, only the positive square root of the expression for $r_{y.12}{}^2$ is used. No meaning can be given to the negative square root of this expression, contrary to the situation when there is one predictor.

The difference may be explained in this way. When there is one predictor, a positive correlation implies that a positive difference between scores of two individuals on the independent variable is accompanied by a positive difference on the dependent variable, and a negative correlation means that the positive difference in one variable is accompanied by a negative difference in the other.

In the case of two predictors, two individuals A and B may differ in a variety of ways in their scores on the independent variables. Thus A may exceed B on both variables, or A may exceed B on one variable but not on the other, or A may be lower than B on both variables. For one dependent variable the difference between A and B is, of course, uniquely determined. A negative value for $r_{y.12}(= - \sqrt{r_{y.12}{}^2})$ does not indicate the nature of the relationship between the dependent variable and either or both of the independent variables. The positive value of $r_{y.12}$ is a measure of the degree of approximation of y_{12} to y as r_{xy} is a measure of the approximation of y_x to y.

The two definitions of $r_{y.12}{}^2$ given above result in equivalent values, but neither provides a good computing procedure. By algebraic manipulation either definition yields the following computational formula

$$(18\text{-}9) \qquad r_{y.12}{}^2 = \frac{\hat{b}_1 S_{y1} + \hat{b}_2 S_{y2}}{S_{yy}}$$

Application of the Formulas to Data. The formulas developed in the preceding sections will now be illustrated by data from Appendix Table VIII. Specifically, it is proposed to predict scores on the final examination in the course from scores on the prognostic test in arithmetic and scores on the midterm test.

The following notation will be used for the variables:

$$Y = \text{Final examination}$$
$$X_1 = \text{Prognostic test in arithmetic}$$
$$X_2 = \text{Midterm test}$$

For these variables the sample means and standard deviations shown in Table 18-2 are obtained for the 98 cases in Table VIII.

Table 18-2   Sample Means and Standard Deviations for Y, X_1, X_2 (from Data of Appendix Table VIII)

Variable	Sample Mean	Sample Standard Deviation
Y	49.1531	9.4218
X_1	31.3776	7.2475
X_2	52.4592	9.2827

Using the symbolism of Table 18-1, the following values are obtained:

$$S_{yy} = 8610.703$$
$$S_{11} = 5095.033$$
$$S_{22} = 8358.328$$
$$S_{y1} = 3322.331$$
$$S_{y2} = 5679.108$$
$$S_{12} = 3382.008$$

The normal equations for calculating $\hat{b}_1$ and $\hat{b}_2$ now have the form

$$5095.033\hat{b}_1 + 3382.008\hat{b}_2 = 3322.331$$
$$3382.008\hat{b}_1 + 8358.328\hat{b}_2 = 5679.108$$

These can be solved directly as two simultaneous equations, or formulas (18-4) can be used, yielding

$$\hat{b}_1 = \frac{(3322.331)(8358.328) - (5679.108)(3382.008)}{(5095.033)(8358.328) - (3382.008)^2} = .2749$$

$$\hat{b}_2 = \frac{(5679.108)(5095.033) - (3322.331)(3382.008)}{(5095.033)(8358.328) - (3382.008)^2} = .5682$$

$$\hat{a} = 49.1531 - .2749(31.3776) - .5682(52.4592) = 10.7201$$

The regression formula for estimation of a y score can now be written as

$$y_{12} = 10.7201 + .2749x_1 + .5682x_2$$

The conditional standard deviation for Y can be estimated by formula (18-7) as

$$s_{y.12} = \sqrt{\frac{8610.703 - .2749(3322.331) - (.5682)(5679.108)}{98 - 3}} = 6.86$$

The gain in precision of estimate which can be achieved through the use of regression is an important concept worthy of some examination. We shall look at it in terms of the sum of the squares of the deviations from prediction. The predicted variable here is final examination score.

If no regression equation is used, the sum of squares is $\Sigma(y - \bar{y})^2 = S_{yy} = 8610.7$, usually called the total SS.

If a regression equation is used with prognostic arithmetic test (X_1) as the single predictor variable, the sum of squares about regression is

$$\Sigma(y - y_1)^2 = S_{yy} - \frac{S_{y1}^2}{S_{11}} = 6444.3$$

If a regression equation is used with midterm score (X_2) as the single predictor variable, the sum of squares about regression is

$$\Sigma(y - y_2)^2 = S_{yy} - \frac{S_{y2}^2}{S_{22}} = 4752.0$$

If a regression equation is used with both midterm score and arithmetic score as predictor variables, the sum of squares about regression is

$$\Sigma(y - y_{12})^2 = S_{yy} - \hat{b}_1 S_{y1} - \hat{b}_2 S_{y2} = 4470.5$$

The ratio of any one of these sums of squares about regression to the total sum of squares shows what proportion of the original variation is independent of variation in the predictors. Thus

$$\frac{\Sigma(y - y_{12})^2}{\Sigma(y - \bar{y})^2} = \frac{4470.5}{8610.7} = .5192$$

and

$$1 - \frac{\Sigma(y - y_{12})^2}{\Sigma(y - \bar{y})^2} = .4808 = r_{y.12}^2$$

This statement may be interpreted to mean that 48 percent of the total variation in Y is explained by the regression on X_1 and X_2, although 52 percent of this variation is not explained by such regression. The residual 52 percent is explained by the effect on Y scores of variables other than X_1 and X_2.

The statistic $r_{y.12}^2$, called the *coefficient of determination*, can also be calculated by formula (18-9) as

$$r_{y.12}^2 = \frac{.2749(3322) + .5682(5679)}{8610.7} = .4808$$

The ratio of each sum of squares about regression to the total sum of squares is displayed in Table 18-3. From this table one can see that if an estimate of final examination scores is needed early in the term, before midterm scores are available, a regression equation with the arithmetic test as predictor would be useful. However, after midterm scores are available, the use of an arithmetic score in addition to a midterm score would not reduce the residual sum of squares very much and so would not increase the multiple correlation greatly.

Table 18-3 Coefficients of Determination and Correlation in the Prediction of Final Examination Scores

Predictor	Ratio of Sums of Squares	Coefficient of Determination	Coefficient of Correlation
Arithmetic, x_1	$\dfrac{\Sigma(y - y_1)^2}{\Sigma(y - \bar{y})^2} = \dfrac{6444.3}{8610.7} = .7484$	$r_{y.1}^2 = .2516$	$r_{y.1} = .502$
Midterm test, x_2	$\dfrac{\Sigma(y - y_2)^2}{\Sigma(y - \bar{y})^2} = \dfrac{4752.0}{8610.7} = .5519$	$r_{y.2}^2 = .4481$	$r_{y.2} = .669$
Arithmetic and midterm tests	$\dfrac{\Sigma(y - y_{12})^2}{\Sigma(y - \bar{y})^2} = \dfrac{4470.5}{8610.7} = .5192$	$r_{y.12}^2 = .4808$	$r_{y.12} = .693$

The Beta Coefficients. A limitation of the $\hat{b}$ coefficients in the regression formula is the fact that these involve the units of measurement in addition to a measure of the contribution to estimation of the dependent variable provided by the independent variable. For this reason, one cannot compare two $\hat{b}$ coefficients as an indication of their relative contribution.

There is available, however, an alternative set of coefficients, called the beta (β) coefficients, from which the units of measurement have been eliminated. In parametric form the coefficients are β_1 and β_2.

In terms of sample statistics the regression formula has the form

$$(18\text{-}10) \qquad \frac{y_{12} - \bar{y}}{s_y} = \hat{\beta}_1 \frac{x_1 - \bar{x}_1}{s_1} + \hat{\beta}_2 \frac{x_2 - \bar{x}_2}{s_2}$$

This formula corresponds to formula (13-10) for one predictor.

In comparing the preceding formula with formula (18-2), one will notice that (18-2) is in raw score form whereas (18-10) is in standard score form, so that both independent variables have unit standard deviation, and the coefficients are comparable with respect to units.

The beta coefficients are related to the b coefficients and can be calculated from them by the formulas

$$(18\text{-}11) \qquad \hat{\beta}_1 = \hat{b}_1 \frac{s_1}{s_y}, \qquad \hat{\beta}_2 = \hat{b}_2 \frac{s_2}{s_y}$$

However, the beta coefficients can be calculated directly from the following set of normal equations

$$(18\text{-}12) \qquad \begin{aligned} \hat{\beta}_1 + r_{12}\hat{\beta}_2 &= r_{y1} \\ r_{12}\hat{\beta}_1 + \hat{\beta}_2 &= r_{y2} \end{aligned}$$

The solutions of these equations have the form

$$(18\text{-}13) \qquad \begin{aligned} \hat{\beta}_1 &= \frac{r_{y1} - r_{12}r_{y2}}{1 - r_{12}{}^2} \\ \hat{\beta}_2 &= \frac{r_{y2} - r_{12}r_{y1}}{1 - r_{12}{}^2} \end{aligned}$$

The multiple correlation coefficient can now be written as

$$(18\text{-}14) \qquad r_{y.12} = \sqrt{r_{y1}\hat{\beta}_1 + r_{y2}\hat{\beta}_2}$$

For the problem of predicting the final examination from the midterm test and the arithmetic test, the means and standard deviations appear in Table 18-2. Also, the correlations among the variables are

$$r_{y1} = .5016, \qquad r_{y2} = .6694, \qquad r_{12} = .5183$$

Then, using formulas (18-13)

$$\hat{\beta}_1 = \frac{.5016 - (.5183)(.6694)}{1 - (.5183)^2} = .2115$$

$$\hat{\beta}_2 = \frac{.6694 - (.5183)(.5016)}{1 - (.5183)^2} = .5598$$

The regression equation in terms of the betas now appears as

$$\frac{\bar{y}_{12} - 49.1531}{9.4218} = .2115 \frac{x_1 - 31.3776}{7.2475} + .5598 \frac{x_2 - 52.4592}{9.2827}$$

One may conclude from the comparison of $\hat{\beta}_1$ and $\hat{\beta}_2$ that the variable X_2, namely the midterm test, contributes substantially more than variable X_1 to correct estimation of scores on the final examination.

Significance Test for Regression. The test to be described indicates whether the regression formula contributes significantly to prediction of the criterion variable. This test is based on an F ratio and is closely related to the method of analysis of variance developed in the preceding chapter.

As was the case in Chapter 17, one is now dealing with a set of populations, namely the conditional distributions corresponding to pairs of values of x_1 and x_2. One may ask whether the means of these distributions are equal. If they are, then the means of the conditional distributions are equal to the mean of all y scores combined, and the regression formula provides no additional information. However, if the means are unequal, then the knowledge provided by the regression formula does contribute to improved estimation.

From this point of view an analysis of variance of the same kind as described in Chapter 17 becomes feasible. One should note an important distinction between the present situation and the one in the previous chapter. In the regression model the means are restricted by the regression formula (18-1), so that whatever their values may be they all lie on the same plane. There was no such restriction in the problems of Chapter 17 where the means might differ arbitrarily from each other. This difference in models is reflected in the choice of degrees of freedom which will be described below.

The analogy having been established for the population, one may turn to a corresponding analogy for the sample. Here the regression estimates y_{12} play the role of sample means. Hence there is a subdivision of total sums of squares into components among and about the regression estimates like the one obtained in Chapter 17.

This subdivision can be expressed in words as

Total sum of squares about the mean of all scores = Sum of squares among regression estimates + Sum of squares of deviations of individual scores from their regression estimates.

In symbols this statement can be written as

(18-15) $$\Sigma(y - \bar{y})^2 = \Sigma(y_{12} - \bar{y})^2 + \Sigma(y - y_{12})^2$$

Another symbolism which will be used is obtained by the abbreviation SS for sum of squares, so that the verbal statement becomes

Total SS = SS for regression + SS for deviations from regression
= SS for regression + SS for error

By algebraic manipulation one obtains the following calculating formulas

(18-16a) $$\text{SS for regression} = \hat{b}_1 S_{y1} + \hat{b}_2 S_{y2}$$
(18-17a) $$\text{SS for error} = S_{yy} - \hat{b}_1 S_{y1} - \hat{b}_2 S_{y2}$$

Further algebraic manipulation yields the following equivalent forms for the sum of squares

(18-16b)   SS for regression $= S_{yy}r_{y.12}{}^2$
(18-17b)   SS for error $= S_{yy}(1 - r_{y.12}{}^2)$

For convenience in computation, the following two forms of partition of sums of squares are now available

$$S_{yy} = (\hat{b}_1S_{y1} + \hat{b}_2S_{y2}) + (S_{yy} - \hat{b}_1S_{y1} - \hat{b}_2S_{y2})$$

and

$$S_{yy} = S_{yy}r_{y.12}{}^2 + S_{yy}(1 - r_{y.12}{}^2)$$

For the purpose of obtaining an F ratio it is necessary to calculate mean squares by dividing the sums of squares by the appropriate numbers of degrees of freedom. The symbol MS will be used for mean square. In the following formulas the mean squares will be expressed in the two ways already discussed.

(18-18)   MS for regression $= \dfrac{\hat{b}_1S_{y1} + \hat{b}_2S_{y2}}{2}$ or $\dfrac{S_{yy}r_{y.12}{}^2}{2}$

(18-19)   MS for error $= \dfrac{S_{yy} - \hat{b}_1S_{y1} - \hat{b}_2S_{y2}}{n - 3}$ or $\dfrac{S_{yy}(1 - r_{y.12}{}^2)}{n - 3}$

One can now obtain F as the quotient

(18-20) $F = \dfrac{\text{MS for regression}}{\text{MS for error}} = \dfrac{r_{y.12}{}^2}{1 - r_{y.12}{}^2} \cdot \dfrac{n - 3}{2}$

The calculations using either of these forms can be entered in an analysis of variance table like Table 17-6 on page 301. Table 18-4 contains results based on prediction of a final examination score from scores on an arithmetic test and on the midterm test

Table 18-4 Analysis of Variance for Regression Data from Appendix Table VIII

Source of Variation	Sum of Squares	Degrees of Freedom	Mean Square	F
Regression	4140	2	2070.00	44.0
Error	4470.5	95	47.06	
Total	8610.5			

Notice that the sums of squares can be calculated from S_{yy} and $r_{y.12}$ as follows

$$\text{SS for regression} = 8610.7(.4808) = 4140$$
$$\text{SS for error} = 8610.7(1 - .4808) = 4470.7$$

The value $F = 44$ points to rejection of the null hypothesis that regression is not useful in prediction for the variables in the study.

Partial Correlation. Imagine that for each of our 98 subjects a regression equation is used to predict his midterm score from his score on the prognostic arithmetic test and the residual error computed. Imagine also that in the same way his final examination score is predicted from his arithmetic score and the residual error computed. There would now be available 98 pairs of residuals, and these could be used to compute a correlation coefficient which would measure the residual relation between final examination and mid-term scores over and above the relation they both have to arithmetic score. When first proposed, this correlation was called *net correlation*, but now it is always called *partial correlation*. The symbol applied to the statistic described above is $r_{y1.2}$.

The tedious process of computing residuals and correlating them would not be carried out in practice. It is described here in order to clarify what partial correlation means. Instead, if the algebraic expressions for these residuals are substituted for x and y in formula (13-2) for r_{xy}, the formula

$$r_{y1.2} = \frac{r_{y1} - r_{y2}r_{12}}{\sqrt{1 - r_{y2}^2}\ \sqrt{1 - r_{12}^2}}$$

is finally obtained. Or in general, if the variables are designated as X_1, X_2, and X_3

$$(18\text{-}21) \qquad r_{12.3} = \frac{r_{12} - r_{13}r_{23}}{\sqrt{1 - r_{13}^2}\ \sqrt{1 - r_{23}^2}}$$

Notice that there are two number symbols to the left of the point in the subscript for partial correlation and only one in the subscript for multiple correlation. These number symbols to the left of the point are called *primary subscripts* and name the variables being correlated; those to the right of the point are called *secondary subscripts* and name the independent variables used as predictors in the regression equation. Thus the symbol $r_{1.2345}$ would denote the correlation between X_1 and estimates of X_1 obtained from a regression equation involving X_2, X_3, X_4 and X_5; and the symbol $r_{12.345}$ would denote the partial correlation between X_1 and X_2 over and above their relation to X_3, X_4, and X_5. A partial correlation is termed first order, second order, and so forth, according to the number of secondary subscripts. More importantly, when there are no secondary subscripts, the expression *zero-order* is often used.

Numerical Relations among Correlation Coefficients. There is an interesting relation between multiple r and the various partials, which for 3 predictors can be written in any of the following ways:

$$1 - r_{y.123}{}^2 = (1 - r_{y1}{}^2)(1 - r_{y2.1}{}^2)(1 - r_{y3.12}{}^2)$$
$$1 - r_{y.123}{}^2 = (1 - r_{y1}{}^2)(1 - r_{y3.1}{}^2)(1 - r_{y2.13}{}^2)$$
$$(18\text{-}22) \qquad 1 - r_{y.123}{}^2 = (1 - r_{y2}{}^2)(1 - r_{y1.2}{}^2)(1 - r_{y3.12}{}^2)$$
$$1 - r_{y.123}{}^2 = (1 - r_{y2}{}^2)(1 - r_{y3.2}{}^2)(1 - r_{y1.23}{}^2)$$
$$1 - r_{y.123}{}^2 = (1 - r_{y3}{}^2)(1 - r_{y1.3}{}^2)(1 - r_{y2.13}{}^2)$$
$$1 - r_{y.123}{}^2 = (1 - r_{y3}{}^2)(1 - r_{y2.3}{}^2)(1 - r_{y1.23}{}^2)$$

Formulas (18-22) have little use as a computing routine; their interest lies rather in the fact that they show clearly that the multiple r must be at least as large numerically as the largest partial which has y as one of its primary subscripts. Thus $r_{y.123}$ is at least as large as any of the other r's named above. It has, however, no such necessary relation to r_{12} or $r_{23.4}$.

In the situations commonly encountered by students of the behavioral sciences, a partial r is likely to be smaller than the zero-order r having as its subscripts the primary subscripts of the partial. However, this is not necessarily so. As an illustration,

if $\qquad r_{xy} = .50, \qquad r_{yz} = .60, \qquad$ and $\qquad r_{xz} = .10,$
then $\qquad r_{xy.z} = .55, \qquad r_{yz.x} = .64, \qquad$ and $\qquad r_{xz.y} = -.29$

If one should hastily and incautiously write down a set of zero order correlation coefficients from which to compute a multiple or a partial correlation coefficient, he might find himself in the uncomfortable position of having selected inconsistent values. Real data never produce a partial correlation numerically larger than 1. Each of the following sets of coefficients may at first glance seem all right, but they are inconsistent because, if inserted in formula (18-21), the outcome is a statement that a partial correlation is either less than -1 or more than $+1$, which cannot be true.

Set	r_{12}	r_{13}	r_{23}
(1)	.80	.70	.10
(2)	.90	.40	-.20
(3)	-.50	.40	.70
(4)	.20	.80	.80

Sometimes the logic of a situation appears to call for a partial and sometimes for a zero order correlation, and it is not always easy to decide which is more appropriate. The difference in meaning is nicely illustrated in data gathered many years ago by Ethel Elderton who made body measurements of a large number of school children in Scotland. For girls ranging in age from 5 to 18, the correlation of height and weight was .91, but when the total group was broken up into smaller groups with an age range of only one year, the correlation was much lower. For children whose ages span a wide range, the correlation is influenced by the fact that older children are generally both taller and heavier than younger children, so the correlation $r = .91$ between height and weight actually involves age also.

If the symbols h, w, and a are used for height, weight, and age, repectively, the intercorrelations are

$$r_{hw} = .91, \qquad r_{ha} = .70, \qquad \text{and} \qquad r_{wa} = .70$$

Then the partial correlation is

$$r_{hw.a} = \frac{.91 - (.70)(.70)}{\sqrt{(1 - .49)(1 - .49)}} = .82$$

Obviously, elimination of the effect of age has resulted in a reduction of the relation between height and weight from .91 to .82.

Extension to More than Two Predictors. The formulas developed in this chapter can be extended to include any number of predictors. However, it would be necessary to introduce more complex concepts such as those of a matrix and the inverse of a matrix to deal with increasing number of normal equations. The computational problems also become increasingly weighty.

The reader is referred to more advanced texts for problems involving many predictors.

Tests of Significance and Confidence Intervals. Among the advanced topics which are also deferred to advanced texts are a variety of tests of significance.

Tests based on the t distribution are available for the $\hat{b}$ coefficients. One can also test differences between two coefficients in the same regression formula.

There are also two confidence intervals of interest. One is the confidence interval for the regression estimate which has been denoted as y_{12}. The other is the interval for a y score when scores on the independent variables have been given.

Computer Applications to Regression. A variety of methods have been developed for calculations dealing with regression problems which are suitable for desk calculators. In spite of all the simplifications provided by these methods, calculations with desk calculators are extremely onerous, and people are turning increasingly to electronic computers to obtain results.

The electronic computer not only provides with vastly greater speed the answers which were obtained in the past by laborious hand calculation. The computer actually provides answers which were unthinkable in the past. Entirely new mathematical devices have been developed to make the computer operate effectively.

Some of the outputs or information provided by the computer will be described. These outputs require an input which consists of the scores on both the dependent variable and the independent variables.

To begin with, the output consists of the means and standard deviations of all the variables. This is followed immediately by a matrix of variances and covariances and a matrix of intercorrelations.

Following the preliminary calculations, the regression analysis is given as a stepwise procedure which provides not only a single regression formula but a sequence of such formulas together with related tests of significance.

At the first step of the sequence a regression equation consisting of the one independent variable which contributes most to prediction of the dependent variable is developed.

In succeeding steps of the sequence additional variables are introduced which jointly provide the best prediction. Goodness of prediction is measured by the squared multiple correlation coefficient.

At each step of the sequence the following information is provided by the computer:

> The regression coefficients
> An analysis of variance to test the significance of regression
> A t test for each regression coefficient
> A matrix of values which are needed for comparisons among the coefficients
> The standard error of estimate
> The coefficient of multiple correlation
> The coefficient of partial correlation between each independent variable and the dependent variable when the effects of all other variables have been eliminated.

Upon instruction, the computer will provide the observed and estimated values and the discrepancy between them.

EXERCISE 18-1

1. The problems appearing below are based on statistics which were obtained from the data in Appendix Table VIII. The following variables are used:

$$Y = \text{Midterm test}$$
$$X_1 = \text{Artificial language test}$$
$$X_2 = \text{Arithmetic test}$$

The following statistics were computed:

$\bar{y} = 52.46$	$S_{yy} = 8358$	$r_{y1} = .620$
$\bar{x}_1 = 46.43$	$S_{11} = 11{,}690$	$r_{y2} = .518$
$\bar{x}_2 = 31.38$	$S_{22} = 5095$	$r_{12} = .539$
$s_y = 9.283$	$S_{y1} = 6122$	$n = 98$
$s_1 = 10.978$	$S_{y2} = 3382$	
$s_2 = 7.247$	$S_{12} = 4159$	

(a) Calculate the coefficients $\hat{a}$, $\hat{b}_1$, and $\hat{b}_2$, and write out the formula for determining y_{12} from x_1 and x_2.
(b) Calculate the beta coefficients $\hat{\beta}_1$ and $\hat{\beta}_2$ directly from the correlation coefficients.
(c) Use the relations in formula (18-11) to check the betas against the b's.
(d) Calculate the conditional standard deviation $s_{y.12}$.
(e) Calculate the multiple correlation coefficient.
(f) Use analysis of variance to check the significance of the regression estimate.
(g) Calculate and interpret each of the partial correlations $r_{y1.2}$ and $r_{y2.1}$.

2. We have already seen that the correlation between intelligence quotient and language score for the 109 fourth-grade children was .58. See Figure 13-2 on page 222. This correlation, however, is partly due to the fact that younger children in the fourth grade tend to have higher language scores and higher intelligence quotients than older children. Examine the subgroups based on age in Figure 18-1. Verify the correlations stated for the subgroups.

3. For the 109 subjects listed in Appendix Table IX, let

$$X_1 = \text{Language score}$$
$$X_2 = \text{Intelligence coefficient}$$
$$X_3 = \text{Age in months}$$

The zero-order correlation coefficients are

$$r_{12} = .583, \quad r_{13} = -.396, \quad \text{and} \quad r_{23} = -.723$$

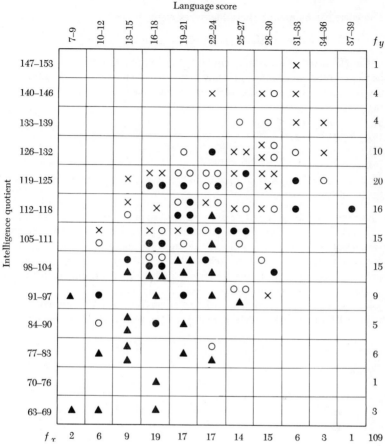

Figure 18-1 Language Score and Intelligence Quotient: Bivariate Distributions for Children in Different Age Groups

✗	Children aged 94–108 months	$n = 27$	$r = .50$
○	Children aged 109–113 months	$n = 31$	$r = .42$
●	Children aged 114–118 months	$n = 26$	$r = .35$
▲	Children aged 119–156 months	$n = 25$	$r = .56$

For entire group $n = 109$ and $r = .58$

What is the value of $r_{12.3}$? How does this value compare with the 4 values of r_{12} for subgroups limited in age range shown in Figure 18-1?

4. In a study of juvenile delinquency, Lander obtained for each of 155 census tracts in Baltimore, Md., a number of measures, among them the following:

$$X_1 = \text{Measure of juvenile delinquency}$$
$$X_2 = \text{Median education of adults}$$
$$X_3 = \text{Overcrowding}$$

The correlations among these measures were

$$r_{12} = -.51, \qquad r_{13} = .73, \qquad \text{and} \qquad r_{23} = -.71$$

If the effect of overcrowding could be eliminated, what do you think would be the correlation between juvenile delinquency and median education? Make the best guess you can, selecting one of the following answers:

(a) $r_{12.3} < -.50$.

(b) $-.50 < r_{12.3} < -.20$.

(c) $-.20 < r_{12.3} < .20$.

(d) $.20 < r_{12.3} < .50$.

(e) $r > .50$.

Now compute $r_{12.3}$, and compare the result with your guess. What does this result mean?

5. The personnel department of an industrial firm has devised a placement test (X_1) given at the time of application and a performance test (X_2) given after two months work on the job. At the end of 6 months work, each employee receives a rating (Y) from his foreman. For 223 employees $r_{y.12} = .45$. What proportion of the variability in foremen's ratings is unrelated to variability in the tests? Do you think these tests provide an effective means of selecting job applicants? Is $r_{y.12}$ significantly different from zero? Would you be willing to assert that there is no relation between the tests and the foremen's ratings?

19

SAMPLES
FROM A FINITE
POPULATION

This chapter will relate to samples from a population which differs in two important respects from those discussed previously: (1) all the individuals in the population are existent at the time the sample is obtained, and (2) the number of individuals in the population is finite.

In an experimental study, it often happens that the only members of the relevant population which actually exist are those which are included in the experiment. Thus, in the tryout of a new operational method, the only individuals upon whom the method is used may be the subjects included in the study. The population to which generalization is made consists of all those individuals from whom the subjects in the experiment might reasonably have been selected as a sample. The population is conceptual but not actually in existence.

The standard error formulas presented in earlier chapters are based on the assumption that sampling has been at random from an infinite population. Sampling made with replacement so that the same individual may be drawn more than once is tantamount to sampling from an infinite population. But in many real situations information is needed concerning a specific, finite set of individuals, and the information is to be obtained by a sample instead of by a complete census. For such sampling, the standard error formulas must take account not only of sample size, but also of population size.

Sampling designed to produce information about some characteristic of an *existent, finite* population is usually called *survey sampling*.

Design of Samples in Surveys. One of the most crucial aspects of a sample survey is the decision as to the number of cases and the methods by which they shall be selected. This decision provides the *design* of the survey.

Avoidance of bias is an obvious requirement for any satisfactory sampling design. The error which causes a statistic to differ from its parameter may arise partly from random errors and partly from bias in the selection of individuals in the sample. The effect of random errors decreases as sample size increases; the effect of bias is as serious in large samples as in small. The early history of opinion polls includes many incidents showing dramatic discrepancy between the predictions made by a poll and the subsequent outcome of an election. Upon analysis the fault was often found to be in a procedure of selection which produced a biased sample.

Another requirement of a "good" sampling procedure is that it provide as much information about the population as can be obtained at a stipulated cost or that it provide a stipulated amount of information for as low a cost as possible. This requirement raises highly technical issues which are outside the scope of this book. Any reader contemplating an important survey should familiarize himself with the technical literature on sampling design such as the references at the end of this chapter, and if possible should secure the assistance of a sampling expert.

A third requirement is that the sample provide not only estimates of population parameters but also information concerning the precision of those estimates. This requirement can be met if, and only if, the sample is a *probability sample*. Before describing a probability sample, it seems essential to clarify the meaning of the terms "sampling unit" and "frame."

Sampling Unit.   The meaning of a sampling unit has been well stated by Yates:

"All rigorous sampling demands a subdivision of the material to be sampled into units, termed *sampling units*, which form the basis of the actual sampling procedure. These units may be natural units of the material, such as individuals in a human population, or natural aggregates of such units, such as households, or they may be artificial units, such as rectangular areas on a map, bearing no relation to the natural subdivisions of the material.

"It is not always necessary to make an actual subdivision of the whole of the material before selection of the sample, provided the selected units can be clearly and unambiguously defined. Thus, with sampling units which are rectangular areas on a map there is no need to demarcate all these areas; they can be defined by co-ordinates, and the selected areas demarcated after selection.

"Clear and unambiguous definition demands the existence or construction of some form of *frame*. In the sampling of a human population, for instance, with households as sampling units, there must be available a list of all households, and this list must be such that any household selected from it can be unambiguously located. In area sampling from maps, the maps must be such that the selected areas can be unambiguously defined on the ground. . . .

"Sampling units may be of the same or differing size. They may contain the same, or approximately the same, number of natural units, or they may contain widely differing numbers. The whole procedure of sampling, including the estimation of the population values and the sampling errors, is simplest when the sampling units are of approximately the same size and contain approximately the same number of natural units. Often, however, the material is such that this condition cannot be conveniently fulfilled. In particular, if the natural units are themselves of widely differing size, variation in size of the sampling units or in the number of natural units they contain is inevitable."

Frame.   This term was adopted in 1948 by the United Nations Subcommittee on Statistical Sampling to mean a description of the sampling units that compose a population. The frame might be a list of names, the cards in a card file, a map, or any other device by which every member of a population can be identified in advance. Cochran once remarked that "sometimes the frame is impossible to construct, as with the population of fish in a lake." The construction of a useful frame may be one of the most important steps in planning a survey. Many helpful suggestions will be found in the references at the end of this chapter.

Probability Sample.   Whenever it is possible to identify every individual in a population by position of a name on a list, of a card in a file, of a point in space or in time, or through the use of some other form of frame, it is possible to draw a sample by a random process with preassigned probability of selection. Such a sample is called a *probability sample*.

The importance of random selection was pointed out in Chapter 9, the use of a table of random numbers was described and an example of its use was given in Exercise 9-1. A modern computer provides a set of random numbers by means of which a random sample can be selected. It is necessary only to indicate the size of population and size of sample required.

Three principal types of probability sample will be described briefly now, in rather general terms, and later the formulas and procedures appropriate for making inferences will be presented for each.

1. *Simple random sampling.* Suppose there are N individuals in a population and n of these are to be selected in a sample. The procedure of drawing each of the n individuals at random from the entire population of N individuals is called *unrestricted random sampling*, or *simple random sampling* and sometimes merely *simple sampling.* Usually it is accomplished by means of a table of random numbers generally without replacement so that no individual enters the sample more than once.

2. *Cluster sampling.* For this method the population is divided into many relatively small groups or clusters of individuals, and the sample consists of a number of these clusters *chosen at random.* For example, the population of all eighth grade pupils in a large city might be sampled by first listing all eighth grades, then selecting at random a number of classes and making observations either on all the pupils in each selected class or on a random sample from each selected class. Cluster sampling is often employed in studies that utilize area sampling, as when small areas like rural townships or city blocks are chosen at random.

Cluster sampling is economical when the cost of measuring an individual is relatively small and the cost of reaching him relatively large. For example, suppose a railroad is making a survey of the ties on which rails rest in order to assess the need for replacements. To take an unrestricted random sample of all the ties in use, would entail heavy costs to travel to the place where the selected ties might be found but very little cost to examine the ties once they are reached. It might cost less and yield more information to measure 100 clusters of 5 ties each than to measure 200 ties chosen individually at random. It would produce more information to measure 100 clusters of 5 ties each than to measure 5 clusters of 100 ties each. The relation of costs to sampling procedure is outside the scope of this text but it is well worth the careful investigation of any one contemplating a large inquiry by sample.

Cluster sampling is most advantageous when there is great heterogeneity within clusters. If all the individuals in each cluster were exactly alike there would be no advantage in observing more than one of them. The larger the variation within clusters, the smaller the number of individuals which will be required to hold $s_{\bar{x}}$ to a predetermined size.

3. *Stratified sampling.* For this method the population is subdivided into a number of groups or strata and a random sample is selected from each stratum. These strata or subpopulations should be so defined that the variance of individuals in each stratum is smaller than the variance of the entire population, and in consequence there is considerable diversity from one stratum to another. Such definition of strata is possible only when there is some prior information about the population. Stratification of a community is often on a geographical basis, as when the residents of a city are classified according to the area in which they live, or residents of a state are classified according to whether they live in a city, in a village, or in open country.

Multistage Sampling.   This term connotes a more complex plan with greater flexibility of design than in the preceding. The population might first be divided into strata and each stratum into clusters, or vice versa. Or in the first stage a sample of clusters might be drawn and in the second stage a sample of individuals from each cluster. Or in the first stage the United States might be divided into five or six geographical regions (strata), and in a second stage the communities in each region might be divided on the basis of density of population (strata). From each such stratum communities might be selected at random (clusters) and from each community 8th grade pupils might be selected at random for a testing program. The possible variations are endless.

Judgment Sampling.   The literature is full of instances showing serious bias in samples composed of individuals chosen because they were presumed to be typical of the population. Even if no bias were present—and there is no way of ascertaining the absence of bias—the method is faulty because the precision of the obtained population estimates cannot be interpreted in probabilistic terms. A probability sample cannot be obtained through the use of conscious selection of cases. Some mechanism for random selection is essential.

Systematic Sampling.   A method of sampling often used is to select one individual at random and then to take other individuals at regular intervals. Thus, if a 5 percent sample is desired, an individual would be chosen at random from the first 20 on the list and every 20th individual thereafter would be included. By this method, every individual in the population has the same chance of being selected but not all samples are equally probable; in fact some samples are impossible.

The great advantage of this method is that it is easier to execute than simple random sampling. The sample is evenly distributed over the population and therefore is usually—but not necessarily—unbiased. If an unnoticed periodicity in the frame of the population should happen to coincide with the pattern of the sample (that is, the selection of every rth case), a serious bias could result. Many sampling experts employ this method with excellent results. The novice may encounter pitfalls. Special methods for appraising the precision of estimates are described in more extensive treatises.

A. SIMPLE RANDOM SAMPLING

Estimation of a Proportion. Methods for making estimates about proportions in infinite populations were developed in Chapter 11. The need for some modification to make these methods applicable to finite populations will now be illustrated.

Suppose that a finite population consisting of N elements can be divided into two classes, one containing A elements and the other $N - A$. Then the population proportion can be taken to be A/N. Hence,

$$(19\text{-}1) \qquad P = \frac{A}{N}$$

Let a sample of n elements be chosen by simple random sampling from the N elements of the population. Suppose the sample contains a of the A elements in the group mentioned above. Then the sample proportion is

$$(19\text{-}2) \qquad p = \frac{a}{n}$$

Here p is the sample estimate of the population proportion P.

As a varies from sample to sample, the value of p varies with it. For example, in drawing samples of 50 each ($n = 50$) from a population of 50 males and 50 females ($N = 100$), a sample may contain as many as 50 males ($a = 50$) or as few as no males ($a = 0$). One should also notice that many samples can have the same value of a when a has values other than 0 or 50. Two samples are regarded as different if they contain different individuals.

Consider now all possible samples of size n. For each sample there is a value of p. The average of these values is P

$$(19\text{-}3) \qquad \text{Average of } p = \mu_p = P$$

The variance of these values is given by the formula

$$(19\text{-}4) \qquad \text{Variance of } p = \sigma_p{}^2 = \frac{N - n}{N - 1} \cdot \frac{P(1 - P)}{n}$$

The standard deviation of p is then

$$(19\text{-}5) \qquad \sigma_p = \sqrt{\frac{N - n}{N - 1} \cdot \frac{P(1 - P)}{n}}$$

The standard deviation in formula (19-5) may be called the standard error. This expression suggests its use in establishing confidence intervals for P, but this cannot be done directly because it contains the unknown value of P. However, substitution of the sample estimate p for P provides a sample estimate of the standard error

$$(19\text{-}6) \qquad s_p = \sqrt{\frac{N - n}{N - 1} \cdot \frac{p(1 - p)}{n}}$$

The expression $\dfrac{N - n}{N - 1}$ which appears in the preceding three formulas is called the *finite population correction*. Without this correction, formula (19-4) gives the variance for samples from an infinite population. The effect of the correction for some special values of sample size n is of interest. When $n = N$, so that the sample includes the whole population, the standard error becomes zero because P is calculated without error. When n is small in comparison with N, then $(N - n)/(N - 1)$ is close to one, and the standard error becomes nearly the same as it is for an infinite population.

One may also comment on the expression $N - 1$ in the denominator of the finite population correction. When the population is not very small, one may use N for $N - 1$ and the correction then becomes $(N - n)/N$ or $1 - n/N$. Here n/N is the fraction of the population contained in the sample, and $1 - n/N$ is the fraction not contained in it.

The standard error, s_p, can be used to construct confidence intervals for P as was done in Chapter 11.

Data obtained from a survey of high school seniors in a given area will be used to illustrate the principles which have been developed. The discussion will concern itself with responses to the question: "Do you plan to enter a four year college program?"

The population of seniors about which information was sought consisted of 51,444 individuals. Of this population, 7,324 were included in the sample. In response to the question regarding planned college attendance, 3,114 stated that they would attend a college providing a four year program. This information can now be fed into the required formulas by use of the numerical equivalents of the appropriate symbols, as follows:

$$N = 51{,}444$$
$$n = 7{,}324$$
$$a = 3{,}114$$

The point estimate, p, of P is

$$p = 3{,}114/7{,}324 = .425$$

The standard error can be estimated as

$$s_p = \sqrt{\frac{51444 - 7324}{51444 - 1} \cdot \frac{(.425)(.575)}{7324}} = .00535$$

One can now calculate confidence limits with any desired confidence coefficient. If the confidence coefficient is set at .99, then the limits are

Upper limit = .425 + 2.58(.00535) = .439
Lower limit = .425 − 2.58(.00535) = .411

The confidence interval indicates that the true fraction of seniors planning to go into four year programs has high probability of being in an interval of 41 to 44 percent.

The normal distribution is applied here since sample size is reasonably great.

Estimation of a Mean. When data measured on a continuous scale are considered, the population parameter to be estimated is a mean. Methods of estimation which were applied to infinite populations in Chapter 10 will be described here in relation to finite populations.

Suppose that the population under study consists of N individuals each of whom is measured on a scaled variable X. The population parameters are the mean

(19-7) $$\mu_x = \frac{\sum\limits_{i}^{N} x_i}{N}$$

and the variance

(19-8) $$\sigma_x^2 = \frac{\sum\limits_{i}^{N} (x_i - \mu_x)^2}{N}$$

Let n elements be chosen from this population by simple random sampling. Then the estimate of μ_x derived from the sample is

(19-9) $$\bar{x} = \frac{\sum\limits_{i}^{n} x_i}{n}$$

The parameter σ_x^2 is estimated as

(19-10)
$$s_x^2 = \frac{\sum_i^n (x_i - \bar{x})^2}{n - 1}$$

as in the case of sampling from an infinite population.

The standard error of $\bar{x}$ takes into account the fact that the population is finite by introduction of the finite population correction. Hence, in parametric terms the standard error is given by the formula

(19-11)
$$\sigma_{\bar{x}} = \sqrt{\frac{N - n}{N - 1}} \frac{\sigma_x}{\sqrt{n}}$$

An estimate of $\sigma_{\bar{x}}$ is given by

(19-12)
$$s_{\bar{x}} = \sqrt{\frac{N - n}{N - 1}} \frac{s_x}{\sqrt{n}}$$

As in the infinite case the mean of the sampling distribution of the sample means is the same as the mean of the scores. Hence,

(19-13)
$$\mu_{\bar{x}} = \mu_x$$

Confidence intervals for μ_x can be obtained by use of tables of the normal distribution.

As an application of the preceding principles, an estimate will be made of the mean change in per-pupil operating expenditures between two successive years for the school districts in an area. There are 70 districts in the area, and a simple random sample of 30 of these is selected. Changes in per-pupil expenditures are obtained for the districts in the sample, and on the basis of the thirty values so obtained the mean and standard deviation are calculated for the sample, using the methods of Chapter 6. The values so obtained are

$$\bar{x} = 42.57 \quad \text{and} \quad s_x = 36.60$$

The standard error of the mean is then calculated as

$$s_{\bar{x}} = \sqrt{\frac{70 - 30}{70 - 1}} \cdot \frac{36.60}{\sqrt{30}} = 5.09$$

A confidence interval with .95 confidence coefficient can be calculated by using $z_{.025} = -1.96$ and $z_{.975} = 1.96$ as indicated by the tables of normal probability, Hence, the limits of the interval are

$$\text{Upper limit} = 42.57 + 1.96(5.09) = 52.55$$
$$\text{Lower limit} = 42.57 - 1.96(5.09) = 32.59$$

One can, therefore, say that the interval 32.59 to 52.55 includes the average change for the districts in the area with probability .95.

Estimation of Totals. In survey sampling one is usually interested not only in means of populations but in their totals as well. In this respect, statistical studies of finite populations differ essentially from studies of infinite populations, for which totals do not exist.

Estimation of totals can be related to estimation of proportions or means by only slight modification of formulas. If the population under consideration consists of two classes, then the confidence interval for the proportion in one class has the form

$$(19\text{-}14) \qquad p + z_{\frac{1-c}{2}} \, s_p < P < p + z_{\frac{1+c}{2}} \, s_p$$

To estimate the total PN in the same class it is only necessary to multiply the terms in the inequality by N, giving the new inequality

$$(19\text{-}15) \qquad pN + z_{\frac{1-c}{2}} \, N s_p < PN < pN + z_{\frac{1+c}{2}} \, N s_p$$

Similarly, estimates of mean and standard error can be converted directly to those for a total. To obtain a point estimate of a total PN, one estimates p from a sample and the total is estimated as

$$(19\text{-}16) \qquad \text{Estimate of population total} = pN$$

The standard error of the estimated total can be obtained directly from the standard error of a proportion by multiplying by N. However, a simplification of the formula is possible by replacing $N - 1$ by N in the finite population correction. The standard error then becomes

$$(19\text{-}17) \qquad s_{pN} = \sqrt{\frac{N(N - n)}{n} \, p(1 - p)}$$

The last two formulas will now be applied to the data previously presented regarding plans of high school seniors. Using these data, one obtains $p = .425$, $n = 7{,}324$ and $N = 51{,}444$. Then the estimated number of students planning to go to four year institutions is

$$.425(51{,}444) = 21{,}864 \text{ or } 21{,}900$$

The standard error is

$$s_{pN} = \sqrt{\frac{51{,}444(51{,}444 - 7{,}324)}{7{,}324}}\,(.425)(.575)$$
$$= 275.194$$

Hence, the .99 confidence limits for number of students entering four year institutions are

Upper limit $= 21{,}864 + 2.58(275.194) = 22{,}574$ or $22{,}600$
Lower limit $= 21{,}864 - 2.58(275.194) = 21{,}154$ or $21{,}200$

Similar formulas for totals can be obtained for measured data. Suppose that the population consists of N individuals, say schools, and for each school an enrollment measure is available, so the total for the population is the total of the enrollments in the schools. This total can be estimated by a simple random sample of enrollments in n schools.
Write

$$x_i = \text{enrollment of a school}$$

Then
$$T = \sum_i^N x_i$$

for the total enrollment of all schools in the population, and

(19-18)
$$x = \sum_i^n x_i$$

for the total enrollment in the sample. Then

(19-19)
$$\hat{T} = N\frac{x}{n} = N\bar{x}$$

is an estimate of total enrollment T. The standard error of $\hat{T}$ is

(19-20)
$$s_{\hat{T}} = s\sqrt{\frac{N(N-n)}{n}}$$

where s^2 is the variance of the scores in the sample,

$$s^2 = \frac{\sum_i^n (x_i - \bar{x})^2}{n-1}$$

EXERCISE 19-1

1. The first part of Table 19-3 on page 346 contains data for stratum A based on a sample of 21 districts. Calculate estimates of the total current year kindergarten enrollment and of the average district enrollment using the entries in the column headed "Current Year." Calculate a point estimate and confidence limits with .90 confidence level. Assume the normal distribution for the means. Note that the entire area has 68 districts as shown in Table 19-2 on page 345.

2. Calculate point and interval estimates for the total change and average change in number of pupils in stratum A using the entries in the Difference Column.

   The total kindergarten enrollment for the prior year is known to be 25,008. Add the upper and lower limits obtained from the calculation using differences to the prior year total to obtain limits for the current year total.

3. Compare the accuracy using the calculations in problems 1 and 2.

B. CLUSTER SAMPLING

In the formulas given here it will be assumed that all the individuals are observed in those clusters which are chosen, but this need not be so. It often happens that a number of clusters is selected at random and then out of those selected clusters a predetermined number of individuals is selected at random.

The formulas must take account of the number of clusters in the population, the number in the sample and the number of individuals in each cluster chosen.

Let M = number of clusters in the population
  m = number of clusters in the sample
  n_i = number of individuals in the ith cluster
  $\bar{x}_i$ = mean of the ith cluster
  $n = \Sigma n_i$ = number of individuals in the sample

Then the sample estimate of the population mean μ is

$$(19\text{-}21) \qquad \bar{x}_c = \frac{\displaystyle\sum_{i=1}^{m} n_i \bar{x}_i}{\displaystyle\sum_{i=1}^{m} n_i} = \frac{\displaystyle\sum_{i}\sum_{\alpha} x_{i\alpha}}{n}$$

and the variance of $\bar{x}_c$ is

$$(19\text{-}22) \qquad s^2_{\bar{x}_c} = \frac{(M - m)m}{M} \cdot \frac{\displaystyle\sum_{i=1}^{m} n_i{}^2(\bar{x}_i - \bar{x}_c)^2}{(m - 1)n^2}$$

or if m is large enough that $m/(m - 1)$ may be treated as 1 this is approximately

$$(19\text{-}23) \qquad s_{\bar{x}_c}{}^2 = \frac{M - m}{M} \cdot \frac{\displaystyle\sum_{i=1}^{m} n_i{}^2(\bar{x}_i - \bar{x}_c)^2}{n^2}$$

In the special case for which $n_1 = n_2 = \cdots = n_m$, the preceding formulas can be much simplified. Then the estimate of μ is merely the mean of the cluster means, or

$$(19\text{-}24) \qquad \bar{x}_c = \frac{\displaystyle\sum_{i=1}^{m} \bar{x}_i}{m}$$

and its variance is

$$(19\text{-}25) \qquad s_{\bar{x}_c}{}^2 = \frac{M - m}{Mm^2} \cdot \sum_{i=1}^{m} (\bar{x}_i - \bar{x}_c)^2$$

In dealing with proportions in samples obtained by cluster sampling we may let

p_i = proportion of cases in the ith sample which have a given characteristic

Then the estimate of the population proportion P is

$$(19\text{-}26) \qquad p_c = \frac{\displaystyle\sum_{i=1}^{m} n_i p_i}{\displaystyle\sum_{i=1}^{m} n_i}$$

and its variance is

$$(19\text{-}27) \qquad s_{p_c}{}^2 = \frac{(M - m)m}{M} \cdot \frac{\displaystyle\sum_{i=1}^{m} n_i{}^2(p_i - p_c)^2}{(m - 1)n^2}$$

These formulas may be used in making interval estimates for μ or P in the usual manner.

The use of cluster sampling in relation to mental testing is discussed by Marks. (See the list of references at the end of this chapter.)

Application of Cluster Sampling. Consider the problem of estimating average fifth grade reading scores of all the children in an area consisting of 60 school districts. To test all the children is a large undertaking, and it is proposed to use a sample.

If a simple random sample of all children were to be used, it would mean visiting all 60 districts and taking some children out of their classes for testing leaving other children behind. To save this inconvenience, the 60 districts are regarded as clusters and a random sample of 15 districts is selected. Within each district all fifth grade children are tested. In this way the data are in accord with the theory of cluster sampling previously developed, since there is a random sample of 15 clusters from a population of 60 clusters, each cluster being fully tested.

Data for the sample of districts appear in Table 19-1. The first column of this table identifies the district. The next two columns show the basic data, namely the number of cases and the mean score of the pupils in the district. The last two columns show portions of the calculations which are needed in the formulas for the combined mean of all samples and for the standard error of the mean. The reader should note that the calculations in the last two columns need not be displayed in full detail in actual practice, but can be combined on a modern calculator. They are exhibited here to clarify the concepts.

From the entries in Table 19-1 the estimate of the mean of all 60 districts is

$$\bar{x}_c = \frac{370,420.80}{7117} = 52.05$$

To obtain the variance of the mean note that $M = 60$, $m = 15$ and $\Sigma n_i^2(\bar{x}_i - \bar{x}_c)^2 = 30,578,482.2437$ as shown in the total of the last column in Table 19-1. Hence,

$$s_{\bar{x}c}^2 = \frac{(60 - 15)(15)}{60}\left(\frac{30,578,482}{(15 - 1)(7117)^2}\right) = .4850$$

and the standard error is

$$s_{\bar{x}_c} = \sqrt{.4850} = .696$$

Now, using a .95 confidence interval the limits for the district mean are

$$\text{Upper limit} = 52.05 + (1.96)(.696) = 53.4$$
$$\text{Lower limit} = 52.05 - (1.96)(.696) = 50.7$$

Table 19-1   Numbers of Cases and Mean Fifth-grade
Reading Scores in Fifteen School
Districts ($\bar{x}_c = 52.05$)

District	Number of Fifth Grade Pupils n_i	Mean Reading Score $\bar{x}_i$	$n_i\bar{x}_i$	$n_i^2(\bar{x}_i - \bar{x}_c)^2$
1	550	53.64	29,502.00	764,750.2500
2	627	46.06	28,879.62	14,105,507.8329
3	313	51.56	16,138.28	23,522.3569
4	373	45.84	17,098.32	5,365,384.6689
5	544	49.48	26,917.12	1,954,627.6864
6	1032	53.12	54,819.84	1,219,345.9776
7	526	53.25	28,009.50	398,413.4400
8	356	52.60	18,725.60	38,337.6400
9	769	54.55	41,948.95	3,696,006.2500
10	197	49.87	9,824.39	184,435.8916
11	542	53.12	28,791.04	336,330.4036
12	441	54.94	24,228.54	1,624,324.7601
13	421	53.90	22,691.90	606,607.3225
14	352	53.38	18,789.76	219,173.7856
15	74	54.81	4,055.94	41,713.9776
Total	7117		370,420.80	30,578,482.2437

EXERCISE 19-2

Calculate limits for the pupil mean using the first 10 of the districts in Table 19-1. Compare the accuracy in using 15 districts with that in using 10.

C.   STRATIFIED SAMPLING

Sampling Procedure.   In stratified sampling the population is subdivided into a number of groups or "strata." A simple random sample is then selected from each stratum and these samples jointly make up the sample from which one obtains estimates of parameters relating to the whole population.

The use of stratified sampling is likely to provide increased accuracy in sample estimates if the strata are so selected that the units within a stratum are less variable than they are in the population as a whole.

Notation.   The following notation will be used in formulas for stratified sampling:

L = number of strata

x_{hi} = the score for i'th case in the h'th stratum

N_h = number of cases in a stratum

$$N = \sum_{h=1}^{L} N_h = \text{number of cases in the entire population}$$

$$T_h = \sum_{i=1}^{N_i} x_{hi} = \text{total of all scores in a stratum}$$

$$T = \sum_{h=1}^{L} T_h = \sum_{h=1}^{L} \sum_{i=1}^{N_i} x_{hi} = \text{total of all scores in the population}$$

n_h = number of cases in the sample from a stratum

$$n = \sum_{h=1}^{L} n_h = \text{number of cases in the entire sample}$$

$$x_h = \sum_{i=1}^{n_h} x_{hi} = \text{total of scores in the sample from a stratum}$$

$$\bar{x}_h = \frac{x_h}{n_h} = \text{mean of scores in a sample from a stratum}$$

$$\mu_h = \frac{T_h}{N_h} = \text{mean of all scores in a stratum}$$

$$\mu = \frac{T}{N} = \text{mean of all scores in the population}$$

$\sigma_h{}^2$ is the variance of all scores in a stratum

σ^2 is the variance of all scores in the population.

Additional notation will be introduced below.

Estimation of Totals and Means.   In the estimation procedure to be described it will be assumed that the numbers of individuals are known or can be estimated, for all strata, so that numbers $N_1, N_2, \ldots, N_L$ are known.

The sample mean $\bar{x}_h$ is calculated for each stratum, and this mean is multiplied by the number of individuals, N_h, in the stratum. The resulting value

$$(19\text{-}28) \qquad \hat{T}_h = N_h \bar{x}_h$$

is a sample estimate of the total in a stratum and is likely to be a number of interest in itself.

As a next step, the estimated totals for the strata are added together to give an estimated total for the entire population,

$$(19\text{-}29) \qquad \hat{T}_{st} = \sum_{h=1}^{L} \hat{T}_h = \sum_{h=1}^{L} N_h \bar{x}_h$$

To obtain a standard error of the total a slight simplification which results from substitution of N_i for $N_i - 1$ provides the following formula

$$(19\text{-}30) \qquad s\hat{_{T_{st}}}^2 = \sum_{h=1}^{L} \frac{N_h(N_h - n_h)}{n_h} s_h^2$$

where s_h^2 is the variance of the sample in a stratum.

Similar formulas can be written for estimation of the mean of a population. The point estimate of a mean is

$$(19\text{-}31) \qquad \bar{x}_{st} = \frac{\displaystyle\sum_{h=1}^{L} N_h \bar{x}_h}{N}$$

The variance of $\bar{x}_{st}$ is

$$(19\text{-}32) \qquad s_{\bar{x}_{st}}^2 = \frac{1}{N^2} \sum_{h=1}^{L} \frac{N_h(N_h - n_h)}{n_h} s_h^2$$

The variances in formulas (19-30) and (19-32) can be used for establishing confidence limits for the total T and the mean μ.

Allocation of Cases to Strata.   A very important problem in stratified sampling is the decision as to the number of cases to be selected from each stratum. The discussion will be limited here to the situation where the total number of cases n in the sample has been agreed upon. The problem reduces itself, therefore, to allocation of the n cases among the strata. Such allocation depends on the knowledge available when the sample is designed. As a minimum, it is assumed that actual values, or estimates, of total numbers of cases in the strata are available. In symbols, one assumes knowledge of the N_h and of $N = \Sigma N_h$.

Important considerations in deciding upon the number of sampling units to be observed in each stratum are cost, size, and variability of stratum.

Sometimes the cost of observing a single sampling unit is much greater in one stratum than in another. Thus, to reach a dwelling in a thinly populated rural area is almost certain to cost more than to reach one in an urban area. As one consideration in efficient sample design is to obtain as much information as possible for a stipulated cost, it is obviously wise to take a larger sample in those units where cost is lowest. This topic, too large to be discussed here, should be carefully explored before any sizable survey is undertaken.

Other things being equal, a large stratum deserves and requires a larger sample than a small one. Recognition of this general principle often causes a research worker to take the same proportion of cases from all strata. Thus the sample size for any particular stratum by proportionate allocation is

$$(19\text{-}33) \qquad n_h = \left(\frac{N_h}{N} \right) n$$

However proportionate allocation is not the best method. For a fixed total sample size, n, the variance of the estimated mean is smallest when the values of n_h are made proportional to $N_h \sigma_h^2$. This is called *optimum allocation*. Then

$$(19\text{-}34) \qquad n_h = \left(\frac{N_h \sigma_h^2}{\sum\limits_{i=1}^{m} N_h \sigma_h^2} \right) n$$

This formula amounts to the rule: Take a larger sample in a stratum which is large (that is, N_h is large) and is more variable (that is, σ_h^2 is large).

If no estimate is available for the variances of the several strata, this may sometimes be obtained from a small preliminary sample. The reader should explore the topics of "optimum allocation" and "multistage sampling" in the references listed.

The reader should note that by either formula (19-33) or (19-34) the total of the n_h values is n, as is required.

Application of Stratified Sampling.   Assume that at the beginning of a school year it is desired to estimate kindergarten enrollment in an area consisting of 121 school districts, and that a sample of 30 districts is convenient as a basis for the estimate.

Since data are available from the previous year, only changes from last year need to be used in the estimates. An average or total based on changes can be estimated with less error than the actual current enrollment.

An examination of the changes for a previous period shows that a sub-area A has less variability from year to year than the remaining area B. Consequently, it is proposed to subdivide the area into two strata consisting of areas A and B. Table 19-2 shows the way in which the entire sample of 30 districts is allocated between the two strata.

Table 19-2   Prior Data on Strata A and B to Be Used in Allocation of Cases in Stratified Sample of 30 Districts

Strata	No. of Districts N_h	Variance of Measures σ_h^2	$N_h\sigma_h^2$	$\dfrac{N_h\sigma_h^2}{\Sigma N_h\sigma_h^2}$	Allocation for $n = 30$
Stratum A	68	1304	88,672	.70	21
Stratum B	53	705	37,365	.30	9
Total	121		126,037	1.00	30

Table 19-3 contains the results of drawing random samples of the required size from the two strata A and B. For each district, kindergarten enrollments are recorded for two successive years, together with their differences. Here the districts are the cases and the differences are the scores which are used in the analysis.

Summary calculations based on Table 19-3 are presented in Table 19-4. The entries in this table follow the notation which was developed in connection with stratified sampling. Of special interest are the entries in the total column when only those entries which are of interest are combined.

The totals in lines 1 and 2 of Table 19-4 are evidently N and n. The total in line 5 is given by formula (19-29) and is an estimate of the sum of the differences for all 121 districts in the area. The total in line 7 is the variance of the estimated total, namely $s^2_{\hat{T}_{st}}$ given by formula (19-30). The entries in lines 8 and 9 are combined forms rather than totals, these being the estimated mean and its variance given by formulas (19-31) and (19-32) respectively.

Table 19-3   Kindergarten Enrollment in Two Strata
of a Total Area for Two Successive Years,
and Their Differences

District	Prior Year	Current Year	Difference
Stratum A			
1	402	359	−43
2	635	616	−19
3	38	45	7
4	130	101	−29
5	559	574	15
6	44	40	−4
7	647	687	40
8	314	379	65
9	82	105	23
10	695	655	−40
11	1050	970	−80
12	1045	1164	119
13	38	36	−2
14	853	911	58
15	121	100	−21
16	19	18	−1
17	666	810	144
18	179	167	−12
19	1001	1004	3
20	121	131	10
21	630	598	−32
Stratum B			
1	317	281	−36
2	809	748	−61
3	507	514	7
4	273	283	10
5	332	338	6
6	332	353	21
7	661	601	−60
8	389	403	14
9	699	675	−24

Confidence limits with coefficient .95 can now be found for both the
total and the mean. These follow.

For the total of all 121 districts:

$$\text{Upper limit} = -73.75 + 1.96 \sqrt{694{,}197} = 1559$$
$$\text{Lower limit} = -73.75 - 1.96 \sqrt{694{,}197} = -1707$$

Table 19-4 Summary Measures Based on Differences in Table 19-3

Line No.	Kind of Data	Stratum A	Stratum B	Total or Combination
1	Cases in stratum, N_h	68	53	121
2	Cases in sample, n_h	21	9	30
3	Sum of enrollment differences in sample, x_h	201	−123	
4	Mean of differences $\bar{x}_h = x_h/n_h$	9.57	−13.67	
5	$N_h\bar{x}_h = \hat{T}_h$	650.76	−724.51	−73.75
6	s_h^2	2787.76	1041.75	
7	$\dfrac{N_h(N_h - n_h)}{n_h}\, s_h^2$	424,269	269,928	694,197
8	$\bar{x}_{st} = \dfrac{\Sigma N_h\bar{x}_h}{N}$			−.610
9	$s_{\bar{x}_{st}}^2$			47.41

For the district mean:

$$\text{Upper limit} = -.610 + 1.96 \sqrt{47.41} = 12.9$$
$$\text{Lower limit} = -.610 - 1.96 \sqrt{47.41} = -14.1$$

The limits which have been obtained to this point apply to the enrollment difference between the current year and the preceding year. These limits can be translated into limits for actual enrollment for the current year.

Such a translation is possible, because at the time of the present study complete data are available for the prior year. For that year and for all 121 districts in the area the following is known.

Total kindergarten enrollment: 53,141
Average kindergarten enrollment: 439.2

One can now use the formula:

current enrollment equals
prior year enrollment
plus the difference between years.

Since prior year's enrollment is known exactly, it is only necessary to insert the limits for the difference to obtain the limits for the total and mean.

The following limits are, therefore, available for current enrollment.

For total enrollment:

$$\text{Upper limit} = 53{,}141 + 1{,}559 = 54{,}700$$
$$\text{Lower limit} = 53{,}141 - 1{,}707 = 51{,}434$$

For average enrollment:

$$\text{Upper limit} = 439.2 + 12.9 = 452.1$$
$$\text{Lower limit} = 439.2 - 14.1 = 425.1$$

EXERCISE 19-3

Calculate limits for current year kindergarten enrollment using the entries for both stratum A and stratum B in the column headed "Current Year."

Compare the accuracy in this calculation with that obtained by using the entries in the "Difference" column.

D. ADDITIONAL COMMENTS ON SURVEY SAMPLING

Further Applications. The applications of survey sampling described in previous pages should not be taken as limiting the scope of the use of the theory. Actually, many additional applications to measurement of opinion and attitude are available. A very important area of application is illustrated by the prediction and opinion polls commonly presented in newspapers. Market surveys conducted by advertising firms also use survey sampling methods.

The important feature of survey sampling theory is its application to measurement of a finite population, as has already been pointed out.

Additional Methods. The methods described in the preceding sections constitute only an elementary introduction to a very extensive theory. Some extensions of the theory will be discussed briefly.

In addition to simple random sampling where all items have equal probability of being chosen, it is possible to introduce sampling with variable probability of selection for different items. Commonly this is done by selection with probability proportional to the size of the item. Thus, if one area has ten times the population of another, it is assigned ten times as many numbers as the other and, therefore, has ten times the chance of being selected. Sampling with varying probability of selection leads to increased accuracy of estimate.

Cluster sampling may be extended to include subsampling within clusters, in contrast to complete sampling of clusters, as was described above. Such subsampling is called two-stage sampling. Actually, subsampling can be extended to involve several stages.

Methods of choosing sample sizes within strata, once total sample sizes have been agreed upon, were described above. However, methods for deciding upon total sample size are also available whether or not stratified sampling is used.

The choice of sample size can be made on two alternative bases. One basis is to decide upon the degree of precision which is desired, that is, the size of the difference between the upper and lower limits. Another basis is provided by the funds which are available for the survey. Sample sizes are then determined so as to obtain the maximum precision possible within the restriction of available funds.

These and other extensions of the elementary theory can be found in books devoted to survey sampling.

E. REFERENCES ON SURVEY SAMPLING

COCHRAN, W. G., "Modern Methods in the Sampling of Human Populations," *American Journal of Public Health,* **41** (1951), 647–653.

———, *Sampling Techniques* (3d ed.). New York: John Wiley & Sons, Inc., 1963.

DEMING, W. EDWARDS, *Some Theory of Sampling.* New York: John Wiley & Sons, Inc., 1950.

———, *Sample Design in Business Research.* New York: John Wiley & Sons, Inc., 1960.

HANSEN, M. H., AND P. M. HAUSER, "Area Sampling—Some Principles of Sample Design," *Public Opinion Quarterly,* Summer 1945, 183–193.

———, W. N. HOROWITZ, AND W. G. MADOW, *Sample Survey Methods and Theory.* New York: John Wiley & Sons, Inc., 1953.

HESS, I., D. C. RIEDEL, AND T. B. FITZPATRICK, *Probability Sampling of Hospitals and Patients.* Ann Arbor: University of Michigan Press, 1961.

KING, A. J., AND R. T. JESSEN, "The Master Sample of Agriculture," *Journal of the American Statistical Association,* **40** (1945), 38–56.

KISH, L., *Survey Sampling.* New York: John Wiley & Sons, Inc., 1965.

MARKS, E. S., "Selective Sampling in Psychological Research," *Psychological Bulletin,* **44** (1947), 267–275.

SLONIM, M. J., *Sampling in a Nutshell.* New York: Simon and Schuster, Inc., 1960.

SNEDECOR G. AND W. G. COCHRAN, *Statistical Methods* (6th ed.). Ames: Iowa State University Press, 1967, Chapter 17, pp. 504–539.

STEPHAN, F. F., "Stratification in Representative Sampling," *Journal of Marketing,* **6** (1941), 38–47.

————, "History of the Uses of Modern Sampling Procedures," *Journal of the American Statistical Association*, **43** (1948), 12–39.

YATES, F., "A Review of Recent Statistical Developments in Sampling and Sampling Surveys," *Journal of the Royal Statistical Society*, **109** (1946), 12–43.

————, *Sampling Methods for Censuses and Surveys* (2d ed.). London: Charles Griffin & Co., Ltd., 1960.

APPENDIX
TABLES

LIST OF TABLES

IN APPENDIX

Table I   Ordinates and Areas of the Standard Normal Curve*

z	Area	Ordinate	z	Area	Ordinate	z	Area	Ordinate
.00	.0000	.3989	.50	.1915	.3521	1.00	.3413	.2420
.01	.0040	.3989	.51	.1950	.3503	1.01	.3438	.2396
.02	.0080	.3989	.52	.1985	.3485	1.02	.3461	.2371
.03	.0120	.3988	.53	.2019	.3467	1.03	.3485	.2347
.04	.0160	.3986	.54	.2054	.3448	1.04	.3508	.2323
.05	.0199	.3984	.55	.2088	.3429	1.05	.3531	.2299
.06	.0239	.3982	.56	.2123	.3410	1.06	.3554	.2275
.07	.0279	.3980	.57	.2157	.3391	1.07	.3577	.2251
.08	.0319	.3977	.58	.2190	.3372	1.08	.3599	.2227
.09	.0359	.3973	.59	.2224	.3352	1.09	.3621	.2203
.10	.0398	.3970	.60	.2257	.3332	1.10	.3643	.2179
.11	.0438	.3965	.61	.2291	.3312	1.11	.3665	.2155
.12	.0478	.3961	.62	.2324	.3292	1.12	.3686	.2131
.13	.0517	.3956	.63	.2357	.3271	1.13	.3708	.2107
.14	.0557	.3951	.64	.2389	.3251	1.14	.3729	.2083
.15	.0596	.3945	.65	.2422	.3230	1.15	.3749	.2059
.16	.0636	.3939	.66	.2454	.3209	1.16	.3770	.2036
.17	.0675	.3932	.67	.2486	.3187	1.17	.3790	.2012
.18	.0714	.3925	.68	.2517	.3166	1.18	.3810	.1989
.19	.0753	.3918	.69	.2549	.3144	1.19	.3830	.1965
.20	.0793	.3910	.70	.2580	.3123	1.20	.3849	.1942
.21	.0832	.3902	.71	.2611	.3101	1.21	.3869	.1919
.22	.0871	.3894	.72	.2642	.3079	1.22	.3888	.1895
.23	.0910	.3885	.73	.2673	.3056	1.23	.3907	.1872
.24	.0948	.3876	.74	.2703	.3034	1.24	.3925	.1849
.25	.0987	.3867	.75	.2734	.3011	1.25	.3944	.1826
.26	.1026	.3857	.76	.2764	.2989	1.26	.3962	.1804
.27	.1064	.3847	.77	.2794	.2966	1.27	.3980	.1781
.28	.1103	.3836	.78	.2823	.2943	1.28	.3997	.1758
.29	.1141	.3825	.79	.2852	.2920	1.29	.4015	.1736
.30	.1179	.3814	.80	.2881	.2897	1.30	.4032	.1714
.31	.1217	.3802	.81	.2910	.2874	1.31	.4049	.1691
.32	.1255	.3790	.82	.2939	.2850	1.32	.4066	.1669
.33	.1293	.3778	.83	.2967	.2827	1.33	.4082	.1647
.34	.1331	.3765	.84	.2995	.2803	1.34	.4099	.1626
.35	.1368	.3752	.85	.3023	.2780	1.35	.4115	.1604
.36	.1406	.3739	.86	.3051	.2756	1.36	.4131	.1582
.37	.1443	.3725	.87	.3078	.2732	1.37	.4147	.1561
.38	.1480	.3712	.88	.3106	.2709	1.38	.4162	.1539
.39	.1517	.3697	.89	.3133	.2685	1.39	.4177	.1518
.40	.1554	.3683	.90	.3159	.2661	1.40	.4192	.1497
.41	.1591	.3668	.91	.3186	.2637	1.41	.4207	.1476
.42	.1628	.3653	.92	.3212	.2613	1.42	.4222	.1456
.43	.1664	.3637	.93	.3238	.2589	1.43	.4236	.1435
.44	.1700	.3621	.94	.3264	.2565	1.44	.4251	.1415
.45	.1736	.3605	.95	.3289	.2541	1.45	.4265	.1394
.46	.1772	.3589	.96	.3315	.2516	1.46	.4279	.1374
.47	.1808	.3572	.97	.3340	.2492	1.47	.4292	.1354
.48	.1844	.3555	.98	.3365	.2468	1.48	.4306	.1334
.49	.1879	.3538	.99	.3389	.2444	1.49	.4319	.1315
.50	.1915	.3521	1.00	.3413	.2420	1.50	.4332	.1295

Table I Ordinates and Areas of the Standard Normal Curve
(*Concluded*)

z	Area	Ordinate	z	Area	Ordinate	z	Area	Ordinate
1.50	.4332	.1295	2.00	.4772	.0540	2.50	.4938	.0175
1.51	.4345	.1276	2.01	.4778	.0529	2.51	.4940	.0171
1.52	.4357	.1257	2.02	.4783	.0519	2.52	.4941	.0167
1.53	.4370	.1238	2.03	.4788	.0508	2.53	.4943	.0163
1.54	.4382	.1219	2.04	.4793	.0498	2.54	.4945	.0158
1.55	.4394	.1200	2.05	.4798	.0488	2.55	.4946	.0154
1.56	.4406	.1182	2.06	.4803	.0478	2.56	.4948	.0151
1.57	.4418	.1163	2.07	.4808	.0468	2.57	.4949	.0147
1.58	.4429	.1145	2.08	.4812	.0459	2.58	.4951	.0143
1.59	.4441	.1127	2.09	.4817	.0449	2.59	.4952	.0139
1.60	.4452	.1109	2.10	.4821	.0440	2.60	.4953	.0136
1.61	.4463	.1092	2.11	.4826	.0431	2.61	.4955	.0132
1.62	.4474	.1074	2.12	.4830	.0422	2.62	.4956	.0129
1.63	.4484	.1057	2.13	.4834	.0413	2.63	.4957	.0126
1.64	.4495	.1040	2.14	.4838	.0404	2.64	.4959	.0122
1.65	.4505	.1023	2.15	.4842	.0395	2.65	.4960	.0119
1.66	.4515	.1006	2.16	.4846	.0387	2.66	.4961	.0116
1.67	.4525	.0989	2.17	.4850	.0379	2.67	.4962	.0113
1.68	.4535	.0973	2.18	.4854	.0371	2.68	.4963	.0110
1.69	.4545	.0957	2.19	.4857	.0363	2.69	.4964	.0107
1.70	.4554	.0940	2.20	.4861	.0355	2.70	.4965	.0104
1.71	.4564	.0925	2.21	.4864	.0347	2.71	.4966	.0101
1.72	.4573	.0909	2.22	.4868	.0339	2.72	.4967	.0099
1.73	.4582	.0893	2.23	.4871	.0332	2.73	.4968	.0096
1.74	.4591	.0878	2.24	.4875	.0325	2.74	.4969	.0093
1.75	.4599	.0863	2.25	.4878	.0317	2.75	.4970	.0091
1.76	.4608	.0848	2.26	.4881	.0310	2.76	.4971	.0088
1.77	.4616	.0833	2.27	.4884	.0303	2.77	.4972	.0086
1.78	.4625	.0818	2.28	.4887	.0297	2.78	.4973	.0084
1.79	.4633	.0804	2.29	.4890	.0290	2.79	.4974	.0081
1.80	.4641	.0790	2.30	.4893	.0283	2.80	.4974	.0079
1.81	.4649	.0775	2.31	.4896	.0277	2.81	.4975	.0077
1.82	.4656	.0761	2.32	.4898	.0270	2.82	.4976	.0075
1.83	.4664	.0748	2.33	.4901	.0264	2.83	.4977	.0073
1.84	.4671	.0734	2.34	.4904	.0258	2.84	.4977	.0071
1.85	.4678	.0721	2.35	.4906	.0252	2.85	.4978	.0069
1.86	.4686	.0707	2.36	.4909	.0246	2.86	.4979	.0067
1.87	.4693	.0694	2.37	.4911	.0241	2.87	.4979	.0065
1.88	.4699	.0681	2.38	.4913	.0235	2.88	.4980	.0063
1.89	.4706	.0669	2.39	.4916	.0229	2.89	.4981	.0061
1.90	.4713	.0656	2.40	.4918	.0224	2.90	.4981	.0060
1.91	.4719	.0644	2.41	.4920	.0219	2.91	.4982	.0058
1.92	.4726	.0632	2.42	.4922	.0213	2.92	.4982	.0056
1.93	.4732	.0620	2.43	.4925	.0208	2.93	.4983	.0055
1.94	.4738	.0608	2.44	.4927	.0203	2.94	.4984	.0053
1.95	.4744	.0596	2.45	.4929	.0198	2.95	.4984	.0051
1.96	.4750	.0584	2.46	.4931	.0194	2.96	.4985	.0050
1.97	.4756	.0573	2.47	.4932	.0189	2.97	.4985	.0048
1.98	.4761	.0562	2.48	.4934	.0184	2.98	.4986	.0047
1.99	.4767	.0551	2.49	.4936	.0180	2.99	.4986	.0046
2.00	.4772	.0540	2.50	.4938	.0175	3.00	.4987	.0044

* SOURCE: This table is reproduced from J. E. Wert, *Educational Statistics*, by courtesy of McGraw-Hill Book Company, New York.

Table II Percentile Values of the Standard Normal Distribution*

Area to the Left of z	z	Area to the Left of z	z	Area to the Left of z	z	Area to the Left of z	z	Area to the Left of z	z
.0001	−3.719	.045	−1.695	.280	−.583	.700	.524	.950	1.645
.0002	−3.540	.050	−1.645	.300	−.524	.720	.583	.955	1.695
.0003	−3.432	.055	−1.598	.320	−.468	.740	.643	.960	1.751
.0004	−3.353	.060	−1.555	.340	−.412	.750	.6745	.965	1.812
.0005	−3.291	.065	−1.514	.360	−.358	.760	.706	.970	1.881
.001	−3.090	.070	−1.476	.380	−.305	.780	.772	.975	1.960
.002	−2.878	.075	−1.440	.400	−.253	.800	.842	.980	2.054
.003	−2.748	.080	−1.405	.420	−.202	.820	.915	.985	2.170
.004	−2.652	.085	−1.372	.440	−.151	.840	.994	.990	2.326
.005	−2.576	.090	−1.341	.460	−.100	.860	1.080	.991	2.366
.006	−2.512	.095	−1.311	.480	−.050	.880	1.175	.992	2.409
.007	−2.457	.100	−1.282	.500	.000	.900	1.282	.993	2.457
.008	−2.409	.120	−1.175	.520	.050	.905	1.311	.994	2.512
.009	−2.366	.140	−1.080	.540	.100	.910	1.341	.995	2.576
.010	−2.326	.160	−.994	.560	.151	.915	1.372	.996	2.652
.015	−2.170	.180	−.915	.580	.202	.920	1.405	.997	2.748
.020	−2.054	.200	−.842	.600	.253	.925	1.440	.998	2.878
.025	−1.960	.220	−.772	.620	.305	.930	1.476	.999	3.090
.030	−1.881	.240	−.706	.640	.358	.935	1.514	.9995	3.291
.035	−1.812	.250	−.6745	.660	.412	.940	1.555	.9996	3.353
.040	−1.751	.260	−.643	.680	.468	.945	1.598	.9999	3.719

* SOURCE: Entries in this table are taken from *The Kelley Statistical Tables*, Harvard University Press, 1938, revised 1948, by permission of the author, Truman Lee Kelley and the publisher.

z_α is the percentile value named by the area to the left of z. Thus $z_{.30} = −.524$ is the 30th percentile. For an interpretation of the values of z as percentiles, see page 117.

Table III Percentile Values of "Student's" Distribution*

ν	$t_{.75}$	$t_{.80}$	$t_{.90}$	$t_{.95}$	$t_{.975}$	$t_{.99}$	$t_{.995}$	$t_{.9995}$	ν
1	1.00	1.38	3.08	6.31	12.71	31.82	63.66	636.62	1
2	.82	1.06	1.89	2.92	4.30	6.96	9.92	31.60	2
3	.76	.98	1.64	2.35	3.18	4.54	5.84	12.94	3
4	.74	.94	1.53	2.13	2.78	3.75	4.60	8.61	4
5	.73	.92	1.48	2.02	2.57	3.36	4.03	6.86	5
6	.72	.91	1.44	1.94	2.45	3.14	3.71	5.96	6
7	.71	.90	1.42	1.89	2.36	3.00	3.50	5.40	7
8	.71	.89	1.40	1.86	2.31	2.90	3.36	5.04	8
9	.70	.88	1.38	1.83	2.26	2.82	3.25	4.78	9
10	.70	.88	1.37	1.81	2.23	2.76	3.17	4.59	10
11	.70	.88	1.36	1.80	2.20	2.72	3.11	4.44	11
12	.70	.87	1.36	1.78	2.18	2.68	3.05	4.32	12
13	.69	.87	1.35	1.77	2.16	2.65	3.01	4.22	13
14	.69	.87	1.34	1.76	2.14	2.62	2.98	4.14	14
15	.69	.87	1.34	1.75	2.13	2.60	2.95	4.07	15
16	.69	.87	1.34	1.75	2.12	2.58	2.92	4.02	16
17	.69	.86	1.33	1.74	2.11	2.57	2.90	3.96	17
18	.69	.86	1.33	1.73	2.10	2.55	2.88	3.92	18
19	.69	.86	1.33	1.73	2.09	2.54	2.86	3.88	19
20	.69	.86	1.32	1.72	2.09	2.53	2.85	3.85	20
21	.69	.86	1.32	1.72	2.08	2.52	2.83	3.82	21
22	.69	.86	1.32	1.72	2.07	2.51	2.82	3.79	22
23	.69	.86	1.32	1.71	2.07	2.50	2.81	3.77	23
24	.68	.86	1.32	1.71	2.06	2.49	2.80	3.74	24
25	.68	.86	1.32	1.71	2.06	2.48	2.79	3.72	25
26	.68	.86	1.32	1.71	2.06	2.48	2.78	3.71	26
27	.68	.86	1.31	1.70	2.05	2.47	2.77	3.69	27
28	.68	.85	1.31	1.70	2.05	2.47	2.76	3.67	28
29	.68	.85	1.31	1.70	2.04	2.46	2.76	3.66	29
30	.68	.85	1.31	1.70	2.04	2.46	2.75	3.65	30
40	.68	.85	1.30	1.68	2.02	2.42	2.70	3.55	40
60	.68	.85	1.30	1.67	2.00	2.39	2.66	3.46	60
120	.68	.85	1.29	1.66	1.98	2.36	2.62	3.37	120
∞	.6745	.842	1.282	1.645	1.960	2.326	2.576	3.291	∞
	$-t_{.25}$	$-t_{.20}$	$-t_{.10}$	$-t_{.05}$	$-t_{.025}$	$-t_{.01}$	$-t_{.005}$	$-t_{.0005}$	

* SOURCE: Reprinted abridged from R. A. Fisher and F. Yates, *Statistical Tables for Biological Agricultural, and Medical Research*, published by Oliver & Boyd Ltd., Edinburgh, 1963, by permission of the authors and publishers.

Table IV   Percentile Values of the Chi-square Distribution*

ν	$\chi^2_{.999}$	$\chi^2_{.995}$	$\chi^2_{.99}$	$\chi^2_{.98}$	$\chi^2_{.975}$	$\chi^2_{.95}$	$\chi^2_{.90}$	$\chi^2_{.75}$	$\chi^2_{.50}$	$\chi^2_{.25}$	$\chi^2_{.10}$	$\chi^2_{.05}$	$\chi^2_{.025}$	$\chi^2_{.02}$	$\chi^2_{.01}$	$\chi^2_{.005}$	ν
1	10.8	7.9	6.6	5.4	5.0	3.8	2.7	1.3	.46	.10	.02	—	—	—	—	—	1
2	13.8	10.6	9.2	7.8	7.4	6.0	4.6	2.8	1.4	.58	.21	.10	.05	.04	.02	.01	2
3	16.3	12.8	11.3	9.8	9.4	7.8	6.3	4.1	2.4	1.21	.58	.35	.22	.18	.11	.07	3
4	18.5	14.9	13.3	11.7	11.1	9.5	7.8	5.4	3.4	1.92	1.1	.71	.48	.43	.30	.21	4
5	20.5	16.7	15.1	13.4	12.8	11.1	9.2	6.6	4.4	2.7	1.6	1.1	.83	.75	.55	.41	5
6	22.5	18.5	16.8	15.0	14.4	12.6	10.6	7.8	5.4	3.5	2.2	1.6	1.2	1.13	.87	.68	6
7	24.3	20.3	18.5	16.6	16.0	14.1	12.0	9.0	6.4	4.3	2.8	2.2	1.7	1.56	1.24	.99	7
8	26.1	22.0	20.1	18.2	17.5	15.5	13.4	10.2	7.3	5.1	3.5	2.7	2.2	2.03	1.65	1.3	8
9	27.9	23.6	21.7	19.7	19.0	16.9	14.7	11.4	8.3	5.9	4.2	3.3	2.7	2.53	2.09	1.7	9
10	29.6	25.2	23.2	21.2	20.5	18.3	16.0	12.5	9.3	6.7	4.9	3.9	3.2	3.06	2.55	2.2	10
11	31.3	26.8	24.7	22.6	21.9	19.7	17.3	13.7	10.3	7.6	5.6	4.6	3.8	3.61	3.05	2.6	11
12	32.9	28.3	26.2	24.1	23.3	21.0	18.5	14.8	11.3	8.4	6.3	5.2	4.4	4.18	3.57	3.1	12
13	34.5	29.8	27.7	25.5	24.7	22.4	19.8	16.0	12.3	9.3	7.0	5.9	5.0	4.76	4.11	3.6	13
14	36.1	31.3	29.1	26.9	26.1	23.7	21.1	17.1	13.3	10.2	7.8	6.6	5.6	5.37	4.66	4.1	14
15	37.7	32.8	30.6	28.3	27.5	25.0	22.3	18.2	14.3	11.0	8.5	7.3	6.3	5.98	5.23	4.6	15
16	39.3	34.3	32.0	29.6	28.8	26.3	23.5	19.4	15.3	11.9	9.3	8.0	6.9	6.61	5.81	5.1	16
17	40.8	35.7	33.4	31.0	30.2	27.6	24.8	20.5	16.3	12.8	10.1	8.7	7.6	7.26	6.41	5.7	17
18	42.3	37.2	34.8	32.3	31.5	28.9	26.0	21.6	17.3	13.7	10.9	9.4	8.2	7.91	7.02	6.3	18
19	43.8	38.6	36.2	33.7	32.9	30.1	27.2	22.7	18.3	14.6	11.7	10.1	8.9	8.57	7.63	6.9	19
20	45.3	40.0	37.6	35.0	34.2	31.4	28.4	23.8	19.3	15.5	12.4	10.9	9.6	9.24	8.26	7.4	20
21	46.8	41.4	38.9	36.3	35.5	32.7	29.6	24.9	20.3	16.3	13.2	11.6	10.3	9.9	8.9	8.0	21
22	48.3	42.8	40.3	37.7	36.8	33.9	30.8	26.0	21.3	17.2	14.0	12.3	11.0	10.6	9.5	8.6	22
23	49.7	44.2	41.6	39.0	38.1	35.2	32.0	27.1	22.3	18.1	14.8	13.1	11.7	11.3	10.2	9.3	23
24	51.2	45.6	43.0	40.3	39.4	36.4	33.2	28.2	23.3	19.0	15.7	13.8	12.4	12.0	10.9	9.9	24
25	52.6	46.9	44.3	41.6	40.6	37.7	34.4	29.3	24.3	19.9	16.5	14.6	13.1	12.7	11.5	10.5	25
26	54.0	48.3	45.6	42.9	41.9	38.9	35.6	30.4	25.3	20.8	17.3	15.4	13.8	13.4	12.2	11.2	26
27	55.5	49.6	47.0	44.1	43.2	40.1	36.7	31.5	26.3	21.7	18.1	16.2	14.6	14.1	12.9	11.8	27
28	56.9	51.0	48.3	45.4	44.5	41.3	37.9	32.6	27.3	22.7	18.9	16.9	15.3	14.8	13.6	12.5	28
29	58.3	52.3	49.6	46.7	45.7	42.6	39.1	33.7	28.3	23.6	19.8	17.7	16.0	15.6	14.3	13.1	29
30	59.7	53.7	50.9	48.0	47.0	43.8	40.3	34.8	29.3	24.5	20.6	18.5	16.8	16.3	15.0	13.8	30
40	73.5	66.8	63.7	60.4	59.3	55.8	51.8	45.6	39.3	33.7	29.1	26.5	24.4	23.8	22.2	20.7	40
60	99.7	92.0	88.4	84.6	83.3	79.1	74.4	67.0	59.3	52.3	46.5	43.2	40.5	39.7	37.5	35.5	60
100	149.5	140.2	135.8	131.1	129.6	124.3	118.5	109.1	99.3	90.1	82.4	77.9	74.2	73.1	70.0	67.3	100

* SOURCE: Abridged from table in *Biometrika*, Vol. 32 (1941), and published with the permission of the author, Catherine M. Thompson, and the editor of *Biometrika*. Columns $\chi^2_{.02}$, $\chi^2_{.98}$, and $\chi^2_{.999}$ are reprinted abridged from R. A. Fisher and F. Yates, *Statistical Tables for Biological, Agricultural, and Medical Research*, published by Oliver & Boyd Ltd., Edinburgh, 1963, by permission of the authors and publishers.

Table V   95th and 99th Percentile

95th Percentile in Light-Face Type,

$\nu_1 = $ *degrees of freedom*

ν_2	1	2	3	4	5	6	7	8	9	10	11	12
1	161	200	216	225	230	234	237	239	241	242	243	244
	4,052	4,999	5,403	5,625	5,764	5,859	5,928	5,981	6,022	6,056	6,082	6,106
2	18.51	19.00	19.16	19.25	19.30	19.33	19.36	19.37	19.38	19.39	19.40	19.41
	98.49	99.01	99.17	99.25	99.30	99.33	99.34	99.36	99.38	99.40	99.41	99.42
3	10.13	9.55	9.28	9.12	9.01	8.94	8.88	8.84	8.81	8.78	8.76	8.74
	34.12	30.81	29.46	28.71	28.24	27.91	27.67	27.49	27.34	27.23	27.13	27.05
4	7.71	6.94	6.59	6.39	6.26	6.16	6.09	6.04	6.00	5.96	5.93	5.91
	21.20	18.00	16.69	15.98	15.52	15.21	14.98	14.80	14.66	14.54	14.45	14.37
5	6.61	5.79	5.41	5.19	5.05	4.95	4.88	4.82	4.78	4.74	4.70	4.68
	16.26	13.27	12.06	11.39	10.97	10.67	10.45	10.27	10.15	10.05	9.96	9.89
6	5.99	5.14	4.76	4.53	4.39	4.28	4.21	4.15	4.10	4.06	4.03	4.00
	13.74	10.92	9.78	9.15	8.75	8.47	8.26	8.10	7.98	7.87	7.79	7.72
7	5.59	4.74	4.35	4.12	3.97	3.87	3.79	3.73	3.68	3.63	3.60	3.57
	12.25	9.55	8.45	7.85	7.46	7.19	7.00	6.84	6.71	6.62	6.54	6.47
8	5.32	4.46	4.07	3.84	3.69	3.58	3.50	3.44	3.39	3.34	3.31	3.28
	11.26	8.65	7.59	7.01	6.63	6.37	6.19	6.03	5.91	5.82	5.74	5.67
9	5.12	4.26	3.86	3.63	3.48	3.37	3.29	3.23	3.18	3.13	3.10	3.07
	10.56	8.02	6.99	6.42	6.06	5.80	5.62	5.47	5.35	5.26	5.18	5.11
10	4.96	4.10	3.71	3.48	3.33	3.22	3.14	3.07	3.02	2.97	2.94	2.91
	10.04	7.56	6.55	5.99	5.64	5.39	5.21	5.06	4.95	4.85	4.78	4.71
11	4.84	3.98	3.59	3.36	3.20	3.09	3.01	2.95	2.90	2.86	2.82	2.79
	9.65	7.20	6.22	5.67	5.32	5.07	4.88	4.74	4.63	4.54	4.46	4.40
12	4.75	3.88	3.49	3.26	3.11	3.00	2.92	2.85	2.80	2.76	2.72	2.69
	9.33	6.93	5.95	5.41	5.06	4.82	4.65	4.50	4.39	4.30	4.22	4.16
13	4.67	3.80	3.41	3.18	3.02	2.92	2.84	2.77	2.72	2.67	2.63	2.60
	9.07	6.70	5.74	5.20	4.86	4.62	4.44	4.30	4.19	4.10	4.02	3.96
14	4.60	3.74	3.34	3.11	2.96	2.85	2.77	2.70	2.65	2.60	2.56	2.53
	8.86	6.51	5.56	5.03	4.69	4.46	4.28	4.14	4.03	3.94	3.86	3.80
15	4.54	3.68	3.29	3.06	2.90	2.79	2.70	2.64	2.59	2.55	2.51	2.48
	8.68	6.36	5.42	4.89	4.56	4.32	4.14	4.00	3.89	3.80	3.73	3.67
16	4.49	3.63	3.24	3.01	2.85	2.74	2.66	2.59	2.54	2.49	2.45	2.42
	8.53	6.23	5.29	4.77	4.44	4.20	4.03	3.89	3.78	3.69	3.61	3.55
17	4.45	3.59	3.20	2.96	2.81	2.70	2.62	2.55	2.50	2.45	2.41	2.38
	8.40	6.11	5.18	4.67	4.34	4.10	3.93	3.79	3.68	3.59	3.52	3.45
18	4.41	3.55	3.16	2.93	2.77	2.66	2.58	2.51	2.46	2.41	2.37	2.34
	8.28	6.01	5.09	4.58	4.25	4.01	3.85	3.71	3.60	3.51	3.44	3.37
19	4.38	3.52	3.13	2.90	2.74	2.63	2.55	2.48	2.43	2.38	2.34	2.31
	8.18	5.93	5.01	4.50	4.17	3.94	3.77	3.63	3.52	3.43	3.36	3.30
20	4.35	3.49	3.10	2.87	2.71	2.60	2.52	2.45	2.40	2.35	2.31	2.28
	8.10	5.85	4.94	4.43	4.10	3.87	3.71	3.56	3.45	3.37	3.30	3.23
21	4.32	3.47	3.07	2.84	2.68	2.57	2.49	2.42	2.37	2.32	2.28	2.25
	8.02	5.78	4.87	4.37	4.04	3.81	3.65	3.51	3.40	3.31	3.24	3.17
22	4.30	3.44	3.05	2.82	2.66	2.55	2.47	2.40	2.35	2.30	2.26	2.23
	7.94	5.72	4.82	4.31	3.99	3.76	3.59	3.45	3.35	3.26	3.18	3.12
23	4.28	3.42	3.03	2.80	2.64	2.53	2.45	2.38	2.32	2.28	2.24	2.20
	7.88	5.66	4.76	4.26	3.94	3.71	3.54	3.41	3.30	3.21	3.14	3.07
24	4.26	3.40	3.01	2.78	2.62	2.51	2.43	2.36	2.30	2.26	2.22	2.18
	7.82	5.61	4.72	4.22	3.90	3.67	3.50	3.36	3.25	3.17	3.09	3.03
25	4.24	3.38	2.99	2.76	2.60	2.49	2.41	2.34	2.28	2.24	2.20	2.16
	7.77	5.57	4.68	4.18	3.86	3.63	3.46	3.32	3.21	3.13	3.05	2.99
26	4.22	3.37	2.98	2.74	2.59	2.47	2.39	2.32	2.27	2.22	2.18	2.15
	7.72	5.53	4.64	4.14	3.82	3.59	3.42	3.29	3.17	3.09	3.02	2.96

$\nu_2 = $ degrees of freedom for denominator

* SOURCE: Reproduced by permission from, *Statistical Methods*, 6th edition, by George W. Snedecor

Values of the F Distribution*

99th Percentile in Bold-Face Type
for numerator

14	16	20	24	30	40	50	75	100	200	500	∞	ν_2
245 **6,142**	246 **6,169**	248 **6,208**	249 **6,234**	250 **6,258**	251 **6,286**	252 **6,302**	253 **6,323**	253 **6,334**	254 **6,352**	254 **6,361**	254 **6,366**	1
19.42 **99.43**	19.43 **99.44**	19.44 **99.45**	19.45 **99.46**	19.46 **99.47**	19.47 **99.48**	19.47 **99.48**	19.48 **99.49**	19.49 **99.49**	19.49 **99.49**	19.50 **99.50**	19.50 **99.50**	2
8.71 **26.92**	8.69 **26.83**	8.66 **26.69**	8.64 **26.60**	8.62 **26.50**	8.60 **26.41**	8.58 **26.35**	8.57 **26.27**	8.56 **26.23**	8.54 **26.18**	8.54 **26.14**	8.53 **26.12**	3
5.87 **14.24**	5.84 **14.15**	5.80 **14.02**	5.77 **13.93**	5.74 **13.83**	5.71 **13.74**	5.70 **13.69**	5.68 **13.61**	5.66 **13.57**	5.65 **13.52**	5.64 **13.48**	5.63 **13.46**	4
4.64 **9.77**	4.60 **9.68**	4.56 **9.55**	4.53 **9.47**	4.50 **9.38**	4.46 **9.29**	4.44 **9.24**	4.42 **9.17**	4.40 **9.13**	4.38 **9.07**	4.37 **9.04**	4.36 **9.02**	5
3.96 **7.60**	3.92 **7.52**	3.87 **7.39**	3.84 **7.31**	3.81 **7.23**	3.77 **7.14**	3.75 **7.09**	3.72 **7.02**	3.71 **6.99**	3.69 **6.94**	3.68 **6.90**	3.67 **6.88**	6
3.52 **6.35**	3.49 **6.27**	3.44 **6.15**	3.41 **6.07**	3.38 **5.98**	3.34 **5.90**	3.32 **5.85**	3.29 **5.78**	3.28 **5.75**	3.25 **5.70**	3.24 **5.67**	3.23 **5.65**	7
3.23 **5.56**	3.20 **5.48**	3.15 **5.36**	3.12 **5.28**	3.08 **5.20**	3.05 **5.11**	3.03 **5.06**	3.00 **5.00**	2.98 **4.96**	2.96 **4.91**	2.94 **4.88**	2.93 **4.86**	8
3.02 **5.00**	2.98 **4.92**	2.93 **4.80**	2.90 **4.73**	2.86 **4.64**	2.82 **4.56**	2.80 **4.51**	2.77 **4.45**	2.76 **4.41**	2.73 **4.36**	2.72 **4.33**	2.71 **4.31**	9
2.86 **4.60**	2.82 **4.52**	2.77 **4.41**	2.74 **4.33**	2.70 **4.25**	2.67 **4.17**	2.64 **4.12**	2.61 **4.05**	2.59 **4.01**	2.56 **3.96**	2.55 **3.93**	2.54 **3.91**	10
2.74 **4.29**	2.70 **4.21**	2.65 **4.10**	2.61 **4.02**	2.57 **3.94**	2.53 **3.86**	2.50 **3.80**	2.47 **3.74**	2.45 **3.70**	2.42 **3.66**	2.41 **3.62**	2.40 **3.60**	11
2.64 **4.05**	2.60 **3.98**	2.54 **3.86**	2.50 **3.78**	2.46 **3.70**	2.42 **3.61**	2.40 **3.56**	2.36 **3.49**	2.35 **3.46**	2.32 **3.41**	2.31 **3.38**	2.30 **3.36**	12
2.55 **3.85**	2.51 **3.78**	2.46 **3.67**	2.42 **3.59**	2.38 **3.51**	2.34 **3.42**	2.32 **3.37**	2.28 **3.30**	2.26 **3.27**	2.24 **3.21**	2.22 **3.18**	2.21 **3.16**	13
2.48 **3.70**	2.44 **3.62**	2.39 **3.51**	2.35 **3.43**	2.31 **3.34**	2.27 **3.26**	2.24 **3.21**	2.21 **3.14**	2.19 **3.11**	2.16 **3.06**	2.14 **3.02**	2.13 **3.00**	14
2.43 **3.56**	2.39 **3.48**	2.33 **3.36**	2.29 **3.29**	2.25 **3.20**	2.21 **3.12**	2.18 **3.07**	2.15 **3.00**	2.12 **2.97**	2.10 **2.92**	2.08 **2.89**	2.07 **2.87**	15
2.37 **3.45**	2.33 **3.37**	2.28 **3.25**	2.24 **3.18**	2.20 **3.10**	2.16 **3.01**	2.13 **2.96**	2.09 **2.89**	2.07 **2.86**	2.04 **2.80**	2.02 **2.77**	2.01 **2.75**	16
2.33 **3.35**	2.29 **3.27**	2.23 **3.16**	2.19 **3.08**	2.15 **3.00**	2.11 **2.92**	2.08 **2.86**	2.04 **2.79**	2.02 **2.76**	1.99 **2.70**	1.97 **2.67**	1.96 **2.65**	17
2.29 **3.27**	2.25 **3.19**	2.19 **3.07**	2.15 **3.00**	2.11 **2.91**	2.07 **2.83**	2.04 **2.78**	2.00 **2.71**	1.98 **2.68**	1.95 **2.62**	1.93 **2.59**	1.92 **2.57**	18
2.26 **3.19**	2.21 **3.12**	2.15 **3.00**	2.11 **2.92**	2.07 **2.84**	2.02 **2.76**	2.00 **2.70**	1.96 **2.63**	1.94 **2.60**	1.91 **2.54**	1.90 **2.51**	1.88 **2.49**	19
2.23 **3.13**	2.18 **3.05**	2.12 **2.94**	2.08 **2.86**	2.04 **2.77**	1.99 **2.69**	1.96 **2.63**	1.92 **2.56**	1.90 **2.53**	1.87 **2.47**	1.85 **2.44**	1.84 **2.42**	20
2.20 **3.07**	2.15 **2.99**	2.09 **2.88**	2.05 **2.80**	2.00 **2.72**	1.96 **2.63**	1.93 **2.58**	1.89 **2.51**	1.87 **2.47**	1.84 **2.42**	1.82 **2.38**	1.81 **2.36**	21
2.18 **3.02**	2.13 **2.94**	2.07 **2.83**	2.03 **2.75**	1.98 **2.67**	1.93 **2.58**	1.91 **2.53**	1.87 **2.46**	1.84 **2.42**	1.81 **2.37**	1.80 **2.33**	1.78 **2.31**	22
2.14 **2.97**	2.10 **2.89**	2.04 **2.78**	2.00 **2.70**	1.96 **2.62**	1.91 **2.53**	1.88 **2.48**	1.84 **2.41**	1.82 **2.37**	1.79 **2.32**	1.77 **2.28**	1.76 **2.26**	23
2.13 **2.93**	2.09 **2.85**	2.02 **2.74**	1.98 **2.66**	1.94 **2.58**	1.89 **2.49**	1.86 **2.44**	1.82 **2.36**	1.80 **2.33**	1.76 **2.27**	1.74 **2.23**	1.73 **2.21**	24
2.11 **2.89**	2.06 **2.81**	2.00 **2.70**	1.96 **2.62**	1.92 **2.54**	1.87 **2.45**	1.84 **2.40**	1.80 **2.32**	1.77 **2.29**	1.74 **2.23**	1.72 **2.19**	1.71 **2.17**	25
2.10 **2.86**	2.05 **2.77**	1.99 **2.66**	1.95 **2.58**	1.90 **2.50**	1.85 **2.41**	1.82 **2.36**	1.78 **2.28**	1.76 **2.25**	1.72 **2.19**	1.70 **2.15**	1.69 **2.13**	26

ν_2 = degrees of freedom for denominator

and William G. Cochran, © 1967 by the Iowa State University Press, Ames, Iowa.

Table V   95th and 99th Percentile

95th Percentile in Light-Face Type,
$\nu_1 = degrees\ of\ freedom$

ν_2 = degrees of freedom for denominator

ν_2	1	2	3	4	5	6	7	8	9	10	11	12
27	4.21	3.35	2.96	2.73	2.57	2.46	2.37	2.30	2.25	2.20	2.16	2.13
	7.68	**5.49**	**4.60**	**4.11**	**3.79**	**3.56**	**3.39**	**3.26**	**3.14**	**3.06**	**2.98**	**2.93**
28	4.20	3.34	2.95	2.71	2.56	2.44	2.36	2.29	2.24	2.19	2.15	2.12
	7.64	**5.45**	**4.57**	**4.07**	**3.76**	**3.53**	**3.36**	**3.23**	**3.11**	**3.03**	**2.95**	**2.90**
29	4.18	3.33	2.93	2.70	2.54	2.43	2.35	2.28	2.22	2.18	2.14	2.10
	7.60	**5.42**	**4.54**	**4.04**	**3.73**	**3.50**	**3.33**	**3.20**	**3.08**	**3.00**	**2.92**	**2.87**
30	4.17	3.32	2.92	2.69	2.53	2.42	2.34	2.27	2.21	2.16	2.12	2.09
	7.56	**5.39**	**4.51**	**4.02**	**3.70**	**3.47**	**3.30**	**3.17**	**3.06**	**2.98**	**2.90**	**2.84**
32	4.15	3.30	2.90	2.67	2.51	2.40	2.32	2.25	2.19	2.14	2.10	2.07
	7.50	**5.34**	**4.46**	**3.97**	**3.66**	**3.42**	**3.25**	**3.12**	**3.01**	**2.94**	**2.86**	**2.80**
34	4.13	3.28	2.88	2.65	2.49	2.38	2.30	2.23	2.17	2.12	2.08	2.05
	7.44	**5.29**	**4.42**	**3.93**	**3.61**	**3.38**	**3.21**	**3.08**	**2.97**	**2.89**	**2.82**	**2.76**
36	4.11	3.26	2.86	2.63	2.48	2.36	2.28	2.21	2.15	2.10	2.06	2.03
	7.39	**5.25**	**4.38**	**3.89**	**3.58**	**3.35**	**3.18**	**3.04**	**2.94**	**2.86**	**2.78**	**2.72**
38	4.10	3.25	2.85	2.62	2.46	2.35	2.26	2.19	2.14	2.09	2.05	2.02
	7.35	**5.21**	**4.34**	**3.86**	**3.54**	**3.32**	**3.15**	**3.02**	**2.91**	**2.82**	**2.75**	**2.69**
40	4.08	3.23	2.84	2.61	2.45	2.34	2.25	2.18	2.12	2.07	2.04	2.00
	7.31	**5.18**	**4.31**	**3.83**	**3.51**	**3.29**	**3.12**	**2.99**	**2.88**	**2.80**	**2.73**	**2.66**
42	4.07	3.22	2.83	2.59	2.44	2.32	2.24	2.17	2.11	2.06	2.02	1.99
	7.27	**5.15**	**4.29**	**3.80**	**3.49**	**3.26**	**3.10**	**2.96**	**2.86**	**2.77**	**2.70**	**2.64**
44	4.06	3.21	2.82	2.58	2.43	2.31	2.23	2.16	2.10	2.05	2.01	1.98
	7.24	**5.12**	**4.26**	**3.78**	**3.46**	**3.24**	**3.07**	**2.94**	**2.84**	**2.75**	**2.68**	**2.62**
46	4.05	3.20	2.81	2.57	2.42	2.30	2.22	2.14	2.09	2.04	2.00	1.97
	7.21	**5.10**	**4.24**	**3.76**	**3.44**	**3.22**	**3.05**	**2.92**	**2.82**	**2.73**	**2.66**	**2.60**
48	4.04	3.19	2.80	2.56	2.41	2.30	2.21	2.14	2.08	2.03	1.99	1.96
	7.19	**5.08**	**4.22**	**3.74**	**3.42**	**3.20**	**3.04**	**2.90**	**2.80**	**2.71**	**2.64**	**2.58**
50	4.03	3.18	2.79	2.56	2.40	2.29	2.20	2.13	2.07	2.02	1.98	1.95
	7.17	**5.06**	**4.20**	**3.72**	**3.41**	**3.18**	**3.02**	**2.88**	**2.78**	**2.70**	**2.62**	**2.56**
55	4.02	3.17	2.78	2.54	2.38	2.27	2.18	2.11	2.05	2.00	1.97	1.93
	7.12	**5.01**	**4.16**	**3.68**	**3.37**	**3.15**	**2.98**	**2.85**	**2.75**	**2.66**	**2.59**	**2.53**
60	4.00	3.15	2.76	2.52	2.37	2.25	2.17	2.10	2.04	1.99	1.95	1.92
	7.08	**4.98**	**4.13**	**3.65**	**3.34**	**3.12**	**2.95**	**2.82**	**2.72**	**2.63**	**2.56**	**2.50**
65	3.99	3.14	2.75	2.51	2.36	2.24	2.15	2.08	2.02	1.98	1.94	1.90
	7.04	**4.95**	**4.10**	**3.62**	**3.31**	**3.09**	**2.93**	**2.79**	**2.70**	**2.61**	**2.54**	**2.47**
70	3.98	3.13	2.74	2.50	2.35	2.23	2.14	2.07	2.01	1.97	1.93	1.89
	7.01	**4.92**	**4.08**	**3.60**	**3.29**	**3.07**	**2.91**	**2.77**	**2.67**	**2.59**	**2.51**	**2.45**
80	3.96	3.11	2.72	2.48	2.33	2.21	2.12	2.05	1.99	1.95	1.91	1.88
	6.96	**4.88**	**4.04**	**3.56**	**3.25**	**3.04**	**2.87**	**2.74**	**2.64**	**2.55**	**2.48**	**2.41**
100	3.94	3.09	2.70	2.46	2.30	2.19	2.10	2.03	1.97	1.92	1.88	1.85
	6.90	**4.82**	**3.98**	**3.51**	**3.20**	**2.99**	**2.82**	**2.69**	**2.59**	**2.51**	**2.43**	**2.36**
125	3.92	3.07	2.68	2.44	2.29	2.17	2.08	2.01	1.95	1.90	1.86	1.83
	6.84	**4.78**	**3.94**	**3.47**	**3.17**	**2.95**	**2.79**	**2.65**	**2.56**	**2.47**	**2.40**	**2.33**
150	3.91	3.06	2.67	2.43	2.27	2.16	2.07	2.00	1.94	1.89	1.85	1.82
	6.81	**4.75**	**3.91**	**3.44**	**3.14**	**2.92**	**2.76**	**2.62**	**2.53**	**2.44**	**2.37**	**2.30**
200	3.89	3.04	2.65	2.41	2.26	2.14	2.05	1.98	1.92	1.87	1.83	1.80
	6.76	**4.71**	**3.88**	**3.41**	**3.11**	**2.90**	**2.73**	**2.60**	**2.50**	**2.41**	**2.34**	**2.28**
400	3.86	3.02	2.62	2.39	2.23	2.12	2.03	1.96	1.90	1.85	1.81	1.78
	6.70	**4.66**	**3.83**	**3.36**	**3.06**	**2.85**	**2.69**	**2.55**	**2.46**	**2.37**	**2.29**	**2.23**
1,000	3.85	3.00	2.61	2.38	2.22	2.10	2.02	1.95	1.89	1.84	1.80	1.76
	6.66	**4.62**	**3.80**	**3.34**	**3.04**	**2.82**	**2.66**	**2.53**	**2.43**	**2.34**	**2.26**	**2.20**
∞	3.84	2.99	2.60	2.37	2.21	2.09	2.01	1.94	1.88	1.83	1.79	1.75
	6.64	**4.60**	**3.78**	**3.32**	**3.02**	**2.80**	**2.64**	**2.51**	**2.41**	**2.32**	**2.24**	**21.8**

Values of the F Distribution (*Concluded*)

99th Percentile in Bold-Face Type
for numerator

14	16	20	24	30	40	50	75	100	200	500	∞	ν_2
2.08	2.03	1.97	1.93	1.88	1.84	1.80	1.76	1.74	1.71	1.68	1.67	27
2.83	**2.74**	**2.63**	**2.55**	**2.47**	**2.38**	**2.33**	**2.25**	**2.21**	**2.16**	**2.12**	**2.10**	
2.06	2.02	1.96	1.91	1.87	1.81	1.78	1.75	1.72	1.69	1.67	1.65	28
2.80	**2.71**	**2.60**	**2.52**	**2.44**	**2.35**	**2.30**	**2.22**	**2.18**	**2.13**	**2.09**	**2.06**	
2.05	2.00	1.94	1.90	1.85	1.80	1.77	1.73	1.71	1.68	1.65	1.64	29
2.77	**2.68**	**2.57**	**2.49**	**2.41**	**2.32**	**2.27**	**2.19**	**2.15**	**2.10**	**2.06**	**2.03**	
2.04	1.99	1.93	1.89	1.84	1.79	1.76	1.72	1.69	1.66	1.64	1.62	30
2.74	**2.66**	**2.55**	**2.47**	**2.38**	**2.29**	**2.24**	**2.16**	**2.13**	**2.07**	**2.03**	**2.01**	
2.02	1.97	1.91	1.86	1.82	1.76	1.74	1.69	1.67	1.64	1.61	1.59	32
2.70	**2.62**	**2.51**	**2.42**	**2.34**	**2.25**	**2.20**	**2.12**	**2.08**	**2.02**	**1.98**	**1.96**	
2.00	1.95	1.89	1.84	1.80	1.74	1.71	1.67	1.64	1.61	1.59	1.57	34
2.66	**2.58**	**2.47**	**2.38**	**2.30**	**2.21**	**2.15**	**2.08**	**2.04**	**1.98**	**1.94**	**1.91**	
1.98	1.93	1.87	1.82	1.78	1.72	1.69	1.65	1.62	1.59	1.56	1.55	36
2.62	**2.54**	**2.43**	**2.35**	**2.26**	**2.17**	**2.12**	**2.04**	**2.00**	**1.94**	**1.90**	**1.87**	
1.96	1.92	1.85	1.80	1.76	1.71	1.67	1.63	1.60	1.57	1.54	1.53	38
2.59	**2.51**	**2.40**	**2.32**	**2.22**	**2.14**	**2.08**	**2.00**	**1.97**	**1.90**	**1.86**	**1.84**	
1.95	1.90	1.84	1.79	1.74	1.69	1.66	1.61	1.59	1.55	1.53	1.51	40
2.56	**2.49**	**2.37**	**2.29**	**2.20**	**2.11**	**2.05**	**1.97**	**1.94**	**1.88**	**1.84**	**1.81**	
1.94	1.89	1.82	1.78	1.73	1.68	1.64	1.60	1.57	1.54	1.51	1.49	42
2.54	**2.46**	**2.35**	**2.26**	**2.17**	**2.08**	**2.02**	**1.94**	**1.91**	**1.85**	**1.80**	**1.78**	
1.92	1.88	1.81	1.76	1.72	1.66	1.63	1.58	1.56	1.52	1.50	1.48	44
2.52	**2.44**	**2.32**	**2.24**	**2.15**	**2.06**	**2.00**	**1.92**	**1.88**	**1.82**	**1.78**	**1.75**	
1.91	1.87	1.80	1.75	1.71	1.65	1.62	1.57	1.54	1.51	1.48	1.46	46
2.50	**2.42**	**2.30**	**2.22**	**2.13**	**2.04**	**1.98**	**1.90**	**1.86**	**1.80**	**1.76**	**1.72**	
1.90	1.86	1.79	1.74	1.70	1.64	1.61	1.56	1.53	1.50	1.47	1.45	48
2.48	**2.40**	**2.28**	**2.20**	**2.11**	**2.02**	**1.96**	**1.88**	**1.84**	**1.78**	**1.73**	**1.70**	
1.90	1.85	1.78	1.74	1.69	1.63	1.60	1.55	1.52	1.48	1.46	1.44	50
2.46	**2.39**	**2.26**	**2.18**	**2.10**	**2.00**	**1.94**	**1.86**	**1.82**	**1.76**	**1.71**	**1.68**	
1.88	1.83	1.76	1.72	1.67	1.61	1.58	1.52	1.50	1.46	1.43	1.41	55
2.43	**2.35**	**2.23**	**2.15**	**2.06**	**1.96**	**1.90**	**1.82**	**1.78**	**1.71**	**1.66**	**1.64**	
1.86	1.81	1.75	1.70	1.65	1.59	1.56	1.50	1.48	1.44	1.41	1.39	60
2.40	**2.32**	**2.20**	**2.12**	**2.03**	**1.93**	**1.87**	**1.79**	**1.74**	**1.68**	**1.63**	**1.60**	
1.85	1.80	1.73	1.68	1.63	1.57	1.54	1.49	1.46	1.42	1.39	1.37	65
2.37	**2.30**	**2.18**	**2.09**	**2.00**	**1.90**	**1.84**	**1.76**	**1.71**	**1.64**	**1.60**	**1.56**	
1.84	1.79	1.72	1.67	1.62	1.56	1.53	1.47	1.45	1.40	1.37	1.35	70
2.35	**2.28**	**2.15**	**2.07**	**1.98**	**1.88**	**1.82**	**1.74**	**1.69**	**1.62**	**1.56**	**1.53**	
1.82	1.77	1.70	1.65	1.60	1.54	1.51	1.45	1.42	1.38	1.35	1.32	80
2.32	**2.24**	**2.11**	**2.03**	**1.94**	**1.84**	**1.78**	**1.70**	**1.65**	**1.57**	**1.52**	**1.49**	
1.79	1.75	1.68	1.63	1.57	1.51	1.48	1.42	1.39	1.34	1.30	1.28	100
2.26	**2.19**	**2.06**	**1.98**	**1.89**	**1.79**	**1.73**	**1.64**	**1.59**	**1.51**	**1.46**	**1.43**	
1.77	1.72	1.65	1.60	1.55	1.49	1.45	1.39	1.36	1.31	1.27	1.25	125
2.23	**2.15**	**2.03**	**1.94**	**1.85**	**1.75**	**1.68**	**1.59**	**1.54**	**1.46**	**1.40**	**1.37**	
1.76	1.71	1.64	1.59	1.54	1.47	1.44	1.37	1.34	1.29	1.25	1.22	150
2.20	**2.12**	**2.00**	**1.91**	**1.83**	**1.72**	**1.66**	**1.56**	**1.51**	**1.43**	**1.37**	**1.33**	
1.74	1.69	1.62	1.57	1.52	1.45	1.42	1.35	1.32	1.26	1.22	1.19	200
2.17	**2.09**	**1.97**	**1.88**	**1.79**	**1.69**	**1.62**	**1.53**	**1.48**	**1.39**	**1.33**	**1.28**	
1.72	1.67	1.60	1.54	1.49	1.42	1.38	1.32	1.28	1.22	1.16	1.13	400
2.12	**2.04**	**1.92**	**1.84**	**1.74**	**1.64**	**1.57**	**1.47**	**1.42**	**1.32**	**1.24**	**1.19**	
1.70	1.65	1.58	1.53	1.47	1.41	1.36	1.30	1.26	1.19	1.13	1.08	1,000
2.09	**2.01**	**1.89**	**1.81**	**1.71**	**1.61**	**1.54**	**1.44**	**1.38**	**1.28**	**1.19**	**1.11**	
1.69	1.64	1.57	1.52	1.46	1.40	1.35	1.28	1.24	1.17	1.11	1.00	∞
2.07	**1.99**	**1.87**	**1.79**	**1.69**	**1.59**	**1.52**	**1.41**	**1.36**	**1.25**	**1.15**	**1.00**	

ν_2 = degrees of freedom for denominator

Table VI Percentile Values of r for ν Degrees of Freedom when $\rho = 0$*

ν	$r_{.95}$	$r_{.975}$	$r_{.99}$	$r_{.995}$	$r_{.9995}$	ν	$r_{.95}$	$r_{.975}$	$r_{.99}$	$r_{.995}$	$r_{.9995}$
1	.988	.997	.9995	.9999	1.000	30	.296	.349	.409	.449	.554
2	.900	.950	.980	.990	.999	35	.275	.325	.381	.418	.519
3	.805	.878	.934	.959	.991	40	.257	.304	.358	.393	.490
4	.729	.811	.882	.917	.974	45	.243	.288	.338	.372	.465
5	.669	.754	.833	.874	.951	50	.231	.273	.322	.354	.443
6	.622	.707	.789	.834	.925	55	.220	.261	.307	.338	.424
7	.582	.666	.750	.798	.898	60	.211	.250	.295	.325	.408
8	.550	.632	.716	.765	.872	65	.203	.240	.284	.312	.393
9	.521	.602	.685	.735	.847	70	.195	.232	.274	.302	.380
10	.497	.576	.658	.708	.823	75	.189	.224	.264	.292	.368
11	.476	.553	.634	.684	.801	80	.183	.217	.256	.283	.357
12	.458	.532	.612	.661	.780	85	.178	.211	.249	.275	.347
13	.441	.514	.592	.641	.760	90	.173	.205	.242	.267	.338
14	.426	.497	.574	.623	.742	95	.168	.200	.236	.260	.329
15	.412	.482	.558	.606	.725	100	.164	.195	.230	.254	.321
16	.400	.468	.542	.590	.708	125	.147	.174	.206	.228	.288
17	.389	.456	.528	.575	.693	150	.134	.159	.189	.208	.264
18	.378	.444	.516	.561	.679	175	.124	.148	.174	.194	.248
19	.369	.433	.503	.549	.665	200	.116	.138	.164	.181	.235
20	.360	.423	.492	.537	.652	300	.095	.113	.134	.148	.188
22	.344	.404	.472	.515	.629	500	.074	.088	.104	.115	.148
24	.330	.388	.453	.496	.607	1000	.052	.062	.073	.081	.104
25	.323	.381	.445	.487	.597	2000	.037	.044	.051	.058	.074
	$-r_{.05}$	$-r_{.025}$	$-r_{.01}$	$-r_{.005}$	$-r_{.0005}$		$-r_{.05}$	$-r_{.025}$	$-r_{.01}$	$-r_{.005}$	$-r_{.0005}$

* SOURCE: Reprinted abridged from R. A. Fisher and F. Yates, *Statistical Tables for Biological, Agricultural, and Medical Research,* published by Oliver & Boyd Ltd. Edinburg, 1963 by permission of the authors and publishers.

Table VII   Values for Transforming r into $z_r = \frac{1}{2} \log_e (1 + r)/(1 - r)$*

	.00	.01	.02	.03	.04	.05	.06	.07	.08	.09
.0	.0000	.0100	.0200	.0300	.0400	.0500	.0599	.0699	.0798	.0898
.1	.0997	.1096	.1194	.1293	.1391	.1489	.1587	.1684	.1781	.1878
.2	.1974	.2070	.2165	.2260	.2355	.2449	.2543	.2636	.2729	.2821
.3	.2913	.3004	.3095	.3185	.3275	.3364	.3452	.3540	.3627	.3714
.4	.3800	.3885	.3969	.4053	.4136	.4219	.4301	.4382	.4462	.4542
.5	.4621	.4700	.4777	.4854	.4930	.5005	.5080	.5154	.5227	.5299
.6	.5370	.5441	.5511	.5581	.5649	.5717	.5784	.5850	.5915	.5980
.7	.6044	.6107	.6169	.6231	.6291	.6352	.6411	.6469	.6527	.6584
.8	.6640	.6696	.6751	.6805	.6858	.6911	.6963	.7014	.7064	.7114
.9	.7163	.7211	.7259	.7306	.7352	.7398	.7443	.7487	.7531	.7574
1.0	.7616	.7658	.7699	.7739	.7779	.7818	.7857	.7895	.7932	.7969
1.1	.8005	.8041	.8076	.8110	.8144	.8178	.8210	.8243	.8275	.8306
1.2	.8337	.8367	.8397	.8426	.8455	.8483	.8511	.8538	.8565	.8591
1.3	.8617	.8643	.8668	.8693	.8717	.8741	.8764	.8787	.8810	.8832
1.4	.8854	.8875	.8896	.8917	.8937	.8957	.8977	.8996	.9015	.9033
1.5	.9052	.9069	.9087	.9104	.9121	.9138	.9154	.9170	.9186	.9202
1.6	.9217	.9232	.9246	.9261	.9275	.9289	.9302	.9316	.9329	.9342
1.7	.9354	.9367	.9379	.9391	.9402	.9414	.9425	.9436	.9447	.9458
1.8	.9468	.9478	.9488	.9498	.9508	.9518	.9527	.9536	.9545	.9554
1.9	.9562	.9571	.9579	.9587	.9595	.9603	.9611	.9619	.9626	.9633
2.0	.9640	.9647	.9654	.9661	.9668	.9674	.9680	.9687	.9693	.9699
2.1	.9705	.9710	.9716	.9722	.9727	.9732	.9738	.9743	.9748	.9753
2.2	.9757	.9762	.9767	.9771	.9776	.9780	.9785	.9789	.9793	.9797
2.3	.9801	.9805	.9809	.9812	.9816	.9820	.9823	.9827	.9830	.9834
2.4	.9837	.9840	.9843	.9846	.9849	.9852	.9855	.9858	.9861	.9863
2.5	.9866	.9869	.9871	.9874	.9876	.9879	.9881	.9884	.9886	.9888
2.6	.9890	.9892	.9895	.9897	.9899	.9901	.9903	.9905	.9906	.9908
2.7	.9910	.9912	.9914	.9915	.9917	.9919	.9920	.9922	.9923	.9925
2.8	.9926	.9928	.9929	.9931	.9932	.9933	.9935	.9936	.9937	.9938
2.9	.9940	.9941	.9942	.9943	.9944	.9945	.9946	.9947	.9949	.9950
3.0	.9951									
4.0	.9993									
5.0	.9999									

* SOURCE: Reprinted abridged from R. A. Fisher and F. Yates, *Statistical Tables for Biological, Agricultural, and Medical Research*, published by Oliver & Boyd Ltd., Edinburgh, 1963 by permission of the authors and publishers. The figures in the body of the table are values of r corresponding to z values read from the scales on the left and top of the table.

Table VIII   Scores Made by Students in a First Course in Statistical
Term, Two Examinations in Subject Matter of the

Code Number of Student	Section	Prognostic Test Score			Criterion Score		
		Reading	Artificial Language	Arithmetic Test	Mid-term Test	Final Exam	Semester Grade
1	III	33	51	33	58	65	62
2	III	39	50	36	53	51	52
3	I	38	38	39	52	53	53
4	I	23	27	29	47	42	45
*5	I	44	40	41	61	50	56
*6	I	45	54	38	47	53	50
7	I	48	57	42	64	64	64
8	II	38	39	35	64	54	59
9	II	32	46	23	58	50	54
*10	II	28	45	37	44	45	45
11	II	33	53	28	62	63	63
12	II	40	43	25	52	50	51
13	II	34	55	38	50	56	58
14	II	34	57	36	71	68	70
15	II	35	46	37	62	65	64
16	I	32	43	39	52	59	56
17	II	34	44	38	59	57	58
18	II	42	51	35	53	60	57
19	I	32	55	36	62	56	59
*20	III	24	34	17	37	43	40
21	I	39	52	34	55	53	54
22	I	34	25	29	30	42	36
23	II	44	58	35	61	56	59
24	III	27	29	26	49	40	45
25	II	34	52	30	43	38	41
26	II	40	53	29	61	54	58
27	I	37	34	26	47	57	52
28	III	22	40	33	47	50	49
29	I	30	54	40	56	47	52
*30	I	36	58	25	52	39	45
31	I	35	43	40	49	37	43
32	I	39	58	32	49	37	43
33	II	37	28	14	43	36	40
34	II	38	51	26	40	42	41
35	II	36	55	31	59	52	56
36	I	46	52	39	61	63	62
37	II	44	57	32	46	53	50
38		40	56	32	64	49	57
39	III	38	30	18	25	38	32
40	I	50	47	32	58	52	55
41	II	16	49	26	52	36	43
42	I	46	60	43	61	57	59
43	I	42	56	40	59	54	57
*44	III	27	53	26	62	48	55
45	III	33	55	25	55	57	56
46	II	40	54	34	52	57	55
47	I	46	53	37	62	49	56
*48	I	46	58	39	58	56	57
*49		25	57	24	46	49	48
*50	III	26	45	18	47	25	36

* A woman student.

Methods on Three Prognostic Tests Given at the Beginning of the Course, and the Semester Grade

Code Number of Student	Section	Prognostic Test Score			Criterion Score		
		Reading	Artificial Language	Arithmetic Test	Mid-term Test	Final Exam	Semester Grade
*51	I	42	46	31	58	56	57
*52	I	34	60	33	62	64	63
53	III	28	37	8	47	46	47
54	II	21	24	16	44	41	43
55	I	44	43	35	58	55	57
56	I	38	47	28	58	56	57
57	III	16	50	27	47	47	47
58	III	40	48	35	56	38	47
59	III	33	42	25	52	46	49
*60	II	35	59	25	52	49	51
61	III	33	40	28	49	47	48
62	III	30	41	30	56	59	58
63	I	38	47	39	62	63	63
*64	III	36	43	26	61	50	56
*65	I	45	57	35	53	46	50
66	III	32	32	21	50	41	46
67	III	36	53	27	49	34	43
*68	III	21	46	39	44	33	39
69	II	30	54	34	47	46	47
70	II	34	49	27	50	44	48
71	II	41	58	39	53	57	55
72	I	44	55	38	47	46	47
*73	II	36	14	20	30	40	35
74	I	58	50	38	68	55	62
75	II	36	41	30	62	56	59
76	III	31	14	18	27	30	29
77	III	40	46	36	58	50	54
78	I	44	60	43	71	45	58
*79	III	36	36	23	49	43	46
80	I	44	51	36	53	47	50
*81	I	34	46	37	43	49	46
*82	I	38	56	36	64	61	63
*83	I	38	52	33	50	56	53
84	I	38	51	33	50	48	49
85	I	46	51	40	62	54	58
86	III	30	42	22	37	33	34
*87	III	26	44	28	58	48	53
*88	I	36	58	42	59	69	64
89	I	36	43	42	46	44	45
*90	III	32	59	36	61	55	58
*91	III	30	23	28	38	38	38
*92	I	31	58	31	44	51	48
93	II	44	34	32	55	49	52
*94	III	14	16	18	40	33	37
95	III	44	26	30	37	19	28
*96	I	30	58	37	61	56	59
97	I	39	34	33	41	46	44
98	I	32	56	30	65	51	58

Table IX Scores for 109 Fourth-Grade Pupils on the
Modern School Achievement Test

Case Number	Age to Nearest Month	IQ	Score on Test of				
			Arithmetic Computation	Arithmetic Reasoning	Reading Speed	Language	History-Civics
1*	121	99	12	11	27	17	24
2	124	83	13	4	12	15	13
3*	103	117	5	8	30	26	10
4*	127	83	8	6	30	12	21
5	115	109	7	4	26	27	13
6	108	111	6	3	17	21	20
7*	106	92	9	9	25	30	12
8*	115	95	5	3	28	12	16
9	140	79	8	3	28	23	11
10	111	115	10	8	20	26	4
11	109	120	5	5	26	19	10
12	119	101	6	4	24	14	20
13	119	87	10	5	8	14	17
14	108	106	7	7	10	16	5
15	131	100	6	5	21	20	8
16*	112	120	13	7	21	19	8
17	118	99	4	4	26	16	10
18	96	119	7	7	20	17	12
19	148	66	10	2	16	11	9
20	132	77	5	3	27	19	21
21*	108	124	8	6	29	29	9
22*	96	125	10	5	30	18	16
23	118	104	9	4	18	23	21
24	123	89	6	5	13	19	13
25*	109	97	8	5	28	26	2
26	106	126	11	11	29	30	18
27*	142	68	5	2	7	17	10
28	109	83	6	3	16	23	15
29	150	64	6	1	11	7	7
30	113	96	4	4	25	26	19

Table IX   Scores on Achievement Test *(Continued)*

Case Number	Age to Nearest Month	IQ	Score on Test of				
			Arithmetic Computation	Arithmetic Reasoning	Reading Speed	Language	History-Civics
31	117	91	6	4	30	19	2
32*	156	75	12	5	30	18	15
33*	119	103	8	5	14	22	18
34*	129	86	9	4	8	14	6
35	114	118	9	3	24	19	10
36	106	140	13	9	40	31	10
37	114	120	12	9	17	21	18
38	134	93	9	3	17	24	12
39*	109	112	7	2	31	24	18
40	115	116	15	5	41	33	22
41	109	123	10	4	34	24	23
42	109	105	11	5	22	25	13
43*	101	121	5	1	35	28	22
44*	108	114	5	4	24	16	10
45*	116	113	13	6	18	20	12
46	111	111	8	4	25	19	12
47*	120	114	9	8	36	23	23
48*	118	99	6	2	33	15	18
49*	109	101	12	2	23	16	18
50*	111	112	12	6	25	15	16
51*	116	125	10	3	33	16	23
52	114	105	13	3	44	20	24
53	129	100	8	6	22	18	20
54	106	144	16	5	40	28	20
55	110	101	10	7	14	17	13
56	125	108	8	4	30	22	18
57*	105	112	6	3	30	14	16
58	124	98	9	7	28	21	12
59	111	105	12	1	24	22	13
60*	118	123	7	3	35	24	22

Table IX   Scores on Achievement Test (*Continued*)

Case Number	Age to Nearest Month	IQ	Score on Test of				
			Arithmetic Computation	Arithmetic Reasoning	Reading Speed	Language	History-Civics
61*	103	148	9	8	43	32	18
62*	129	93	10	2	24	17	12
63	104	129	8	3	39	25	17
64*	116	106	6	5	45	26	20
65	112	111	8	4	35	17	13
66*	111	121	7	2	45	25	17
67*	109	115	12	7	44	20	20
68*	104	130	8	9	39	29	23
69	107	114	15	7	42	30	21
70	117	118	16	12	49	39	18
71*	114	103	6	7	26	30	19
72	108	108	3	2	21	12	3
73	114	123	1	1	26	31	9
74	112	142	2	8	45	30	5
75*	107	120	2	2	26	25	7
76*	109	136	10	5	34	29	10
77*	113	131	5	7	24	28	8
78	108	141	4	9	23	24	10
79*	116	106	8	5	26	22	3
80*	109	126	4	9	24	19	2
81	110	125	7	7	24	23	2
82	107	131	6	5	40	36	17
83*	113	104	7	3	28	29	19
84*	109	128	10	9	42	30	18
85	109	133	8	7	31	26	9
86	114	127	9	9	28	24	7
87	107	138	3	6	46	36	4
88*	112	132	4	1	44	31	12
89*	107	136	9	2	29	32	5
90*	113	119	3	7	30	22	9

Table IX   Scores on Achievement Test (*Concluded*)

Case Number	Age to Nearest Month	IQ	Score on Test of				
			Arithmetic Computation	Arithmetic Reasoning	Reading Speed	Language	History-Civics
91	115	106	9	4	24	17	3
92*	108	124	1	5	50	29	23
93	108	124	9	4	48	14	7
94*	122	102	6	2	26	20	5
95*	124	83	8	3	10	15	6
96*	126	93	7	6	16	26	14
97	104	113	8	8	30	23	8
98	115	90	0	2	24	18	5
99*	117	103	8	5	27	17	5
100*	111	88	12	5	10	11	6
101*	115	111	9	7	16	17	10
102	103	130	2	2	41	25	2
103	122	91	4	8	32	9	5
104*	113	106	7	3	15	10	6
105	111	115	6	2	43	28	7
106	114	120	1	3	27	26	8
107*	118	116	11	4	40	19	5
108	109	120	9	2	50	35	4
109	118	123	10	6	41	18	4

* All case numbers starred refer to boys, unstarred numbers to girls.

Table **X**   Pre-registration Scores of **447** College Students on
the Cooperative Test Service English Test
(Data supplied by Dr. Irving Lorge)

Subject	Score	Subject	Score	Subject	Score	Subject	Score	Subject	Score
1	141	41	133	81	102	121	156	161	093
2	120	42	146	82	138	122	098	162	094
3	150	43	138	83	132	123	085	163	119
4	156	44	138	84	118	124	115	164	139
5	122	45	128	85	135	125	129	165	166
6	128	46	127	86	100	126	143	166	108
7	112	47	142	87	111	127	108	167	103
8	178	48	173	88	169	128	087	168	092
9	120	49	109	89	124	129	121	169	175
10	160	50	131	90	134	130	187	170	160
11	104	51	157	91	135	131	179	171	152
12	088	52	111	92	180	132	140	172	140
13	100	53	114	93	102	133	136	173	177
14	191	54	117	94	117	134	086	174	133
15	137	55	115	95	085	135	175	175	111
16	108	56	155	96	169	136	120	176	114
17	147	57	190	97	127	137	175	177	106
18	127	58	159	98	128	138	133	178	147
19	156	59	141	99	131	139	107	179	090
20	201	60	100	100	167	140	119	180	167
21	131	61	125	101	118	141	135	181	156
22	174	62	116	102	142	142	100	182	130
23	096	63	148	103	143	143	102	183	143
24	140	64	169	104	094	144	138	184	142
25	102	65	139	105	117	145	129	185	151
26	090	66	145	106	114	146	134	186	168
27	177	67	084	107	091	147	145	187	134
28	125	68	180	108	091	148	148	188	120
29	164	69	139	109	120	149	155	189	124
30	150	70	166	110	119	150	124	190	176
31	181	71	159	111	135	151	109	191	170
32	118	72	126	112	158	152	103	192	120
33	192	73	177	113	138	153	113	193	165
34	169	74	165	114	123	154	135	194	102
35	117	75	174	115	199	155	117	195	109
36	152	76	173	116	111	156	123	196	116
37	176	77	149	117	105	157	101	197	145
38	089	78	173	118	100	158	107	198	136
39	151	79	093	119	158	159	108	199	099
40	148	80	118	120	103	160	093	200	163

Table **X**   Pre-registration Scores of 447 College Students on
the Cooperative Test Service English Test
(Data supplied by Dr. Irving Lorge) (*Concluded*)

Subject	Score	Subject	Score	Subject	Score	Subject	Score	Subject	Score
201	108	251	098	301	130	351	095	401	068
202	144	252	104	302	140	352	185	402	078
203	117	253	128	303	113	353	095	403	063
204	166	254	131	304	189	354	160	404	077
205	115	255	099	305	152	355	133	405	032
206	089	256	178	306	105	356	120	406	058
207	128	257	119	307	111	357	149	407	063
208	165	258	114	308	109	358	137	408	070
209	176	259	090	309	139	359	137	409	058
210	148	260	129	310	110	360	115	410	053
211	127	261	148	311	156	361	087	411	080
212	143	262	096	312	156	362	097	412	072
213	143	263	108	313	103	363	124	413	052
214	183	264	162	314	165	364	128	414	078
215	151	265	121	315	121	365	158	415	069
216	144	266	154	316	105	366	157	416	083
217	139	267	178	317	153	367	149	417	066
218	146	268	168	318	132	368	137	418	055
219	151	269	089	319	114	369	079	419	036
220	086	270	147	320	184	370	068	420	051
221	136	271	154	321	118	371	058	421	070
222	148	272	087	322	105	372	081	422	080
223	107	273	177	323	186	373	045	423	072
224	132	274	154	324	141	374	072	424	073
225	180	275	140	325	173	375	060	425	076
226	118	276	093	326	124	376	050	426	083
227	106	277	148	327	140	377	075	427	080
228	086	278	088	328	103	378	044	428	082
229	143	279	157	329	165	379	076	429	081
230	158	280	136	330	172	380	039	430	057
231	122	281	094	331	185	381	058	431	078
232	096	282	155	332	094	382	071	432	052
233	143	283	163	333	094	383	029	433	063
234	151	284	123	334	091	384	072	434	065
235	163	285	110	335	164	385	063	435	052
236	132	286	125	336	172	386	080	436	081
237	168	287	147	337	174	387	083	437	056
238	125	288	092	338	136	388	050	438	081
239	164	289	091	339	164	389	070	439	058
240	162	290	169	340	101	390	082	440	083
241	106	291	180	341	131	391	078	441	072
242	085	292	149	342	129	392	058	442	016
243	159	293	115	343	144	393	080	443	069
244	136	294	155	344	178	394	069	444	076
245	152	295	142	345	088	395	040	445	063
246	144	296	157	346	183	396	046	446	028
247	122	297	104	347	146	397	067	447	082
248	185	298	086	348	099	398	047		
249	159	299	109	349	116	399	047		
250	136	300	132	350	108	400	065		

Table XI Random Numbers*

Line\Col.	(1)	(2)	(3)	(4)	(5)	(6)	(7)	(8)	(9)	(10)	(11)	(12)	(13)	(14)
1	10480	15011	01536	02011	81647	91646	69179	14194	62590	36207	20969	99570	91291	90700
2	22368	46573	25595	85393	30995	89198	27982	53402	93965	34095	52666	19174	39615	99505
3	24130	48360	22527	97265	76393	64809	15179	24830	49340	32081	30680	19655	63348	58629
4	42167	93093	06243	61680	07856	16376	39440	53537	71341	57004	00849	74917	97758	16379
5	37570	39975	81837	16656	06121	91782	60468	81305	49684	60672	14110	06927	01263	54613
6	77921	06907	11008	42751	27756	53498	18602	70659	90655	15053	21916	81825	44394	42880
7	99562	72905	56420	69994	98872	31016	71194	18738	44013	48840	63213	21069	10634	12952
8	96301	91977	05463	07972	18876	20922	94595	56869	69014	60045	18425	84903	42508	32307
9	89579	14342	63661	10281	17453	18103	57740	84378	25331	12566	58678	44947	05585	56941
10	85475	36857	53342	53988	53060	59533	38867	62300	08158	17983	16439	11458	18593	64952
11	28918	69578	88231	33276	70997	79936	56865	05859	90106	31595	01547	85590	91610	78188
12	63553	40961	48235	03427	49626	69445	18663	72695	52180	20847	12243	90511	33703	90322
13	09429	93969	52636	92737	88974	33488	36320	17617	30015	08272	84115	27156	30613	74952
14	10365	61129	87529	85689	48237	52267	67689	93394	01511	26358	85104	20285	29975	89868
15	07119	97336	71048	08178	77233	13916	47564	81056	97735	85977	29372	74461	28551	90707
16	51085	12765	51821	51259	77452	16308	60756	92144	49442	53900	70960	63990	75601	40719
17	02368	21382	52404	60268	89368	19885	55322	44819	01188	65255	64835	44919	05944	55157
18	01011	54092	33362	94904	31273	04146	18594	29852	71585	85030	51132	01915	92747	64951
19	52162	53916	46369	58586	23216	14513	83149	98736	23495	64350	94738	17752	35156	35749
20	07056	97628	33787	09998	42698	06691	76988	13602	51851	46104	88916	19509	25625	58104
21	48663	91245	85828	14346	09172	30168	90229	04734	59193	22178	30421	61666	99904	32812
22	54164	58492	22421	74103	47070	25306	76468	26384	58151	06646	21524	15227	96909	44592
23	32639	32363	05597	24200	13363	38005	94342	28728	35806	06912	17012	64161	18296	22851

24	29334	27001	87637	87308	58731	00256	45834	15398	46557	41135	10367	07684	36188	18510
25	02488	33062	28834	07351	19731	92420	60952	61280	50001	67658	32586	86679	50720	94953
26	81525	72295	04839	96423	24878	82651	66566	14778	76797	14780	13300	87074	79666	95725
27	29676	20591	68086	26432	46901	20849	89768	81536	86645	12659	92259	57102	80428	25280
28	00742	57392	39064	66432	84673	40027	32832	61362	98947	96067	64760	64584	96096	98253
29	05366	04213	25669	26422	44407	44048	37937	63904	45766	66134	75470	66520	34693	90449
30	91921	26418	64117	94305	26766	25940	39972	22209	71500	64568	91402	42416	07844	69618
31	00582	04711	87917	77341	42206	35126	74087	99547	81817	42607	43808	76655	62028	76630
32	00725	69884	62797	56170	86324	88072	76222	36086	84637	93161	76038	65855	77919	88006
33	69011	65795	95876	55293	18988	27354	26575	08615	40801	59920	29841	80150	12777	48501
34	25976	57948	29888	88604	67917	48708	18912	82271	65424	69774	33611	54262	85963	03547
35	09763	83473	73577	12908	30883	18317	28290	35797	05998	41688	34952	37888	38917	88050
36	91567	42595	27958	30134	04024	86385	29880	99730	55536	84855	29080	09250	79656	73211
37	17955	56349	90999	49127	20044	59931	06115	20542	18059	02008	73708	83517	36103	42791
38	46503	18584	18845	49618	02304	51038	20655	58727	28168	15475	56942	53389	20562	87338
39	92157	89634	94824	78171	84610	82834	09922	25417	44137	84813	25555	21246	35509	20468
40	14577	62765	35605	81263	39667	47358	56873	56307	61607	49518	89656	20103	77490	18062
41	98427	07523	33362	64270	01638	92477	66969	98420	04880	45585	46565	04102	46880	45709
42	34914	63976	88720	82765	34476	17032	87589	40836	32427	70002	70663	88863	77775	69348
43	70060	28277	39475	46473	23219	53416	94970	25832	69975	94884	19661	72828	00102	66794
44	53976	54914	06990	67245	68350	82948	11398	42878	80287	88267	47363	46634	06541	97809
45	76072	29515	40980	07391	58745	25774	22987	80059	39911	96189	41151	14222	60697	59583
46	90725	52210	83974	29992	65831	38857	50490	83765	55657	14361	31720	57375	56228	41546
47	64364	67412	33339	31926	14883	24413	59744	92351	97473	89286	35931	04110	23726	51900
48	08962	00358	31662	25388	61642	34072	81249	35648	56891	69352	48373	45578	78547	81788
49	95012	68379	93526	70765	10592	04542	76463	54328	02349	17247	28865	14777	62730	92277
50	15664	10493	20492	38391	91132	21999	59516	81652	27195	48223	46751	22923	32261	85653

Table XI Random Numbers (Concluded)

Line \ Col.	(1)	(2)	(3)	(4)	(5)	(6)	(7)	(8)	(9)	(10)	(11)	(12)	(13)	(14)
51	16408	81899	04153	53381	79401	21438	83035	92350	36693	31238	59649	91754	72772	02338
52	18629	81953	05520	91962	04739	13092	97662	24822	94730	06496	35090	04822	86774	98289
53	73115	35101	47498	87637	99016	71060	88824	71013	18735	20286	23153	72924	35165	43040
54	57491	16703	23167	49323	45021	33132	12544	41035	80780	45393	44812	12515	98931	91202
55	30405	83946	23792	14422	15059	45799	22716	19792	09983	74353	68668	30429	70735	25499
56	16631	35006	85900	98275	32388	52390	16815	69298	82732	38480	73817	32523	41961	44437
57	96773	20206	42559	78985	05300	22164	24369	54224	35083	19687	11052	91491	60383	19746
58	38935	64202	14349	82674	66523	44133	00697	35552	35970	19124	63318	29686	03387	59846
59	31624	76384	17403	53363	44167	64486	64758	75366	76554	31601	12614	33072	60332	92325
60	78919	19474	23632	27889	47914	02584	37680	20801	72152	39339	34806	08930	85001	87820

* SOURCE: Taken from the 30-page table of 105,000 random digits prepared by the Bureau of Transport Economics and Statistics of the Interstate Commerce Commission, Washington, D.C., Mr. W. H. S. Stevens, Director. It is used in this text with their permission.

TABLE XII   Values of $k = np$ Leading to Rejection of a Hypothesis about a Proportion at .05 Level of Significance

$n = $ *Number of cases in sample*
$k = $ *Number of cases with given characteristic*
$P = $ *Proportion under hypothesis tested*
$p = k/n$

A. Two-Tail Region (.025 in each tail)

For $P \leqq .5$, reject hypothesis if observed k is less than or equal to left value in cell or is greater than or equal to right value.
For $P > .5$, substitute $1 - P$ for P and $n - k$ for k.

n	.05	.10	.15	.20	.25	.30	.35	.40	.45	.50	n
5	* 2	* 3	* 4	* 4	* 4	* 5	* 5	* 5	* 5	* *	5
6	* 3	* 3	* 4	* 4	* 5	* 5	* 5	* 6	* 6	0 6	6
7	* 3	* 4b	* 4	* 5	* 5	* 6	* 6	* 6	0 7	0 7	7
8	* 3	* 4	* 4	* 5	* 6	* 6	* 7b	0 7	0 7	0 8	8
9	* 3	* 4	* 5	* 5	* 6	* 7b	0 7	0 8b	0 8	1 8	9
10	* 3	* 4	* 5	* 6	* 6	* 7	0 8b	0 8	1 9	1 9	10
11	* 3	* 4	* 5	* 6	* 7	0 7	0 8	0 9	1 9	1 10	11
12	* 3	* 5b	* 5	* 6	* 7	0 8	0 9b	1 9	1 10	2 10	12
13	* 3	* 5	* 6	* 7	0 7	0 8	0 9	1 10	2 10	2 11	13
14	* 4	* 5	* 6	* 7	0 8	0 9	1 9	1 10	2 11	2 12	14
15	* 4	* 5	* 6	* 7	0 8	0 9	1 10	1 11	2 12b	3 12	15
16	* 4	* 5	* 6	* 8	0 9	0 10b	1 11	2 11	2 12	3 13	16
17	* 4	* 5	* 7	0 8	0 9	1 10	1 11	2 12	3 13	4 13	17
18	* 4	* 6	* 7	0 8	0 9	1 10	2 11	2 12	3 13	4 14	18
19	* 4	* 6	* 7	0 8	0 10	1 11	2 12	3 13	3 14	4 15	19
20	* 4	* 6	* 8	0 9	1 10	1 11	2 12	3 13	4 14	5 15	20
21	* 4	* 6	* 8	0 9	1 10	1 12	2 13	3 14	4 15	5 16	21
22	* 4	* 6	* 8	0 9	1 11	2 12	3 13	3 14	4 15	5 17	22
23	* 5b	* 6	0 8	0 10	1 11	2 12	3 14	4 15	5 16	6 17	23
24	* 5	* 7	0 8	0 10	1 11	2 13	3 14	4 15	5 17	6 18	24
25	* 5	* 7	0 9b	0 10	1 12	2 13	3 15b	4 16	5a 17	7 18	25
26	* 5	* 7	0 9	1 10	1a 12	2a 14b	4a 15	5 16	6 18	7 19	26
27	* 5	* 7	0 9	1 11	2 12	3 14	4 15	5 17	6 18	7a 20b	27
28	* 5	* 7	0 9	1 11	2 13	3 14	4 16	5 17	7 19	8 20	28
29	* 5	* 7	0 9	1 11	2 13	3 15	4 16	6 18	7 19	8 21	29
30	* 5	* 8b	0 10	1 12b	2 13	3 15	5 17	6 18	7 20	9 21	30
31	* 5	* 8	0 10	1 12	2 14	4 15	5 17	6 19	8 20	9 22	31
32	* 5	* 8	0 10	1 12	2a 14	4 16	5 18	7 19	8 21	9a 23b	32
33	* 5	* 8	0 10	1 12	3 14	4 16	5 18	7 20	8 21	10 23	33
34	* 6b	* 8	0 10	2 13	3 15	4 17	6 18	7 20	9 22	10 24	34
35	* 6	*c 8	1 11	2 13	3 15	4 17	6 19	7 21b	9 23b	11 24	35

* There is no region of rejection in the indicated tail.
a Increasing this number by 1 will not increase the lower tail to more than .026.
b Decreasing this number by 1 will not increase the upper tail to more than .026.
c Changing * to 0 will not increase the lower tail to more than .026.

(Table XII, *Continued* **) B. Upper-Tail Region (.05 in upper tail)**

For $P \leq .5$, reject hypothesis if observed k is equal to or greater than cell value.
For $P > .5$, substitute $1 - P$ for P and $n - k$ for k; use Table C.

n \ P	.05	.10	.15	.20	.25	.30	.35	.40	.45	.50	n
5	2	3	3	4	4	4	5	5	5	5	5
6	2	3	3	4	4	5	5	5	6	6	6
7	2	3	4	4	5	5	6	6	6	7	7
8	3	3	4	5	5	6	6	6	7	7	8
9	3	4	4	5	5	6	7	7	7	8	9
10	3	4	4	5	6	6	7	8	8	9	10
11	3	4	5	6b	6	7	8	8	9	9	11
12	3	4	5	6	7	7	8	9	9	10	12
13	3	4	5	6	7	8	8	9	10	10	13
14	3	4	5	6	7	8	9	10	10	11	14
15	3	5	6	7	8	9b	9	10	11	12	15
16	3	5	6	7	8	9	10	11	11	12	16
17	4b	5	6	7	8	9	10	11	12	13	17
18	4	5	6	8	9	10	11	12	13	13	18
19	4	5	7	8	9	10	11	12	13	14	19
20	4	5	7	8	9	10	12	13	14	15	20
21	4	6	7	8	10	11	12	13	14	15	21
22	4	6	7	9	10	11	12	14	15	16	22
23	4	6	7	9	10	12	13	14	15	16	23
24	4	6	8	9	11	12	13	15	16	17	24
25	4	6	8	9	11	12	14	15	16	18	25
26	4	6	8	10	11	13	14	16	17	18	26
27	4	6	8	10	12	13	15	16	17	19	27
28	4	7	8	10	12	13	15	16	18	19	28
29	5	7	9	10	12	14	15	17	18	20	29
30	5	7	9	11	13b	14	16	17	19	20	30
31	5	7	9	11	13	15	16	18	20b	21	31
32	5	7	9	11	13	15	17	18	20	22	32
33	5	7	9	12b	13	15	17	19	21	22	33
34	5	7	10	12	14	16	18	19	21	23	34
35	5	8	10	12	14	16	18	20	22	23	35

b Decreasing this number by 1 will not increase the upper tail to more than .051.

(Table XII, *Continued* **) C. Lower-Tail Region (.05 in lower tail)**

For $P \leq .5$, reject hypothesis if observed k is equal to or less than cell value.
For $P > .5$, substitute $1 - P$ for P and $n - k$ for k; use Table B.

n \ P	.05	.10	.15	.20	.25	.30	.35	.40	.45	.50	n
5	*	*	*	*	*	*	*	*	*	0	5
6	*	*	*	*	*	*	*	*	0	0	6
7	*	*	*	*	*	*	*	0	0	0	7
8	*	*	*	*	*	*	0	0	0	1	8
9	*	*	*	*	*	0	0	0	1	1	9
10	*	*	*	*	*	0	0	1	1	1	10
11	*	*	*	*	0	0	0	1	1	2	11
12	*	*	*	*	0	0	1	1	2	2	12
13	*	*	*	*	0	0	1	1	2	3	13
14	*	*	*	0	0	1	1	2	2	3	14
15	*	*	*	0	0	1	1	2	3	3	15
16	*	*	*	0	0	1	2	2	3	4	16
17	*	*	*	0	0a	1	2	3	3	4	17
18	*	*	*	0	1	1	2	3	4	5	18
19	*	*	0	0	1	2	2	3	4	5	19
20	*	*	0	0	1	2	3	3a	4	5	20
21	*	*	0	0	1	2	3	4	5	6	21
22	*	*	0	1	1	2	3	4	5	6	22
23	*	*	0	1	2	2	3	4	5a	7	23
24	*	*	0	1	2	3	4	5	6	7	24
25	*	*	0	1	2	3	4	5	6	7	25
26	*	*	0	1	2	3	4	5a	7	8	26
27	*	*	0	1	2	3	4a	6	7	8	27
28	*	*	0	1	2	4	5	6	7	9	28
29	*	0	0	1a	3	4	5	6	8	9	29
30	*	0	1	2	3	4	5	7	8	10	30
31	*	0	1	2	3	4	6	7	8	10	31
32	*	0	1	2	3	4a	6	7	9	10	32
33	*	0	1	2	3	5	6	8	9	11	33
34	*	0	1	2	4	5	6	8	10	11	34
35	*	0	1	2	4	5	7	8	10	12	35

* No region of rejection is available.
a Increasing this number by 1 will not increase the lower tail to more than .051.

Table XIII Transformation of Ranks to Modified Standard Scores* ($T = 50 + 10z$)

Number of Persons Ranked

Rank	5	6	7	8	9	10	11	12	13	14	15	16	17	18	19	20	21	22	23	24	25	26	27	28	29	30	Rank
1	63	64	65	65	66	66	67	67	68	68	68	69	69	69	69	70	70	70	70	70	71	71	71	71	71	71	1
2	55	57	58	59	60	60	61	62	62	62	63	63	64	64	64	64	65	65	65	65	66	66	66	66	66	66	2
3	50	52	54	55	56	57	57	58	59	59	60	60	60	61	61	62	62	62	62	63	63	63	63	63	64	64	3
4	45	48	50	52	53	54	55	55	56	57	57	58	58	59	59	59	60	60	60	61	61	61	61	62	62	62	4
5	37	43	46	48	50	51	52	53	54	55	55	56	56	57	57	58	58	58	59	59	59	59	60	60	60	60	5
6		36	42	45	47	49	50	51	52	53	53	54	55	55	56	56	56	57	57	57	58	58	58	59	59	59	6
7			35	41	44	46	48	49	50	51	52	52	53	54	54	55	55	55	56	56	56	57	57	57	58	58	7
8				35	40	43	45	47	48	49	50	51	51	52	53	53	54	54	55	55	55	56	56	56	56	57	8
9					34	40	43	45	46	47	48	49	50	51	51	52	52	53	53	54	54	54	55	55	55	56	9
10						34	39	42	44	45	47	48	49	49	50	51	51	52	52	53	53	53	54	54	54	55	10
11							33	38	41	43	45	46	47	48	49	49	50	51	51	52	52	52	53	53	54	54	11
12								33	38	41	43	44	45	46	47	48	49	49	50	51	51	51	52	52	53	53	12
13									32	38	40	42	44	45	46	47	48	48	49	49	50	50	51	51	52	52	13
14										32	37	40	42	43	44	45	46	47	48	48	49	50	50	50	51	51	14
15											32	37	40	41	43	44	45	46	47	47	48	49	49	50	50	50	15
16												31	36	39	41	42	44	45	45	46	47	48	48	49	49	50	16
17													31	36	39	41	42	43	44	45	46	47	47	48	48	49	17
18														31	36	38	40	42	43	44	45	46	46	47	47	48	18
19															31	36	38	40	41	43	44	44	45	46	46	47	19
20																30	35	38	40	41	42	43	44	45	45	46	20
21																	30	35	38	39	41	42	43	44	45	45	21
22																		30	35	37	39	41	42	43	44	44	22
23																			30	35	37	39	40	41	42	43	23
24																				30	34	37	39	40	41	42	24
25																					29	34	37	38	40	41	25
26																						29	34	37	38	40	26
27																							29	34	36	38	27
28																								29	34	36	28
29																									29	34	29
30																										29	30

* From *Improvement of Grading Practices for Air Training Command Schools*, ATRC Manual 50–900–9.

Table XIV   Squares, Square Roots, and Reciprocals

n	n^2	$\sqrt{n}$	$\sqrt{10n}$	$1/n$	n	n^2	$\sqrt{n}$	$\sqrt{10n}$	$1/n$
1	1	1.000	3.162	1.00000	36	1296	6.000	18.974	.02778
2	4	1.414	4.472	.50000	37	1369	6.083	19.235	.02703
3	9	1.732	5.477	.33333	38	1444	6.164	19.494	.02632
4	16	2.000	6.325	.25000	39	1521	6.245	19.748	.02564
5	25	2.236	7.071	.20000	40	1600	6.325	20.000	.02500
6	36	2.449	7.746	.16667	41	1681	6.403	20.248	.02439
7	49	2.646	8.367	.14286	42	1764	6.481	20.494	.02381
8	64	2.828	8.944	.12500	43	1849	6.557	20.736	.02326
9	81	3.000	9.487	.11111	44	1936	6.633	20.976	.02273
10	100	3.162	10.000	.10000	45	2025	6.708	21.213	.02222
11	121	3.317	10.488	.09091	46	2116	6.782	21.448	.02174
12	144	3.464	10.954	.08333	47	2209	6.856	21.679	.02128
13	169	3.606	11.402	.07692	48	2304	6.928	21.909	.02083
14	196	3.742	11.832	.07143	49	2401	7.000	22.136	.02041
15	225	3.873	12.247	.06667	50	2500	7.071	22.361	.02000
16	256	4.000	12.649	.06250	51	2601	7.141	22.583	.01961
17	289	4.123	13.038	.05882	52	2704	7.211	22.804	.01923
18	324	4.243	13.416	.05556	53	2809	7.280	23.022	.01887
19	361	4.359	13.784	.05263	54	2916	7.348	23.238	.01852
20	400	4.472	14.142	.05000	55	3025	7.416	23.452	.01818
21	441	4.583	14.491	.04762	56	3136	7.483	23.664	.01786
22	484	4.690	14.832	.04545	57	3249	7.550	23.875	.01754
23	529	4.796	15.166	.04348	58	3364	7.616	24.083	.01724
24	576	4.899	15.492	.04167	59	3481	7.681	24.290	.01695
25	625	5.000	15.811	.04000	60	3600	7.746	24.495	.01667
26	676	5.099	16.125	.03846	61	3721	7.810	24.698	.01639
27	729	5.196	16.432	.03704	62	3844	7.874	24.900	.01613
28	784	5.292	16.733	.03571	63	3969	7.937	25.100	.01587
29	841	5.385	17.029	.03448	64	4096	8.000	25.298	.01562
30	900	5.477	17.321	.03333	65	4225	8.062	25.495	.01538
31	961	5.568	17.607	.03226	66	4356	8.124	25.690	.01515
32	1024	5.657	17.889	.03125	67	4489	8.185	25.884	.01493
33	1089	5.745	18.166	.03030	68	4624	8.246	26.077	.01471
34	1156	5.831	18.439	.02941	69	4761	8.307	26.268	.01449
35	1225	5.916	18.708	.02857	70	4900	8.367	26.458	.01429

Table XIV   Squares, Square Roots, and Reciprocals (*Concluded*)

n	n^2	$\sqrt{n}$	$\sqrt{10n}$	$1/n$	n	n^2	$\sqrt{n}$	$\sqrt{10n}$	$1/n$
71	5041	8.426	26.646	.01408	86	7396	9.274	29.326	.01163
72	5184	8.485	26.833	.01389	87	7569	9.327	29.496	.01149
73	5329	8.544	27.019	.01370	88	7744	9.381	29.665	.01136
74	5476	8.602	27.203	.01351	89	7921	9.434	29.833	.01124
75	5625	8.660	27.386	.01333	90	8100	9.487	30.000	.01111
76	5776	8.718	27.568	.01316	91	8281	9.539	30.166	.01099
77	5929	8.775	27.749	.01299	92	8464	9.592	30.332	.01087
78	6084	8.832	27.928	.01282	93	8649	9.644	30.496	.01075
79	6241	8.888	28.107	.01266	94	8836	9.695	30.659	.01064
80	6400	8.944	28.284	.01250	95	9025	9.747	30.822	.01053
81	6561	9.000	28.460	.01235	96	9216	9.798	30.984	.01042
82	6724	9.055	28.636	.01220	97	9409	9.849	31.145	.01031
83	6889	9.110	28.810	.01205	98	9604	9.899	31.305	.01020
84	7056	9.165	28.983	.01190	99	9801	9.950	31.464	.01010
85	7225	9.220	29.155	.01176	100	10000	10.000	31.623	.01000

Table XV   Four-place Squares of Numbers

n	0	1	2	3	4	5	6	7	8	9
1.0	1.000	1.020	1.040	1.061	1.082	1.103	1.124	1.145	1.166	1.188
1.1	1.210	1.232	1.254	1.277	1.300	1.323	1.346	1.369	1.392	1.416
1.2	1.440	1.464	1.488	1.513	1.538	1.563	1.588	1.613	1.638	1.664
1.3	1.690	1.716	1.742	1.769	1.796	1.823	1.850	1.877	1.904	1.932
1.4	1.960	1.988	2.016	2.045	2.074	2.103	2.132	2.161	2.190	2.220
1.5	2.250	2.280	2.310	2.341	2.372	2.403	2.434	2.465	2.496	2.528
1.6	2.560	2.592	2.624	2.657	2.690	2.723	2.756	2.789	2.822	2.856
1.7	2.890	2.924	2.958	2.993	3.028	3.063	3.098	3.133	3.168	3.204
1.8	3.240	3.276	3.312	3.349	3.386	3.423	3.460	3.497	3.534	3.572
1.9	3.610	3.648	3.686	3.725	3.764	3.803	3.842	3.881	3.920	3.960
2.0	4.000	4.040	4.080	4.121	4.162	4.203	4.244	4.285	4.326	4.368
2.1	4.410	4.452	4.494	4.537	4.580	4.623	4.666	4.709	4.752	4.796
2.2	4.840	4.884	4.928	4.973	5.018	5.063	5.108	5.153	5.198	5.244
2.3	5.290	5.336	5.382	5.429	5.476	5.523	5.570	5.617	5.664	5.712
2.4	5.760	5.808	5.856	5.905	5.954	6.003	6.052	6.101	6.150	6.200
2.5	6.250	6.300	6.350	6.401	6.452	6.503	6.554	6.605	6.656	6.708
2.6	6.760	6.812	6.864	6.917	6.970	7.023	7.076	7.129	7.182	7.236
2.7	7.290	7.344	7.398	7.453	7.508	7.563	7.618	7.673	7.728	7.784
2.8	7.840	7.896	7.952	8.009	8.066	8.123	8.180	8.237	8.294	8.352
2.9	8.410	8.468	8.526	8.585	8.644	8.703	8.762	8.821	8.880	8.940
3.0	9.000	9.060	9.120	9.181	9.242	9.303	9.364	9.425	9.486	9.548
3.1	9.610	9.672	9.734	9.797	9.860	9.923	9.986	10.05	10.11	10.18
3.2	10.24	10.30	10.37	10.43	10.50	10.56	10.63	10.69	10.76	10.82
3.3	10.89	10.96	11.02	11.09	11.16	11.22	11.29	11.36	11.42	11.49
3.4	11.56	11.63	11.70	11.76	11.83	11.90	11.97	12.04	12.11	12.18
3.5	12.25	12.32	12.39	12.46	12.53	12.60	12.67	12.74	12.82	12.89
3.6	12.96	13.03	13.10	13.18	13.25	13.32	13.40	13.47	13.54	13.62
3.7	13.69	13.76	13.84	13.91	13.99	14.06	14.14	14.21	14.29	14.36
3.8	14.44	14.52	14.59	14.67	14.75	14.82	14.90	14.98	15.05	15.13
3.9	15.21	15.29	15.37	15.44	15.52	15.60	15.68	15.76	15.84	15.92
4.0	16.00	16.08	16.16	16.24	16.32	16.40	16.48	16.56	16.65	16.73
4.1	16.81	16.89	16.97	17.06	17.14	17.22	17.31	17.39	17.47	17.56
4.2	17.64	17.72	17.81	17.89	17.98	18.06	18.15	18.23	18.32	18.40
4.3	18.49	18.58	18.66	18.75	18.84	18.92	19.01	19.10	19.18	19.27
4.4	19.36	19.45	19.54	19.62	19.71	19.80	19.89	19.98	20.07	20.16
4.5	20.25	20.34	20.43	20.52	20.61	20.70	20.79	20.88	20.98	21.07
4.6	21.16	21.25	21.34	21.44	21.53	21.62	21.72	21.81	21.90	22.00
4.7	22.09	22.18	22.28	22.37	22.47	22.56	22.66	22.75	22.85	22.94
4.8	23.04	23.14	23.23	23.33	23.43	23.52	23.62	23.72	23.81	23.91
4.9	24.01	24.11	24.21	24.30	24.40	24.50	24.60	24.70	24.80	24.90
5.0	25.00	25.10	25.20	25.30	25.40	25.50	25.60	25.70	25.81	25.91
5.1	26.01	26.11	26.21	26.32	26.42	26.52	26.63	26.73	26.83	26.94
5.2	27.04	27.14	27.25	27.35	27.46	27.56	27.67	27.77	27.88	27.98
5.3	28.09	28.20	28.30	28.41	28.52	28.62	28.73	28.84	28.94	29.05
5.4	29.16	29.27	29.38	29.48	29.59	29.70	29.81	29.92	30.03	30.14
n	0	1	2	3	4	5	6	7	8	9

Table XV Four-place Squares of Numbers (Concluded)

n	0	1	2	3	4	5	6	7	8	9
5.5	30.25	30.36	30.47	30.58	30.69	30.80	30.91	31.02	31.14	31.25
5.6	31.36	31.47	31.58	31.70	31.81	31.92	32.04	32.15	32.26	32.38
5.7	32.49	32.60	32.72	32.83	32.95	33.06	33.18	33.29	33.41	33.52
5.8	33.64	33.76	33.87	33.99	34.11	34.32	34.34	34.46	34.57	34.69
5.9	34.81	34.93	35.05	35.16	35.28	35.40	35.52	35.64	35.76	35.88
6.0	36.00	36.12	36.24	36.36	36.48	36.60	36.72	36.84	36.97	37.09
6.1	37.21	37.33	37.45	37.58	37.70	37.82	37.95	38.07	38.19	38.32
6.2	38.44	38.56	38.69	38.81	38.94	39.06	39.19	39.31	39.44	39.56
6.3	39.69	39.82	39.94	40.07	40.20	40.32	40.45	40.58	40.70	40.83
6.4	40.96	41.09	41.22	41.34	41.47	41.60	41.73	41.86	41.99	42.12
6.5	42.25	42.38	43.51	42.64	42.77	42.90	43.03	43.16	43.30	43.43
6.6	43.56	43.69	43.82	43.96	44.09	44.22	44.36	44.49	44.62	44.76
6.7	44.89	45.02	45.16	45.29	45.43	45.56	45.70	45.83	45.97	46.10
6.8	46.24	46.38	46.51	46.65	46.79	46.92	47.06	47.20	47.33	47.47
6.9	47.61	47.75	47.89	48.02	48.16	48.30	48.44	48.58	48.72	48.86
7.0	49.00	49.14	49.28	49.42	49.56	49.70	49.84	49.98	50.13	50.27
7.1	50.41	50.55	50.69	50.84	50.98	51.12	51.27	51.41	51.55	51.70
7.2	51.84	51.98	52.13	52.27	52.42	52.56	52.71	52.85	53.00	53.14
7.3	53.29	53.44	53.58	53.73	53.88	54.02	54.17	54.32	54.46	54.61
7.4	54.76	54.91	55.06	55.20	55.35	55.50	55.65	55.80	55.95	56.10
7.5	56.25	56.40	56.55	56.70	56.85	57.00	57.15	57.30	57.46	57.61
7.6	57.76	57.91	58.06	58.22	58.37	58.52	58.68	58.83	58.98	59.14
7.7	59.29	59.44	59.60	59.75	59.91	60.06	60.22	60.37	60.53	60.68
7.8	60.84	61.00	61.15	61.31	61.47	61.62	61.78	61.94	62.09	62.25
7.9	62.41	62.57	62.73	62.88	63.04	63.20	63.36	63.52	63.68	63.84
8.0	64.00	64.16	64.32	64.48	64.64	64.80	64.96	65.12	65.29	65.45
8.1	65.61	65.77	65.93	66.10	66.26	66.42	66.59	66.75	66.91	67.08
8.2	67.24	67.40	67.57	67.73	67.90	68.06	68.23	68.39	68.56	68.72
8.3	68.89	69.06	69.22	69.39	69.56	69.72	69.89	70.06	70.22	70.39
8.4	70.56	70.73	70.90	71.06	71.23	71.40	71.57	71.74	71.91	72.08
8.5	72.25	72.42	72.59	72.76	72.93	73.10	73.27	73.44	73.62	73.79
8.6	73.96	74.13	74.30	74.48	74.65	74.82	75.00	75.17	75.34	75.52
8.7	75.69	75.86	76.04	76.21	76.39	76.56	76.74	76.91	77.08	77.26
8.8	77.44	77.62	77.79	77.97	78.15	78.32	78.50	78.68	78.85	79.03
8.9	79.21	79.39	79.57	79.74	79.92	80.10	80.28	80.46	80.64	80.82
9.0	81.00	81.18	81.36	81.54	81.72	81.90	82.08	82.26	82.45	82.63
9.1	82.81	82.99	83.17	83.36	83.54	83.72	83.91	84.09	84.27	84.46
9.2	84.64	84.82	85.01	85.19	85.38	85.56	85.75	85.93	86.12	86.30
9.3	86.49	86.68	86.86	87.05	87.24	87.42	87.61	87.80	87.98	88.17
9.4	88.36	88.55	88.74	88.92	89.11	89.30	89.49	89.68	89.87	90.06
9.5	90.25	90.44	90.63	90.82	91.01	91.20	91.39	91.58	91.78	91.97
9.6	92.16	92.35	92.54	92.74	92.93	93.12	93.32	93.51	93.70	93.90
9.7	94.09	94.28	94.48	94.67	94.87	95.06	95.26	95.45	95.65	95.84
9.8	96.04	96.24	96.43	96.63	96.83	97.02	97.22	97.42	97.61	97.81
9.9	98.01	98.21	98.41	98.60	98.80	99.00	99.20	99.40	99.60	99.80
n	0	1	2	3	4	5	6	7	8	9

GREEK ALPHABET

*Alpha	A	α	*Nu	N	ν
*Beta	B	β	Xi	Ξ	ξ
Gamma	Γ	γ	Omicron	O	o
Delta	Δ	δ	*Pi	Π	π
Epsilon	E	ε	*Rho	P	ρ
*Zeta	Z	ζ	†Sigma	Σ	σ
Eta	H	η	Tau	T	τ
Theta	Θ	θ	Upsilon	Υ	υ
Iota	I	ι	*Phi	Φ	φ
Kappa	K	κ	*Chi	X	χ
Lambda	Λ	λ	Psi	Ψ	ψ
*Mu	M	μ	Omega	Ω	ω

* The small letter is used in this text.
† Both capital and small letters are used in this text.

LIST OF FORMULAS

NUMBER   **PAGE**

(6-12) $\bar{x} = a + i\left(\dfrac{\Sigma f x'}{n}\right)$ 84

(6-13)   Variance $= i^2\left(\dfrac{n\Sigma f(x')^2 - (\Sigma f x')^2}{n(n-1)}\right)$ 84

(6-14) $\Sigma(x - \bar{x})^2 = \Sigma x^2 - \dfrac{(\Sigma x)^2}{n}$ 85

(6-15) $\Sigma(x - \bar{x})^2 = i^2\left[\Sigma f(x')^2 - \dfrac{(\Sigma f x')^2}{n}\right]$ 85

(6-16) $s^2 = \dfrac{\Sigma x^2 - (\Sigma x)^2/n}{n-1}$ 85

(6-17) $s^2 = i^2\left[\dfrac{\Sigma f(x')^2 - (\Sigma f x')^2/n}{n-1}\right]$ 85

Raising the arbitrary origin by r intervals:
(6-18)   changes $\Sigma x'$ to $\Sigma x' - nr$ 85
(6-19)   changes $\Sigma(x')^2$ to $\Sigma(x')^2 - 2r\Sigma x' + nr^2$ 85
Lowering the arbitrary origin by r intervals:
(6-20)   changes $\Sigma x'$ to $\Sigma x' + nr$ 85
(6-21)   changes $\Sigma(x')^2$ to $\Sigma(x')^2 + 2r\Sigma x' + nr^2$ 85
(6-22) $n_c = n_1 + n_2 + \cdots + n_k$ 89

(6-23) $\bar{x}_c = \dfrac{1}{n_c}(n_1\bar{x}_1 + n_2\bar{x}_2 + \cdots + n_k\bar{x}_k)$ 89

(6-24) $d_i = \bar{x}_i - \bar{x}_c$ 89
(6-25) $(n_c - 1)s_c^2 = (n_1 - 1)s_1^2 + (n_2 - 1)s_2^2 + \cdots$
           $+ (n_k - 1)s_k^2 + n_1 d_1^2 + n_2 d_2^2 + \cdots + n_k d_k^2$ 89

(8-1) $z = \dfrac{x - \mu}{\sigma}$ 112

(8-2) $y = \dfrac{1}{\sigma\sqrt{2\pi}}\, e^{-\frac{(x-\mu)^2}{2\sigma^2}}$ 115

(8-3) $y = \dfrac{1}{\sqrt{2\pi}}\, e^{-\frac{z^2}{2}}$ 115

(8-4) $p = 1 - \dfrac{R - 0.5}{n}$ 130

(9-1) $\sigma_{\bar{x}} = \dfrac{\sigma}{\sqrt{n}}$ 144

(9-2) $z = \dfrac{x - \bar{x}}{s}$ 147

(9-3) $z = \dfrac{x - \mu}{\sigma}$ 148

(9-4) $z = \dfrac{\bar{x} - \mu}{\sigma/\sqrt{n}} = \dfrac{(\bar{x} - \mu)\sqrt{n}}{\sigma}$ 148

(9-5) $t = \dfrac{\bar{x} - \mu}{s/\sqrt{n}} = \dfrac{(\bar{x} - \mu)\sqrt{n}}{s}$ 149

Confidence limits for μ when n is large are:

(9-6) $\bar{x} + z_{\frac{1-c}{2}}\dfrac{s}{\sqrt{n}}$ and $\bar{x} + z_{\frac{1+c}{2}}\dfrac{s}{\sqrt{n}}$ 151

When n is large: $z_{\frac{1-c}{2}} < \dfrac{\bar{x} - \mu}{s/\sqrt{n}} < z_{\frac{1+c}{2}}$ is true for a pro-

(9-7)   portion c of samples   151

Interval estimate for μ when n is large:

(9-8) $\bar{x} + z_{\frac{1-c}{2}}\dfrac{s}{\sqrt{n}} < \mu < \bar{x} + z_{\frac{1+c}{2}}\dfrac{s}{\sqrt{n}}$ 151

Confidence limits for μ when n is small are:

(9-9) $\bar{x} + t_{\frac{1-c}{2}}\dfrac{s}{\sqrt{n}}$ and $\bar{x} + t_{\frac{1+c}{2}}\dfrac{s}{\sqrt{n}}$ 154

(10-1) $s_{\bar{x}_1 - \bar{x}_2} = \sqrt{(s_1{}^2/n_1) + (s_2{}^2/n_2)}$ for large samples when $\sigma_1{}^2$ is not assumed equal to $\sigma_2{}^2$ 164

(10-2) $t = \dfrac{(\bar{x}_1 - \bar{x}_2) - (\mu_1 - \mu_2)}{\sqrt{(s_1{}^2/n_1) + (s_2{}^2/n_2)}}$ 164

(10-3) $s^2 = \dfrac{(n_1 - 1)s_1{}^2 + (n_2 - 1)s_2{}^2}{n_1 + n_2 - 2}$ 165

(10-4) $s_{\bar{x}_1 - \bar{x}_2}{}^2 = \dfrac{s^2(n_1 + n_2)}{n_1 n_2}$ 165

(10-5) $s_{\bar{x}_1 - \bar{x}_2}{}^2 = \dfrac{(n_1 - 1)s_1{}^2 + (n_2 - 1)s_2{}^2}{n_1 + n_2 - 2} \cdot \dfrac{n_1 + n_2}{n_1 n_2}$ 165

(10-6) $t = \dfrac{(\bar{x}_1 - \bar{x}_2) - (\mu_1 - \mu_2)}{s_{\bar{x}_1 - \bar{x}_2}}$ 165

(10-7) $s^2 = \dfrac{\Sigma x_1{}^2 + \Sigma x_2{}^2 - \dfrac{(\Sigma x_1)^2}{n_1} - \dfrac{(\Sigma x_2)^2}{n_2}}{n_1 + n_2 - 2}$ 165

NUMBER   **PAGE**

(13-19) $z_r = \dfrac{1}{2} \log_e \dfrac{1+r}{1-r}$ 228

(13-20) $z_r = 1.1503 \log_{10} \dfrac{1+r}{1-r}$ 228

(13-21) $z_r = 1.1503[\log_{10}(1+r) - \log_{10}(1-r)]$ 228

(13-22) $\sigma_{z_r} = \dfrac{1}{\sqrt{n-3}}$ 228

As confidence interval for ζ:

(13-23)   Upper limit $= z_r + z_{\frac{1+c}{2}}\left(\dfrac{1}{\sqrt{n-3}}\right)$

Lower limit $= z_r + z_{\frac{1-c}{2}}\left(\dfrac{1}{\sqrt{n-3}}\right)$ 229

(13-24) $z = \dfrac{(z_{r_1} - z_{r_2}) - (\zeta_1 - \zeta_2)}{\sqrt{\dfrac{1}{n_1-3} + \dfrac{1}{n_2-3}}}$ 230

(13-25) $1 - r^2 = \dfrac{\Sigma(y - y_x)^2}{\Sigma(y - \bar{y})^2}$ 232

(13-26) $R = \dfrac{2r}{1+r}$ 233

(13-27) $\phi = \dfrac{ad - bc}{\sqrt{(a+b)(a+c)(b+d)(c+d)}}$ 236

(13-28) $R = 1 - \dfrac{6\Sigma d^2}{n(n^2-1)}$ 237

(14-1) $r =$ 245

$$\dfrac{\Sigma x_1 y_1 + \Sigma x_2 y_2 + \Sigma x_3 y_3 - \dfrac{(\Sigma x_1)(\Sigma y_1)}{n_1} - \dfrac{(\Sigma x_2)(\Sigma y_2)}{n_2} - \dfrac{(\Sigma x_3)(\Sigma y_3)}{n_3}}{\sqrt{\Sigma x_1^2 + \Sigma x_2^2 + \Sigma x_3^2 - \dfrac{(\Sigma x_1)^2}{n_1} - \dfrac{(\Sigma x_2)^2}{n_2} - \dfrac{(\Sigma x_3)^2}{n_3}}\sqrt{\Sigma y_1^2 + \Sigma y_2^2 + \Sigma y_3^2 - \dfrac{(\Sigma y_1)^2}{n_1} - \dfrac{(\Sigma y_2)^2}{n_2} - \dfrac{(\Sigma y_3)^2}{n_3}}}$$

(14-2) $R_{xy} \cong \dfrac{r_{xy}S_x}{\sqrt{r_{xy}^2 S_x^2 + s_x^2(1 - r_{xy}^2)}}$ 248

(15-1) $\chi^2 = \dfrac{\Sigma(x - \bar{x})^2}{\sigma^2}$ 249

(15-2) $\chi^2 = \dfrac{(n-1)s^2}{\sigma^2}$ 249

LIST OF FIGURES IN TEXT

LIST OF TABLES IN TEXT

GLOSSARY OF SYMBOLS

The capital letters X and Y (and sometimes Z) are used here to name variables; the corresponding lower case letters, x, y, and z are used to denote particular values of those variables. Constants are usually denoted by letters at the beginning of the alphabet, either upper or lower case. Lower case letters from the middle of the alphabet, especially i and j, are used as subscripts to indicate reference to specific individuals or groups.

Parameters are customarily denoted by Greek letters and sample statistics by English. Some exceptions are made: P and Q are used to denote proportions in a population and F to denote a hypothetical frequency; χ^2 and ϕ are used to denote sample statistics.

A bar over a lower case letter indicates the mean of a sample of observations on the variable named by the corresponding capital letter.

Greek letters and English letters are listed separately according to their respective alphabets. Symbols of operation are listed separately.

Each definition of a symbol is followed by the number of the page on which reference to the symbol first appears in the text.

Symbols for constants which are used only once for a specific purpose and which have no general interest are not listed below.

SYMBOL	DEFINITION
A	Alternative to a hypothesis (169).
a	Arbitrary origin (81).
a, b, c, d	Frequencies in a 2×2 contingency table (236).
$\hat{a}, \hat{a}_{yx}, \hat{a}_{xy}$	Constant term in a regression equation, the first subscript indicating the variable to be predicted and

SYMBOL	DEFINITION

the others indicating the predictors (199, 217).

$\hat{b}, \hat{b}_{yx}, \hat{b}_{xy}$ Regression coefficient, the first subscript indicating the variable to be predicted and the others indicating the predictors (199, 217).

C Correction for ties by ranks (307).

c (1) Confidence coefficient (151).

(2) Subscript for statistics obtained from combined groups (89).

(3) Subscript for statistics obtained in cluster sampling (338).

(4) Number of columns in a contingency table (272).

d Difference between two ranks assigned to the same individual (237).

d_i Difference between the mean of the ith group and the mean of combined groups, $\bar{x}_i - \bar{x}_c$ (89).

$d.f.$ Degrees of freedom (298).

e A constant whose value is approximately 2.71828. It has many uses in mathematics, but in this book is used only in the equation for the normal curve (115) and in the z_r transformation (228).

F (1) The ratio of two independent variances (260).

(2) Statistic for comparing means of several populations (299).

$F_{.95}(\nu_1, \nu_2)$ The 95th percentile of the probability distribution of F with ν_1 and ν_2 degrees of freedom. Other percentiles are indicated by appropriate subscripts (261).

F_i The frequency expected in the ith category on the basis of some theory about the population (268).

F_{ij} The expected frequency in row i and column j of a contingency table (274).

f Observed frequency (36).

f_i The observed frequency in the ith category of a frequency distribution (268).

f_{ij} The observed frequency in row i and column j of a contingency table (273).

$f_{i.}$ The observed marginal frequency for row i of a contingency table (273).

$f_{.j}$ The observed marginal frequency for column j of a contingency table (273).

$f_{..}$ The total frequency in a contingency table (273).

SYMBOL	DEFINITION

H A hypothesis about a parameter (169).

H Statistic for analysis of variance among ranks (306).

h Subscript for a stratum in stratified sampling (342).

i Size of a class interval in scale units (84).

i, j, k Subscripts used for reference to specific individuals or specific subgroups.

k (1) Number of cases with a given characteristic, $k = np$ (Appendix Table XII) (186).

     (2) Number of subgroups in a total sample (89, 303).

L Number of strata in stratified population (342).

M Number of clusters in a finite population (338).

MS   Mean square (298).

MS_B Mean square between means (298).

MS_W Mean square within groups (299).

m Number of clusters in a sample (338).

N Number of cases in a finite population (332).

n Number of cases in a sample (54, 76).

N_h Number of cases in stratum h of a finite population (342).

n_h Number of cases from stratum h in a sample (342).

n_c Number of individuals in a combined sample (89).

n_i (1) Number of individuals in the ith category (279).

     (2) Number of individuals in cluster i in a cluster sample (338).

P The proportion of cases in one class of a dichotomous population (179).

p The proportion of cases in one class of a sample from a dichotomous population (179).

Q $1 - P$ (179).

Q_1, Q_L $x_{.25}$, the first or lower quartile (59).

Q_3, Q_U $x_{.75}$, the third or upper quartile (60).

q $1 - p$ (179).

R (1) Coefficient of rank correlation (237).

     (2) Validity coefficient for an entire group based on data for a restricted range (248).

     (3) Reliability coefficient estimated by Spearman-Brown formula (233).

r, r_{xy} Coefficient of correlation (213).

r Number of rows in a contingency table (272).

r_{y1} Correlation between y and x_1 (317).

SYMBOL   DEFINITION

$r_{y.12}$ Multiple correlation between a variable Y and two predictors X_1 and X_2 (313).

$r_{y1.2}$ Coefficient of partial correlation between Y and X_1 with influence of X_2 eliminated (320).

S Sample standard deviation, used when needed to distinguish from lower case s (247).

SS Sum of squares of deviations (296).

SS_B Sum of squares of deviations of sample means from the total mean; sum of squares between groups (298).

SS_W Sum of all the sums of squares of scores from the mean of their own group; sum of squares within groups (298).

S_{yy} $\Sigma(y - \bar{y})^2$ (311).

S_{11}, S_{22} $\Sigma(x_1 - \bar{x}_1)^2; \Sigma(x_2 - \bar{x}_2)^2$ (311).

S_{y1}, S_{y2} $\Sigma(y - \bar{y})(x_1 - \bar{x}_1); \Sigma(y - \bar{y})(x_2 - \bar{x}_2)$ (311).

S_{12} $\Sigma(x_1 - \bar{x}_1)(x_2 - \bar{x}_2)$ (311).

s^2 Variance of a sample (78).

s Standard deviation of a sample (78).

$s_{x.y}, s_{y.x}$ Standard error of estimate (203).

$s_c{}^2$ Variance of a combined group (89).

$s_{\bar{x}}{}^2$ Sample estimate of $\sigma_{\bar{x}}{}^2$, the variance of the mean.

$s_b{}^2$ Sample estimate of $\sigma_b{}^2$, the variance of a regression coefficient (205).

$s_{\bar{x}_c}{}^2$ Sample estimate of the variance of the mean of a finite population obtained by cluster sampling (339).

$s_p{}^2$ Sample estimate of the variance of a proportion.

$s_{p_c}{}^2$ Sample estimate of the variance of a proportion in cluster sampling from a finite population (339).

$s_{\hat{T}_{st}}{}^2$ Sample estimate of the variance for the estimate of the total for an entire population obtained by stratified sampling (343).

$s_{\bar{x}_1 - \bar{x}_2}$ Standard error of the difference between means (164, 165).

$s_{p_1 - p_2}$ Standard error of the difference between two proportions (188).

$s_{y.12}{}^2$ Sample estimate of conditional variance (312).

st Subscript for statistics in stratified sampling (343).

T (1) $50 + 10z$, where z is a normal deviate inferred from a percentile rank (127).

(2) Total of all scores in a finite population (337).

SYMBOL	DEFINITION
T_h	The total of all scores in stratum h of a finite population (342).
$\hat{T}$	Sample estimate of T (337).
$\hat{T}_h$	Sample estimate of T_h (343).
t	(1) Statistic for testing hypotheses about means (149).
	(2) A scale value of "Student's" distribution or a variable which has that distribution (152).
	(3) The number of tied scores (307).
$t_{.95}$	The 95th percentile of the t distribution, and similarly for other percentiles (153).
X, Y, Z	Name of a variable (38).
x, y, z	Specific score on a variable (38).
$x_{.50}$	Median score on variable X. (Other subscripts indicate other percentile values) (58).
$\bar{x}$	Mean of a sample (76).
$x_{ij}, x_{i\alpha}$	The jth or αth individual in the ith group (291).
x_y	Value of X estimated from known value of Y (216).
x'	Coded score, $ix' = x - a$ (84).
$\bar{x}_c$	(1) Mean of combined samples (89).
	(2) Sampling estimate of μ in cluster sampling (338).
y_x	Value of Y estimated from known value of X (199).
y_{12}	Value of Y estimated from values of X_1 and X_2 (310).
z	(1) A standard score $(x - \bar{x})/s$ (90).
	(2) A standard normal deviate (115).
z_r	Transformation for r (228).
$z_{.05}, z_{.95}$	Percentile value of the standard normal distribution (117).
$z_{\frac{1-c}{2}}$	$(1 - c)/2$th percentile value of standard normal distribution (151).
α	(1) Subscript indicating an individual (291).
	(2) Level of significance (168, 262).
β_1, β_2	Regression coefficients in parametric form for an equation to estimate y from x_1 and x_2 when variables are expressed as standard scores (316).
$\hat{\beta}_1, \hat{\beta}_2$	Sample estimates of β_1 and β_2 (316).
ζ	Transformation of parameter ρ (228).
μ	Population mean (110).
μ_p	Mean of the sampling distribution of p (182).

SYMBOL   DEFINITION

$\mu_{y.12}$ Population mean of distribution of Y scores for specific values of X_1 and X_2 (310).

$\mu_{y.x}$ Population mean of distribution of Y scores for specific values of X (196).

ν Number of degrees of freedom (152).

π Ratio of circumference of circle to its diameter, approximately 3.1416.

ρ Correlation coefficient in a population (226).

Σ The sum of (76).

σ Standard deviation of a population or of a theoretical distribution (78, 110).

$\sigma_{\bar{x}}, \sigma_p, \sigma_{\bar{x}_1-\bar{x}_2}$ Standard error of the sampling distribution of sta-
$\sigma_{p_1-p_2}, \sigma_r, \sigma_{z_r}$ tistic named in subscript.

$\sigma_{y.x}$ Conditional standard deviation of distribution of Y for a specific value of X, parametric form (196).

ϕ Measure of relationship between two dichotomous variables (236).

χ^2 (1) Statistic for comparing an observed variance with a theoretical value (249).

(2) Statistic for comparing observed frequencies with those expected on some theoretical basis (269).

χ_y^2 Chi-square with Yates' correction (282).

$=$ Is equal to.

$\neq$ Is not equal to.

$a < b$ a is less than b.

$a \leq b$ a is less than or equal to b; a is not greater than b.

$a > b$ a is greater than b.

$a \geq b$ a is equal to or greater than b; a is not less than b.

$|a|$ The absolute or numerical value of a, the sign being taken as positive (77).

$\sqrt{a}$ The positive square root of a.

∞ Infinity.

ANSWERS

2. (a) 21 years; 28.6 years; 26.0 years; 33.9 years; greater for females; greater for nonwhites.

(b) Not if all races are combined. In 1920 it was slightly higher for nonwhite males than for nonwhite females.

(c) Discrepancy has increased for both whites and nonwhites.

(d) Discrepancy has decreased for both males and females.

(e) No; no.

(f) (1) 1910–20; 1920–30; 1910–20; 1940–50.

(2) 1920–30; 1910–20.

3. (a) $0 + 70.2 = 70.2$; $20 + 52.7 = 72.7$; $70 + 11.7 = 81.7$.

(b) No. (c) Age 70; age 70.

4.

16	33	49
33	16	49
49	49	

Exercise 4-1, page 40

1. (a) 6.5–7.5; 1. (b) 6–8; 2. (c) 6.75–7.25; .5.

(d) 5.5–8.5; 3. (e) 6.95–7.05; 0.1. (f) $6\frac{7}{8}$–$7\frac{1}{8}$; $\frac{1}{4}$.

2. (b) 19.9, 20.0, 20.1, 20.2, 20.3.

(c) 17.0, 20.0, 23.0, 26.0, 29.0.

(d) 19.99, 20.00, 20.01, 20.02, 20.03.

Exercise 4-2, page 42

1. (b)

Real Limits	Score Limits	Class Index	i
9.25–10.75	9.5–10.5	10.0	1.5
7.75– 9.25	8.0– 9.0	8.5	1.5
6.25– 7.75	6.5– 7.5	7.0	1.5

2. (a)

Scores in Interval	Real Limits	Class Index	i
17, 18, 19	16.5–19.5	18	3
14, 15, 16	13.5–16.5	15	3

(b)

Real Limits	Class Index	i
29.5–39.5	34.5	10
19.5–29.5	24.5	10
9.5–19.5	14.5	10

(c)

Real Limits	Class Index	i
69.5–76.5	73	7
62.5–69.5	66	7
55.5–62.5	59	7

3. 35–37 and 38–40; $i = 3$.

Exercise 5-1, page 65

2. (a) Not possibly.   (b) More variable than I on all measures, more variable than II on some.
3. No; no.
4. III; I; I; I; II.
5. 63.1; a score; *above* not *in*.
6. In I nearer to $x_{.75}$; in II a little nearer to $x_{.25}$; in III nearer to $x_{.25}$; in combined group a little nearer to $x_{.75}$.
9. *B D A C.*   10. *C A B E D.*   11. *A* stands higher.
12. (a) *E* or *OE*.   (b) *D* or *OD*.   (c) *F* or *OF*.
    (d) *C* or *OC*.   (e) *CF*.   (f) area *CFJM*.
    (g) Area to right of *GI* or area *GHI*.
    (h) *B* or *OB*.   (i) *BG*.
13. (a) Vertical.
    (b) *A* is the fifth percentile, or $x_{.05}$; *B* is the tenth percentile, or $x_{.10}$.
    (c) Median, 50th percentile, fifth decile, $x_{.50}$.
    (d) $AE = x_{.95} - x_{.05}$; $BD = x_{.90} - x_{.10}$.
    (e) *FH* is middle 80 percent and *OG* is lower 50 percent of the distribution.

Exercise 5-2, Page 69

1.
Score:	25	21	18	24	20	21	21
Rank:	1	4	7	2	6	4	4

Sum of ranks $= 28 = 7(8)/2$

2.
Score:	6	9	13	6	14	8	12	5	2
Rank:	$6\frac{1}{2}$	4	2	$6\frac{1}{2}$	1	5	3	8	9

Sum of ranks $= 45 = 9(10)/2$

3.
Score:	12	8	7	10	6	7	9	7	13	8	14	5
Rank:	3	$6\frac{1}{2}$	9	4	11	9	5	9	2	$6\frac{1}{2}$	1	12

Sum of ranks $= 78 = 12(13)/2$

4. Score: 5   4   7   9   8   5   3   5   9   12   8   2   13   6

  Rank: 10   12   7 $3\frac{1}{2}$ $5\frac{1}{2}$ 10   13   10 $3\frac{1}{2}$ 2 $5\frac{1}{2}$ 14   1   8

Sum of ranks $= 105 = 14(15)/2$

Exercise 5-3, page 71

1. 7800 is the 57th percentile; 57 is the percentile rank of 7800.
2. 12 is the 87th percentile; 87 is the percentile rank of 12.
3. The 50th percentile is 72.0; the percentile rank of 72.0 is 50.
4. The 95th percentile is 10,850; the percentile rank of 10,850 is 95.

Exercise 6-1, page 76

1. $\bar{x} = 5.7$; $x_{.50} = 5.5$.
2. $\bar{x} = 16$; $x_{.50} = 18$.
3. $\bar{x} = 409$; $x_{.50} = 409.5$.

Exercise 6-2, page 80

2. Score 10.2 has percentile rank 79 because $\dfrac{17 + .7(4)}{25} = .792$.

3. Score 5.4 has percentile rank 19 because $\dfrac{2 + .9(3)}{25} = .188$.

5. Range/s $= (12 - 4)/2.4 = 3.33$.

6. and 7.

	$x_{.50}$	$\bar{x}$	s^2	s	Range/s
(a)	9	9	15.8	3.97	3.3
(b)	26	25.4	5.8	2.41	2.5
(c)	10.5	9.1	27.7	5.26	2.7
(d)	1.2	3.5	18.9	4.34	2.5
(e)	2.0	1.0	50.5	7.11	2.4
(f)	−4.0	−2.0	40.7	6.38	2.8

8. It is possible for the standard deviation to be larger than the mean.

Exercise 6-3, page 87

1. For ungrouped data

	I	II	III	Combined
$\bar{x}$	55.0	52.8	48.2	52.4
s^2	68.3	80.1	95.4	86.1
s	8.3	8.9	9.8	9.3

For grouped data

	I	II	III	Combined
$\bar{x}$	54.9	52.7	48.1	52.3
s^2	67.5	78.3	95.1	85.2
s	8.2	8.8	9.8	9.2

3. Yes.

4.

	I	II	III	Combined
Grouped	−1.0	−.2	−.1	−.8
Ungrouped	−2.6	.4	−.8	−.1

5. III, II, I; III, II, I.

6.
Section	Ungrouped	Grouped
I	$41/8.3 = 4.9$	$45/8.2 = 5.5$
II	$41/8.9 = 4.6$	$45/8.8 = 5.0$
III	$37/9.8 = 3.8$	$40/9.8 = 4.1$
Combined	$46/9.3 = 4.9$	$50/9.2 = 5.4$

7. Mistakes:

(a) $x_{.20} < x_{.40} < x_{.50}$.

(c) $\Sigma(x - 5) = n + \Sigma(x - 6)$.

(d) $\Sigma(x - \bar{x}) = 0$.

(f) Range/s is unreasonably large here.

(i) $\Sigma(x - 11) = n + \Sigma(x - 12)$.

(k) $\Sigma(x - \bar{x})^2 < \Sigma(x - a)^2$ if $a \neq \bar{x}$.

Exercise 6-4, page 93

1.
Range in Symbol Form	Range in Score Form	Number of Students	Percent of Students
Above $\bar{x} + 2s$	Above 71.1	0	0
$\bar{x} + s$ to $\bar{x} + 2s$	61.8–71.1	17	17.3%
$\bar{x}$ to $\bar{x} + s$	52.5–61.8	32	32.7
$\bar{x} - s$ to $\bar{x}$	43.2–52.5	35	35.7
$\bar{x} - 2s$ to $\bar{x} - s$	33.9–43.2	10	10.2
Below $\bar{x} - 2s$	Below 33.9	4	4.1
		98	100.0%

2.
	Jones	Smith	Brown
A	66	53	41
B	52	48	53
C	51	57	52
D	66	57	45
Mean	59	54	48

3.
Test	$(x - \bar{x})/s$
English	$-.095$
Algebra	.57
History	1.28

Exercise 8-1, page 117

1. (a) .23 (c) .45 (e) .48 (g) .34
    (b) .16 (d) .04 (f) .46 (h) .26

2. (a) .3472 (c) .6826 (e) .9898
    (b) .4972 (d) .9500 (f) .6970

3. (a) .5596 (c) .9713 (e) .0322 (g) .7486
    (b) .0934 (d) .9525 (f) .0495 (h) .2514

4. (a) .0500 (c) .0204 (e) .4532
    (b) .1010 (d) .0098 (f) .0050

Exercise 8-2, page 119

1. (a) .07   (b) .96   (c) .55   (d) .31   (e) .45
2. (a) .01   (b) .95   (c) .60   (d) .005   (e) .06
3. $z_{.07}, z_{.31}, z_{.45}, z_{.05}, z_{.40}$
4. (a) .98   (c) .50   (e) .95
   (b) .99   (d) .06   (f) .11
6. (a) .02   (c) .05   (e) .10
   (b) .01   (d) .04   (f) .20
8. (a) $z_{.005}$ and $z_{.995}$.   (e) $z_{.05}$ and $z_{.95}$.
   (b) $z_{.01}$ and $z_{.99}$.   (f) $z_{.10}$ and $z_{.90}$.
   (c) $z_{.25}$ and $z_{.75}$.   (g) $z_{.375}$ and $z_{.625}$.
   (d) $z_{.025}$ and $z_{.975}$.   (h) $z_{.0005}$ and $z_{.9995}$.

Exercise 8-4, page 128

2. T scores are 29, 38, 45, 52, 59, 66.
3. With $\bar{x} = 41.6$ and $s = 2.6$, the modified standard scores are 21, 25, 28, 32, 36, 40, 44, 48, 52, 55, 59, 63.

Exercise 9-2, page 145

Distribution of $\bar{x}$

	Mean	Standard Error	Range of Middle 95 Percent of Sample Means	Range of Middle 50 Percent of Sample Means
2.	60	1.5	57.1 to 62.9	59.0 to 61.0
3.	40	2.0	36.1 to 43.9	38.7 to 41.3
4.	40	3.0	34.1 to 45.9	38.0 to 42.0
5.	92	.8	90.4 to 93.6	91.5 to 92.5
6.	53	2.5	48.1 to 57.9	51.3 to 54.7
7.	39	.5	38.0 to 40.0	38.7 to 39.3
8.	39	1.25	36.6 to 41.4	38.2 to 39.8

Exercise 9-3, page 149

1. (a) .50   (b) .07   (c) .84   (d) .68
2. (a) .98   (b) .87
4. (a) 64.7 and 85.3.   (d) 71.9 and 78.1.
   (b) 67.3 and 82.7.   (e) 73.5 and 76.5.
   (c) 69.8 and 80.2.   (f) 74.0 and 76.0.
5. (a) 74.4 to 75.6.   (d) 72.9 to 77.1.
   (b) 73.6 to 76.4.   (e) 72.2 to 77.8.
   (c) 73.3 to 76.7.   (f) 71.7 to 78.3.

Exercise 9-4, page 153

2. (a) (i) 1.83   (iii) − .88   (b) (i) 1.75   (iii) − .87
      (ii) −1.38   (iv) −2.82   (ii) −1.34   (iv) −2.60

(c)   (i) $-$ .72 and .72.   (iv) -2.45 and 2.45.
      (ii) -1.44 and 1.44.   (v) -3.71 and 3.71.
      (iii) -1.94 and 1.94.   (vi) -5.96 and 5.96.

3.   (i) $-$ .69 and .69.   (iv) -2.09 and 2.09.
      (ii) -1.33 and 1.33.   (v) -2.86 and 2.86.
      (iii) -1.73 and 1.73.   (vi) -3.88 and 3.88.

4.   95 percent. Answer depends only on n.

Exercise 9-5, page 156

1.   (a)   14.65 and 32.35.   **2.**   (a)   22.0 and 25.0.
      (b)   17.11 and 29.89.   (b)   22.3 and 24.7.
      (c)   18.25 and 28.75.   (c)   22.5 and 24.5.

3.   (a)   20.6 and 26.4.   (b)   21.4 and 25.6.

Exercise 10-1, page 167

1. $t = .17$; $t_{.975} = 2.26$; not significant.
2. $t = 6.81$; $t_{.99} = 2.43$; significant.
3. $t = 1.96$; $t_{.99} = 2.43$; not significant.
4. $t = .186$; $t_{.99} = 2.43$; not significant.
5. $t = .72$; $t_{.975} = 2.03$; not significant.

Exercise 11-1, page 189

1.   For normal approximation the region of rejection is $p > .81$, so the hypothesis that deaf and hearing are alike is rejected. Hypothesis is also rejected by Table XIIB.
2.   Proportion $D\text{-}H$ positive over $D\text{-}H$ positive or negative equals .4. At .05 level do not reject H if $.32 < p < .68$. Consistent with Exercise 10-1.
3. $z = 1.44$ not significant at .05 level.

Exercise 12-1, page 211

2.   (a) $y_x = 31.6 + .66(25) = 48.1$; $y - y_x = 52 - 48.1 = 3.9$.
      (b) $y_x = 31.6 + .66(39) = 57.3$; $y - y_x = 44 - 57.3 = -13.3$.

Exercise 13-1, page 220

4. $r = 11{,}500/\sqrt{(177{,}400)(212{,}875)} = .059$.
      $y_x = 45.7 + .108x$; $x_y = 15.4 + .032y$.

Exercise 13-2, page 225

1. No.   **2.** 2 were in both groups.   **3.** No.   **4.** No.   **5.** 7.
6. 6.

Exercise 13-3, page 233

1. $\dfrac{2(.52)}{1.52} = .68$; not necessarily; something not far from .52.

2. $\dfrac{2(.45)}{1.45} = .62$; the result is less likely to be achieved.

Exercise 14-1, page 248

2. All decreased; s_x proportionately more than s_y.
3. All decreased; s_y proportionately more than s_x.
4. All increased; s_x proportionately more than s_y.
5. All increased; s_y proportionately more than s_x.

Exercise 15-2, page 259

1. (a) R (b) E (c) R (d) R
   (e) E (f) R (g) R (h) E
2. (a) S (b) C (c) X (d) X (e) S
3. (a) S (b) C (c) C (d) S (e) S (f) C
4. $(n-1)s^2/\sigma^2 = 7.5$; accept at .05 level.
5. $12 < \sigma^2 < 69$; $3.5 < \sigma < 8.3$.
6. Using $\chi_{.25}^2 = 1.92$ and $\chi_{.75}^2 = 5.4$, the intervals for the samples are: A (601, 1691); B (2184, 6143); C (404, 1138); D (199, 561); E (1442, 4055). Only one interval contains $\sigma^2 = 1380.4$, so that $c = .50$ is too low a confidence level.
7. $6.91 < \sigma < 8.23$; $47.75 < \sigma^2 < 67.73$.
8. Reject H: $\sigma^2 = 100$ at .01 level.

Exercise 15-3, page 262

	F	ν_1	ν_2	α
1.	3.7	12	20	$\alpha < .02$
2.	3.7	20	12	$.02 < \alpha < .10$
3.	1.4	24	14	$.10 < \alpha$
4.	1.1	75	40	$.10 < \alpha$
5.	1.3	89	119	$.10 < \alpha$
6.	1.8	34	74	$.02 < \alpha < .10$
7.	2.3	26	51	$\alpha < .02$
8.	2.1	30	30	$.02 < \alpha < .10$
9.	6.3	9	∞	$\alpha < .02$

Exercise 16-1, page 271

1. $\chi^2 = 10.6$; for $\nu = 4$, $\chi_{.95}^2 = 9.5$.

Exercise 16-2, page 275

1. f_{12}; f_{24}; f_{31}; f_{36}; f_{43}.
2. $f_{2.}$; $f_{3.}$; $f_{.2}$; $f_{.3}$.
3. (a) $f_{1.}$; (b) $f_{5.}$; (c) $f_{.5}$.
4. $f_{.3} = f_{13} + f_{23} + f_{33} + f_{43} = D$.
   $f_{3.} = f_{31} + f_{32} + f_{33} + f_{34} + f_{35} + f_{36} = B$.
   $f_{.2} = f_{12} + f_{22} + f_{32} + f_{42} = C$.
5. $\displaystyle\sum_{i=1}^{3} f_{i4} = f_{14} + f_{24} + f_{34}$.

   $\displaystyle\sum_{j=1}^{3} f_{2j} = f_{21} + f_{22} + f_{23}$.
6. $F_{12} = \dfrac{(f_{1.})(f_{.2})}{n}$; $F_{24} = \dfrac{(f_{2.})(f_{.4})}{n}$.

   $F_{31} = \dfrac{(f_{3.})(f_{.1})}{n}$; $F_{36} = \dfrac{(f_{3.})(f_{.6})}{n}$.

   $F_{43} = \dfrac{(f_{4.})(f_{.3})}{n}$.

7.

18	22.5	36	13.5	90
10	12.5	20	7.5	50
12	15	24	9	60
40	50	80	30	200

8. 6 cells.
9. 8 cells, related to r and c only.

Exercise 16-3, page 286

1. $\chi^2 = 26.5$; $\nu = 1$; $z = 5.1$.
2. $\chi^2 = 6.56$; $\nu = 1$.
3.

	Urban	Rural
50+	1	8
less than 50	7	2

$\chi^2 = 5.625$; $t = 3.52$.
$(\nu = 1)$ $(\nu = 16)$

4. $\chi^2 = 28.3$; $\nu = 4$; $z = 5.3$.
5. $\chi^2 = 6.1$; $\nu = 1$.

Exercise 17-1, page 292

1. 5, 4, 6, 5.
2. 19, 21, 10.

3. x_{14}, x_{22}.

4. 535.   **5.** 337.   **6.** 25, 10, 15, sum is 50.

7. 50.   **8.** $\displaystyle\sum_{\alpha=1}^{6} x_{5\alpha}$.

9. $\displaystyle\sum_{\alpha=1}^{4} x_{3\alpha}{}^2$.   **10.** $\displaystyle\sum_{i=1}^{4} x_{i\alpha}$.

11. $\displaystyle\sum_{i=1}^{3}\sum_{\alpha=1}^{n_i} x_{i\alpha}$; $n_1 = 4$; $n_2 = 7$; $n_3 = 2$.

12. $\displaystyle\sum_{i=1}^{3}\sum_{\alpha-1}^{4} x_{i\alpha}{}^2$.   **13.** $x_{41} + x_{42} + x_{43}$.

14. $x_{71} + x_{72} + x_{73} + x_{74}$.

15. $x_{11} + x_{12} + x_{13} + x_{21} + x_{22} + x_{23}$.

16. n.   **17.** $\bar{x}_3$.   **18.** $\displaystyle\sum_{i}\sum_{\alpha} x_{i\alpha}$.

19. $\displaystyle\sum_{\alpha} (x_{2\alpha} - \bar{x}_2)^2$.

20. $\displaystyle\left(\sum_{\alpha} x_{1\alpha}\right)^2$.   **21.** $\displaystyle\sum_{\alpha} x_{3\alpha}{}^2$.   **22.** $\displaystyle\left(\sum_{i}\sum_{\alpha} x_{i\alpha}\right)^2$.   **23.** $\displaystyle\sum_{i}\sum_{\alpha} x_{i\alpha}{}^2$.

Exercise 17-2, page 295

Sample	n	Sum of Scores	Sum of Squared Scores	Mean	Sum of Squared Deviations
1	5	112	3264	22.40	755.20
2	4	41	535	10.25	114.75
3	6	102	1964	17.00	230.00
4	5	82	1464	16.40	119.20
Combined Samples	20	337	7227	16.85	1548.55
Within Samples					1219.15

Exercise 17-3, page 302

1.

Source of Variation	SS	d.f.	MS	F	$F_{.95}$
Between means	329.40	3	109.8	1.44	3.24
Within samples	1219.15	16	76.20		
Total	1548.55	19			

2. $\mathrm{MS}_W = 7396/93 = 79.5$.
   $\mathrm{MS}_B = 798/2 = 399; F = 5.0$.
3. $\mathrm{MS}_B = 2756; \mathrm{MS}_W = 119; F = 23.6$.

Exercise 18-1, page 324

1. (a) $\hat{a} = 23.197; \hat{b}_1 = .4052; \hat{b}_2 = .3330$.
   (b) $\hat{\beta}_1 = .481; \hat{\beta}_2 = .260$.
   (d) 7.07.   (e) .657.
   (f) MS for regression $= 3607/2$.
        MS for error $= 4751/95; F = 36$.
   (g) $r_{y1.2} = .474; r_{y2.1} = .278$.
3. $r_{12.3} = .468$. Mean of 4 correlations $= .458$.
4. $r_{12.3} = .02$.
5. 80% of the variability in ratings is unrelated to variability on the test. The test has some utility. $F = 27.9$, so the hypothesis $\rho_{y.12}{}^2 = 0$ can be rejected.

Exercise 19-1, page 338

1. For total: $\hat{T} = 30,665$; lower limit $= 22,990$; upper limit $= 38,340$.
   For district mean: $\bar{x} = 451$; lower limit $= 337$; upper limit $= 565$.
2. For differences: point estimate of total is 651; lower limit $= -421$, upper limit $= 1727$.
   Point estimate of mean of differences 9.6; lower limit $= -6.2$, upper limit $= 25.4$.
   For total, using prior year: point estimate $= 25,659$; lower limit $= 24,583$, upper limit $= 26,735$.
3. In Problem 1, the difference between limits for total $= 15,350$. In Problem 2, this difference is 2,152.

Exercise 19-2, page 341

For pupil mean, lower limit $= 49.5$, upper limit $= 53.3$. For 10 districts difference between limits $= 3.8$. For 15 districts the difference is 2.7.

Exercise 19-3, page 348

Using current year entries to estimate Total with .95 confidence coefficient, lower limit $= 44,722$, upper limit $= 66,028$, the difference between limits being 21,306.

Using differences, the difference between limits was found to be 3,266.

INDEX